PROBABILISTIC ENGINEERING DESIGN

MECHANICAL ENGINEERING

A Series of Textbooks and Reference Books

EDITORS

L. L. FAULKNER	**S. B. MENKES**
Department of Mechanical Engineering	*Department of Mechanical Engineering*
The Ohio State University	*The City College of the*
Columbus, Ohio	*City University of New York*
	New York, New York

OTHER VOLUMES IN PREPARATION

PROBABILISTIC ENGINEERING DESIGN

PRINCIPLES AND APPLICATIONS

JAMES N. SIDDALL

Faculty of Engineering
McMaster University
Hamilton, Ontario, Canada

MARCEL DEKKER, INC.   New York and Basel

Library of Congress Cataloging in Publication Data

Siddall, James N.
  Probabilistic engineering design.

  (Mechanical engineering; 23)
  Includes bibliographical references and indexes.
  1. Engineering design--Statistical methods.
2. Probabilities. I. Title. II. Series.
TA174.S5734   1983   620'.00425'01519   83-15203
ISBN 0-8247-7022-6

MARCEL DEKKER, INC.
270 Madison Avenue, New York, New York 10016

Current printing (last digit):
10 9 8 7 6 5 4 3 2 1

PRINTED IN THE UNITED STATES OF AMERICA

Innovation means uncertainty;
and the price of progress is trouble.

Probabilistic design is the codification of risk judgment;
and is not the statistical analysis of the design problem.

Preface

This book is an attempt to provide an engineer with the necessary background to apply probability concepts in design. Engineers have, of course, always used probability in their design work. Judgment of risks is one of the most important ingredients in the design process. Until rather recently, however, this has been done almost wholly intuitively, with the risk being codifed in a simple factor of safety. Statistical techniques have been used in engineering in areas such as quality control, reliability, and the control and interpretation of research experiments. But there has been very little application of probability theory in the central role of the engineer—design.

The digital computer has made it possible to apply formal probabilistic techniques in design, which enhance the designer's risk judgment and codify his or her risk decisions in a much more meaningful way. This codification will also promote a more widespread sharing of risk experience, so that we can expect to see handbooks full of standard probability functions associated with different devices and components, and design offices will have catalogs of probability functions for all of their manufactured parts and assemblies, as a data source for their designers.

In writing a book on probabilistic design in engineering, it is unfortunately necessary to take a stand on the meaning of probability in the engineering sense—a stand that could be quite controversial. Much of the theoretical background of engineering is taken from the sciences, but sometimes the material must be interpreted in a different way than scientists do. The concept of probability used may well be one of the more important of the many differences between science and engineering. In engineering design, it seems to me, probability must be defined as unequivocally *subjective* in the strictest sense—a sense that precludes the concept of confidence limits and even the use of

Bayesian statistics. Thus, even Bayesian statisticians, who do define probability as subjective, may need some convincing that engineering is different. This question is discussed further in Sections 3.2 and 3.6 and Appendix C.

I hasten to emphasize, however, that I am not condemning the use of a different concept of probability in other fields. Kendall and Stuart exemplify the best approach: "It seems, however, that every man must choose for himself [the definition of probability] and that his psychological make-up, his experience and his fields of interest all determine the kind of axiomatization which he prefers. In statistics it is a mark of immaturity to argue overmuch about the fundamentals of probability theory."

It would be helpful for the user of this text to have a first course in probability and statistics, although I have found that students with only a traditional course in classical statistics have gained very little insight into the probabilistic concepts that are so important to design applications. Sufficient introductory theory is provided in Chapters 2 to 5 so that the book can be used without reference to elementary texts, although examining other sources is always helpful. In any event, it is desirable that students review these chapters to enhance their understanding of probability theory in an engineering context.

Computer programs are included that were developed to provide the necessary software for applying formal probability techniques in design.

Probabilistic design is closely related to optimization in design; both are aspects of analytical decision making in design. There is in fact some overlap in the topic of probabilistic optimization, and Chapter 10 herein is identical to Chapter 3 in the companion book, *Optimal Engineering Design: Principles and Applications*. Also, much of Chapter 1, which gives the context in general design theory of probabilistic design, is common with the introductory material in the companion book.

James N. Siddall

Contents

PROBABILISTIC
ENGINEERING DESIGN

1

Introduction

The subject of probabilistic design in the context of the general theory of design and identified as one aspect of analytical decision making

1.1 ANALYTICAL DECISION MAKING FOR DESIGN

We are concerned with exploring aspects of the *new theory of engineering design*. Any such theory must explain in some sense the practice of design, or represent a hypothesis for an improved procedure for design.

Any theory for the practice of design should contain the following components:

1. *Taxonomy*: the identification of standard elements such as gears, bearings, and pumps, and the identification of classes of devices and systems
2. *Morphology*: the identification of the steps or components of the total process of design
3. *Creativity*: the explanation of the process of the creation of configurations
4. *Decision making*: procedures for deciding on the best configurations and the proper specifications

This book deals with category 4, decision making. However, we exclude most optimization aspects of decision making, which are covered in a companion volume entitled *Optimal Engineering Design: Principles and Applications*. It is important to distinguish here between

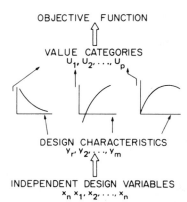

OBJECTIVE FUNCTION

VALUE CATEGORIES
$U_1, U_2, \ldots, U_p$

DESIGN CHARACTERISTICS
$Y_r, Y_2, \ldots, Y_m$

INDEPENDENT DESIGN VARIABLES
$x_n, x_1, x_2, \ldots, x_n$

FIG. 1.1 Hierarchy of design variables.

the theory of engineering design and the theory of engineering modeling. The latter is used to predict performance of components, devices, and systems, and utilizes the many subjects of engineering, such as mechanics, thermodynamics, and electrical science.

Design decisions have been made almost wholly by intuitive means in the past, in an attempt to produce optimal designs. It should be possible, however, to rationalize a significant amount of this decision making, basing it on codification of values and analytical techniques.

Before considering the decision-making process in design, it will be helpful to consider the hierarchy of design variables, shown in Fig. 1.1. At bottom are the independent design variables—the quantities with which the designer directly copes at the lowest level of decision making. What diameter for a shaft? What yield strength for a steel? What viscosity for a lubricant? What smoothness for a bearing surface? From these decisions the designer *hopes* to generate certain measures for design characteristics, such as shaft torque capacity, engine power, or aircraft weight. We shall see in Sec. 1.3 that these in turn generate *values* and the relationship can commonly be codified in curves. The combination of these values, when added up, provides a decision criterion, or objective function, for the overall design. This criterion can be used in different kinds of decision making, as we shall see below.

1.2 CATEGORIES OF DECISION MAKING

It is important next to categorize ways in which decision making is used to pick the best design.

1. *The design option problem*: Two or more "frozen" designs are compared as a means for choosing the best. Numerical values of all the independent design variables have been fixed. So the

decision problem is simply that of determining the total value of each design and choosing the one having the maximum. The problem is complicated by a probabilistic treatment.

2. *The design optimization problem*: The designer has created a *general* configuration in which the numerical values of the independent variables have not been fixed. To obtain the best specific configuration, the problem must be formulated in a general way as follows. An optimization or objective function is set up defining the total value in terms of the independent variables.

$$U(x_1, x_2, \ldots, x_n) = \text{maximum} \qquad (1.2.1)$$

Equality and inequality constraints are developed which define feasibility with respect to all possible modes of failure.

$$\psi_i(x_1, x_2, \ldots, x_n) = 0, \quad i = 1, m \qquad (1.2.2)$$

$$\phi_j(x_1, x_2, \ldots, x_n) \geqslant 0, \quad j = 1, p \qquad (1.2.3)$$

Numerical procedures from the mathematics of operations research can be used to adjust the independent variables so that these expressions are satisfied. The companion volume, *Optimal Engineering Design*, is devoted primarily to this problem, although probabilistic aspects are discussed in this text in Chapter 13.

3. *The specification problem*: Many of the constraint functions referred to previously are based on specifications. All designs are for subsystems, and specifications are junctions with other parts of the system of which the device being designed is a part. A change in a specification will change the optimum design point of both adjoining parts of the system, and *what is best for the whole* must be selected.

4. *The producibility problem*: Many decisions must be made, particularly at the level of detail design, in order to select the best configurations from the point of view of producibility. This requires a good knowledge of materials and production processes, and the decisions are made primarily on cost criteria.

5. *The risk-level problem*: It is a design decision as to the choice of level of risk associated with successful operation of a device. These decisions are subjective. The risk-level decision problem is the primary topic of this book.

1.3 THE ROLE OF VALUE

When talking about optimal design or generating the best possible design, we are implying that there is some value criterion for our choice of best. The criteria that we most commonly use are design *character-*

istics, the second level in the hierarchy of Fig. 1.1. So we tend to choose designs so as to minimize cost, or maximize capacity, or minimize weight, or minimize error, and the like. It is meaningful and worthwhile to do analytical optimal design using these criteria. However, these criteria have two important limitations. They do not permit the designer to use multiple criteria. Thus the engineer may wish not just to minimize cost, but simultaneously to maximize capacity and minimize weight. Yet such characteristics, with different dimensions, cannot be combined directly in a criterion expression. The second limitation is that the use of design characteristics as criteria does not engage the designer with the real or basic criteria in the minds of the users. Humankind has a large and varied spectrum of values that motiviate decision making of all types. Unless the designer systematically considers all of these potential values, there is risk of basing decisions on incorrect criteria. The use of values as criteria also automatically solves the multiple-criteria problem just referred to.

A completely rational approach to probabilistic decision making is not possible without including a theory of values. Such a theory is included in the companion volume, *Optimal Engineering Design*.

1.4 PRACTICAL ASPECTS OF PROBABILISTIC DESIGN

This book is about the *practice* of probabilistic design, and therefore our primary concern is not with the mathematics, but with the application of the mathematics. The level of probability theory required is not high and a first course in probability and statistics provides an adequate background. Chapters 2 to 5 provide a review of the necessary background material for this text. However, a clear and intimate understanding is very important of the basic concepts of an event, a population, a sample, subjective probability, a random variable, a probability density function, and a cumulative distribution function. These are not all easy concepts.

Finally, it must be emphasized that the use of the computer is absolutely essential for the practice of probabilistic design. The computer programs in the Appendixes are set up in such a way that they are easily used, and may be called by short, fully prescribed main programs that require a minimal knowledge of FORTRAN. However, difficulties in execution are commonplace, and the most expert the user is in programming, the most easily he or she can cope with these difficulties.

2

Concepts and Theorems of Probability

An intuitive understanding of the nature of probability and its role in engineering • Some basic aspects of probability theory, including the probability laws, conditions for independence, and Bayes' theorem • A generalization of the probability laws using the concept of basic events

2.1 THE BASIC CONCEPT OF PROBABILITY

Engineers must deal with uncertainty. It is an all-pervading and dominant characteristic of engineering practice. Engineering uncertainty may occur in three basic ways.

Uncertainty occurs when we measure something or make predictions of dependent variables from measured quantities. We are inclined, as engineers, to assume that this kind of uncertainty does not exist at all. We tend to treat problems of this type as determinate, and overdesign to compensate for uncertainty. This must be considered as a rather crude treatment of this kind of uncertainty, and we must learn to bring our judgment to bear on the problem in a more rational way, using the concepts of probability theory. Let us examine the example of heat transfer from a finned tube. The coefficient of heat conductivity for the tube and fin material would commonly be taken from reference sources, with no indication of uncertainty. Even if we measured it we would have little suggestion of uncertainty. We would expect that the measurement could be repeated with very consistent results. This is typical of engineering measurements, in which the discrimination of the instrument's scale corresponds to its inherent variation, and we get the impression that we are making an exact, or almost exact, measurement. However, substantial uncertainty about

the true value of the heat conductivity may really exist. It may be
due to a bias in the measurement technique, or variations in material
from lot to lot. We would also require film coefficients of heat transfer. These would have considerably more uncertainty, being more difficult to measure, and subject to indeterminate variations in surface
roughness, fluid velocities, and fluid conductivity properties. Uncertainty about tube and fin dimensions would also contribute to the
uncertainty about the overall heat transfer. Furthermore, the equation predicting heat transfer may be empirical, with considerable uncertainty about the "correct" value of coefficients. Or the predictive
law may be a poor model for the true behavior, to an uncertain degree.

The second basic type of engineering uncertainty occurs when
we are concerned with an event that may or may not occur, or the
time of its occurrence may be uncertain. Typical of this would be a
certain kind of loading on a structure or machine, or the time of use
of a telephone by any individual, or pressure in a chemical reactor
reaching a critical level.

We may also be uncertain about the validity of a hypothesis or
theory that is to be used to predict performance of an engineering
design. This is similar to the case mentioned in our discussion of the
first type of uncertainty, the question of a poor model for the design
analysis. However, we are concerned there about whether or not a
model, which was known to have good validity in certain circumstances, was applicable to a particular design for which the circumstances
are uncertain. Here we are concerned with the validity of one or perhaps alternative models when the circumstances are well defined or
well controlled.

To deal with uncertainty in a rational, ordered manner, we must
be able, in a sense, to measure it. We shall use the term *event* to
designate all the kinds of things we are uncertain about, as discussed
above. *The measure of our degree of uncertainty about the likelihood
of an event occurring is probability.* In our attempt to measure probability we begin by arbitrarily defining its range as 0 to 1, or 0 to
100%. If we are certain that an event will not occur, we say that it
has a probability of 0; if we are certain that an event will occur, it
has a probability of 1.

Now we are faced with the problem of measuring intermediate
values of probability. We can begin with examples where the measure
is obvious from the situation, or known a priori. Common examples
are tossing a coin, dealing cards, and throwing a die. A head is
equally likely to fall as a tail, the first card dealt is *equally likely* to
be any one of 52, and any of the six faces of a die is *equally likely* to
show. The measure, then, of our uncertainty about whether to expect
a head on a given toss is 0.5; or we may say that the likelihood of getting a head is 0.5; or the probability of getting a head is 0.5. Simi-

larly, the probability of being dealt the ace of spades is 1/52, and the probability of rolling a three is 1/6.

All a priori probabilities are not equal. Consider an urn containing six black balls, four red balls, seven blue balls, and three white balls. If we draw a ball blindly, any ball has an equal chance of being drawn. It seems clear that the probability of drawing a black ball is 6/20, a red ball 4/20, a blue ball 7/20, and a white ball 3/20.

These examples, in which a priori probabilities may be assumed, represent a rather special type of situation quite unrepresentative of engineering problems. It is unfortunate that they are so often used to illustrate probability concepts. They also must be used with caution. We should specify a *balanced* coin, *well-shuffled* cards, a *fair* die, and *well-mixed* balls.

Continuing with these examples for the moment, we must examine the probability concept more closely. What do we really mean by the likelihood of an event occurring? There is no doubt that it is quantitatively meaningful in real life, and its meaning is illustrated by how we use probability measures to guide our actions. There are in fact several meanings. One straightforward meaning uses the frequency concept. We can say that if the probability of getting a head in tossing a coin is 1/2; then in a given number of tosses, say 100, we would expect *about* 50 heads. The "about" is a sticky point here. In some probability situations we can give the "about" some more quantitative meaning; in this case all we can say is that, as the number of tosses increases, the closer the number of heads will approach 1/2.

But suppose that we are concerned about the result of just one toss. Is probability measure meaningful here? It is to a gambler. If he bets a dollar on a head resulting from the toss of a coin, he would expect to win a dollar if a head did fall. Or if he bet a dollar on a blue ball being drawn from the urn, he would expect to win 13/7 dollars. The concept of a fair bet is based on the expectation of no loss or gain in a large number of plays. Thus in 100 draws from the urn the gambler would lose about 13/20 × 100 times or $65.00, and win about 7/20 × 100 times or $35.00. The calculation of winnings is based on the concept of *odds*, which is the ratio of the probability of losing (13/20) to the probability of winning (7/20). Another way of expressing a fair bet is as follows:

Probability of losing × bet = probability of winning × payoff

$$(2.1.1)$$

We might call this the *fair bet equation*.

This is clear enough for repeatable events with a priori probabilities. Probability is a measure of the number of times we expect an event to occur, or it is the measure of a fair bet for a single event. There is still some difficulty even for repeatable events with a priori probabilities. We must introduce the rather difficult concept of *value*.

In gambling, we are implying that the bet and payoff are in money,
but this leads to an inconsistency in the fair bet equation, as we can
see by examples. A poor man would not bet $1000 against 10:1 odds
with a fair payoff because the probability is high that he would lose
the $1000 and be ruined. He would require a much higher payoff than
that calculated by the fair bet equation. On the other hand, he might
bet 25 cents against 1000:1 odds with a less than fair payoff, because
the return would still be very attractive to him. A wealthy man's at-
titude toward these bets would be quite different. We see then that
(2.1.1) is strictly true only if the bet and payoff are in some kind of
value units. However, we can often assume a linear relationship between value and money, so that (2.1.1) can be used. Equation
(2.1.1) has, in fact, been used in an attempt to determine the relationship between value and money for different individuals (Schlaifer,
1961). We are only hinting here at the rather complex theory of values related to decision making under uncertainty, which is discussed
in more detail in Siddall (1982).

Let us leave simple events with a priori probabilities, and examine more complex events. Consider three engineering examples.

EXAMPLE 2.1 Light Bulb

In the first an engineer is concerned with the problem of selecting an electric light bulb to have a desired design life. She requests from a manufacturer a quotation to meet the specifications.
She is rather disconcerted when the quotation is for three models, for none of which they guarantee to meet the specified life.
Instead, they specify that out of every 1000 bulbs, about X number will have the required life. The value of X is different for
each model and the cost increases with X. The engineer is now
very concerned. It is vital in the design that the bulb be on at
all times during the required life, and the best model of bulb has
an X value of only 850. She therefore goes back to the manufacturer and asks them if they cannot supply a bulb that, for practical purposes, has *no chance* of failure (i.e., X = 1000). They
are reluctant to even consider this, but they finally come up with
a quotation for a special bulb, saying that they *think* that, out
of 1000 bulbs, 999 will have the desired life. The engineer reluctantly decides that she will have to accept this small amount
of *risk*. However, she then realizes that the special bulb is pro-
hibitively large and expensive. She reluctantly concludes that
she must use one of the standard bulbs and begins to search
about for some way around the problem *by design*. One solution
seems to be to have one or more parallel circuits carrying extra
bulbs that are always burning. It seem intuitively clear that the
probability of at least one bulb not failing is much improved. After considerable thought she develops a formula predicting rela-

tive frequency of combined survival X_c from the number of bulbs n and the relative frequency of survival for a single bulb X (we shall derive it later):

$$X_c = 1 - (1 - X)^n \qquad\qquad (2.1.2)$$

By considerations of costs and available space she decides final-ly on a combined frequency of survival of 995 out of 1000, using three of the second-best bulbs. She is later approached by a customer who is interested in the device, who asks if the device is completely reliable. The engineer rather reluctantly admits that about 5 in every 1000 will have a premature failure. This does not satisfy the customer; he wants to know if the particular machine *he buys* will meet specifications. The engineer then tells the customer that he must accept a 0.5% *risk* of failure, or probability of failure, to meet specifications. The customer still does not find this too meaningful, so the engineer, who by now has thought a lot about the concept of uncertainty, goes on to explain that it is somewhat like a gamble. The odds are 199:1 that the customer will get a reliable device. If he wants better odds, he must be prepared to find another source and pay a higher price.

All engineering components present this uncertainty about their performance. Other examples are purity of a chemical, yield stress of a metal, resistance of a resistor, and life of a ball bearing. Very often the cost of obtaining a very high pro-bability of a component meeting specifications is relatively small, and may be obtained with adequate accuracy by applying engi-neering judgment to the selection of a conservative factor of safety. Using this approach, probability measures are not used at all; all quantities are assumed determinate. As engineering becomes increasingly more sophisticated, this "factor of igno-rance" approach becomes correspondingly unsatisfactory, *and we must learn to apply measure to our uncertainty.*

EXAMPLE 2.2 Wind Loading

In our second example a team of engineers are designing a very large wind-powered generator for the Canadian Northland. The maximum design wind load is of concern to the mechanical engi-neers in the design of the propeller and drive system, and of concern to the civil engineers in predicting the wind load on the supporting structure. Wind velocities have been recorded at a weather station 300 miles away for the previous 10 years; but this is the only information available for predicting the maximum wind loading on the wind generator during it design life of 35 years. In this problem the uncertainty is not so neatly tied to

well-controlled frequencies as it was in the first example. The
engineers might be tempted simply to use the maximum recorded
wind velocity, but they cannot be sure that it will not be exceeded in the next 35 years. Also, they are uncertain as to how well
wind velocities at the site correspond to those at the nearest
weather station. The designers are next tempted to apply a factor of safety based on judgment to the maximum recorded wind
velocity at the weather station. However, when they begin their
cost estimating, they realize that the provision for maximum wind
speed is very expensive, and the factor of safety must be pared
as low as possible. They begin to look for methods of handling
the uncertainty in a more rational way, and discover *extreme
value theory*. This provides a technique for predicting the maximum wind speed that is likely to occur, by analysis of recorded
wind speeds. In this case they find that the maximum recorded
wind speed was 78 mph. Analysis indicates, however, that the
maximum speed in the next 30 years would be 123 mph, with a
probability of 0.999 that this would not be exceeded. The probability here is again a *measure of risk*. However, the engineers
must now use judgment to decide how valid these figures are for
the site. They examine similar data from other Northland sites
and taking into consideration differences in terrain, they finally
conclude that the figures are reasonably good.

EXAMPLE 2.3 Component for a Chemical Process

In this example a company is asked to provide a one-off component for a chemical process. The part is similar to, but different
from, any of their standard line of parts. It is a very critical
part in this application, and the company is asked to provide a
warranty of $10,000 to compensate the purchaser for his expens-
es in the event of a failure. The selling price for a normal pro-
fit would be $185, but the company manager feels they should get
a higher price because of the risk of having to pay the warranty.
We need a sort of insurance premium, she tells the chief engineer,
and asks him to set a proper price. The chief engineer has no
directly applicable frequency data available, and what is available
is rather limited. However, on the basis of this information, and
his own experience and intuition, he sets the probability of no
failure at 0.95. He then set up the fair bet equation as follows:

$$0.95 \times \text{premium} = 0.05 \times 10,000$$

giving a premium of $526 and a price of $711. The manager is
troubled by the fact that the chief engineer cannot justify his
probability number with statistical data. The engineer insists
that it has valid significance, and that *it is a measure of the degree of his belief that the component will not fail*. And, after all,

many of his important decisions are based purely on judgment. The manager then expresses concern about the risk of a serious loss, whereas insurance companies make on one policy what they lose on another; and how could she explain to the board of directors if she lost so much money in a gamble? The engineer replies that all decisions in life are in effect a gamble. However, he adds that if the manager has high aversion to the risk of loss, she should compensate by asking for a higher premium.

In these examples we have seen how uncertainty commonly enters into engineering practice, and how the concept of probability measure is useful in dealing with uncertainty. We have also illustrated several apparently different meanings for probability. The first and rather simple meaning suggests a definition that probability is the relative frequency of occurrence of an event. We must examine this definition a little more closely because of the difficulty we have experienced in being forced to say that the probability measure gives *about* the frequency of occurrence of events in a sample. Experience shows that a series of a given number of trials or tests or measurements of an event will not yield exactly the same frequency of occurrence each time. We are concerned with random or unpredictable events, and we would expect this to be so. Experience also shows that variation in frequency becomes less as the sample increases. *We could define probability in this sense more precisely as the limit of the ratio of the number of occurrences to the number of trials, as the number of trials increases without limit.*

We have also seen in our examples that probability is a measure of risk that an event will or will not occur, a measure of uncertainty about the occurrence of an event, and a measure of the degree of a person's belief that an event will occur. These are closely related concepts, and represent successive increases in generality of the concept of probability. The measure of risk is related to the concept of gambling. In gambling we risk something in the hope of a return. It seems reasonable and meaningful to consider the measure of the degree of risk of a single event occurring the same as the relative frequency with which a whole set of these events occur. It is consistent with the fair bet equation—which is, of course, only a theory itself. The measure of uncertainty is really the same thing as the measure of risk, but it is more subjective. If *I* am uncertain about the occurrence of an event, there is a risk involved, and the measure of the risk is also the measure of *my* uncertainty. We can invert this and say that if I have a measure of my uncertainty, this can also be taken as the measure of risk. This leads us to our most general concept of probability. *The subjective measure of uncertainty can be considered to be the degree of a person's belief that an event will occur.* This is sometimes called *subjective probability.* This is the most general concept because we have seen that it includes uncertainty and risk, and it also encompasses the relative frequency interpretation. Observed relative frequencies can be considered simply as information substantiating to a greater or less degree a person's belief, depending on the extent of the data. A subjective probability choice may accept completely evidence of statistical data, or it may modify it in view of the decision maker's other but less tangible experience.

We may wish to cling to relative frequencies as being more tangible and meaningful as a measure of probability. However, we are faced with the cruel fact that many engineering events occur only once or rarely. It is not possible to obtain relative frequencies or even think in terms of hypothetical relative frequencies; yet we must measure their uncertainty. Consider, for example, the designer who is trying to estimate the probability of the event that a short-circuit overload occurs on the large generator driven by a hydraulic turbine. It is an extremely rare event, yet hypothetically possible, and the designer must decide how much resources he should devote to accommodating it, based on the probability of its occurrence. He may not consciously or explicitly use a probability approach in applying his judgment. However, his work will be more coherent and consistent if he does so.

A subjective probability estimate may be considered objective in the sense that all reasonable people with the same information will arrive at the same probability estimate. The potential information available would be everything related to the events. This potential information is commonly so vast and so dispersed that any two people would rarely agree on the subjective probability.

The extreme case of no information available to predict events is of some interest. If one of four possible events may occur, and we have no knowledge whatsoever to assist in deciding the probability with which each is likely to occur, we can only assume that all are equally likely with a probability of 1/4.

Probability is not a concept that bears too close examination, just as the concepts of force and mass do not in mechanics. The engineer simply develops an intuitive feeling for the significance of the concept, applies it to reality, and finds that it works for him or her. The basic concepts of probability are discussed by Churchman (1961), Lindley (1965), Good (1965), Savage (1954), Tribus (1969), and Ang and Tang (1975).

2.2 THE CONCEPTS OF EVENTS AND POPULATIONS

2.2.1 Sample Space

The analogy to mechanics again gives us insight into what we are trying to illustrate here. In mechanics, real-world things are represented by idealized generalizations, such as a point mass, rigid body,

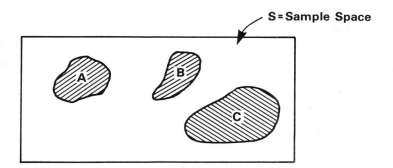

FIG. 2.1 Venn diagram.

vector, or concentrated force. These permit us to generalize the
study of mechanics and conveniently model specific real-world applications. Similarly, in the study of probability and statistics, we set up
idealized generalizations of real-world things. We should note here
that, contrary to the highly standardized notation and terminology of
mechanics, there is wide variation of these in probability theory.

In abstract, nonreal terms, then, we define a *sample space* as a
generalized collection of points representing *elementary events*. The
sample space may be subdivided into *events* or *sets* that are collec-
tions of elementary events. The events are thought of as some kind
of classification of elementary events, so that any elementary event
can be identified with one or more sets or events. It is useful to
represent events by abstract figures, called Venn diagrams. Each
event is assigned an area in the sample space, corresponding to its
proportion of elementary events, as shown in Fig. 2.1. In the diagram the events are A, B, and C.

Two events, A and B, may intersect and the area of intersection,
designated AB, is the collection of all elementary events contained in
both A and B. This is illustrated in Fig. 2.2. The area of an event

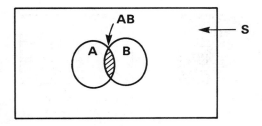

FIG. 2.2 Intersection of events.

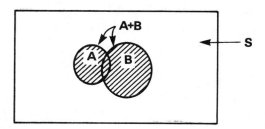

FIG. 2.3 Union of events.

corresponds to the number of elementary events it contains. The
union of two events is the collection of all elementary events contained
in A or B or both. This is expressed as A + B and is shown in Fig.
2.3. The complement of event A, expressed as $\overline{A}$, is the collection of
all elementary events not in A.

Events in a sample space are *mutually exclusive* if none intersect,
or there is no elementary event contained in more than one event.
Events in a sample space are *collectively exhaustive* if every elementary event in the sample space occurs in at least one event. The Venn
diagrams in Fig. 2.4 illustrate these definitions.

Now, having set up some idealized abstractions, let us identify
them with the real-world example of the light bulbs; the sample space
is the 1000 bulbs tested, event A is a bulb not failing, event B is a
bulb failing, and an elementary event is the failing or not failing of
any specific bulb. The events are mutually exclusive and collectively
exhaustive, and $\overline{A}$ = B.

There is an algebra of the logic of events, using the symbolism
introduced above. If we have the occurrence of events A and B or the
occurrence of events C and D, or both, it would be expressed as AB
+ CD; or equally well as DC + BA. If we have the occurrence of
events A or B and C, it would be written as (A + B)C or AC + BC.
Thus the commutative and distributive laws of algebra apply, by definition. Also, the associative law applies, represented by the example

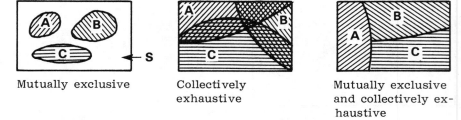

Mutually exclusive   Collectively   Mutually exclusive
                          exhaustive   and collectively ex-
                                                 haustive

FIG. 2.4 Mutually exclusive and collectively exhaustive events.

$$(A + B) + C = A + (B + C)$$

It is clear also that

$$A + A = A \qquad\qquad\qquad (2.2.1)$$

$$AA = A \qquad\qquad\qquad (2.2.2)$$

and if

$$A = BC \quad \text{then} \quad \bar{A} = \bar{B} + \bar{C} \qquad\qquad (2.2.3)$$

The equal sign of this algebra requires some examination. It means that if an elementary event occurs in A [in (2.2.3)], then an elementary events also occurs in B and in C. A somewhat different conceptual approach is to say that if an elementary event occurs in A, then A is *true*. Thus an equal sign also means that both sides of an "equation" are true or both are false. To illustrate, consider the following real events based on the mathematical model, and the significance of $A = BC$.

B = alloy steel is not used for a member

C = the working stress of the member exceeds 75,000 psi

A = the member fails

In the problem where the engineers were concerned with the maximum wind speed, the events are not so obvious. Just what they are is a matter of judgment, and we are not yet conceptually prepared to identify them.

Our third example, where there was just one-off of the component, the events are failing and not failing. The sample space consists of these two mutually exclusive and collectively exhaustive events. In this case the model is not fully applicable, as there is no set of elementary events. However, we can use the true or false concept of events.

Consider a new example where we are concerned with the concentration of a chemical entering a reactor. We set up the following possible events in the sample space:

A = a concentration less than 10 ppm

B = a concentration between 10 and 20 ppm

C = a concentration between 20 and 30 ppm

D = a concentration between 30 and 40 ppm

E = a concentration greater than 40 ppm

These are mutually exclusive and collectively exhaustive.

Recognizing or defining events in a sample space is not always easy or obvious. Consider the following example, where we are concerned with the possible failure of an engineering device. We identify the following events that appear to represent the sample space.

A = failure due to corrosion

B = failure due to overheating

C = failure due to excess pressure

D = failure due to excess wind

E = failure due to earthquake

F = no failure

These are mutually exclusive and collectively exhaustive. This example illustrates that we must use engineering judgment in setting up a mathematical model for probability analysis. There may be some doubt that the event of failure due to earthquake is even remotely likely to occur. We must also be careful not to overlook any possibility of failure.

2.2.2 Conditional Events

We are concerned primarily with the *probability* of occurrence of the events that we have been discussing. There is another type of event closely related to probability. A *conditional event* is one that is true given that another event is true. It is written as A|B, an elementary event occurs in A given that an elementary event occurs in B; or A is true given that B is true. Some authors insist that there are always conditioning events, and they should always be indicated. Thus BC|S might read: the event that alloy steel is not used for a member (B) and the event that the working stress of the member exceeds 75,000 psi (C) given the design of the member and the loadings (S). We shall omit the general conditioning event S when it is common to all events in the problem. However, at this point it is a useful concept. We are saying that S corresponds to the sample space. When we are referring to a conditional event A|B, we are actually referring event A to a more restricted sample space B. The distribution law applies to conditional events. Thus

$$(A + B)|C = A|C + B|C \qquad\qquad (2.2.4)$$

An event A is said to be *independent* of another event B if

$$P(A|B) = P(A)$$

$P(A \mid B)$ is called a *conditional probability*. Real-life events are independent if the occurrence of one has no effect on the occurrence of another; or if the occurrence of B provides no new information that affects our estimate of $P(A)$. The distinction between AB and $A \mid B$ is sometimes somewhat difficult to make. This question is discussed further in Sec. 2.5.

The concept of independence of n events, $A_1, A_2, \ldots, A_n$, is somewhat more complicated. All n events are *mutually independent* if and only if

$$P(A_i \mid A_j A_k \cdots A_m) = P(A_i) \qquad (2.2.5)$$

where $i \neq j, k, \ldots, m$ and $A_j A_k \cdots A_m$ represents any combination of any number of the remaining events. Thus, for complete mutual independence, the occurrence of event A_i is unaffected by the occurrence of any combination of the remaining events. There may be limited independence without complete mutual independence. For example, we may have

$$P(A_1 \mid A_2 A_3) = P(A_1)$$

but

$$P(A_1 \mid A_2 A_3 A_4) \neq P(A_1)$$

Actually, all possible variations of (2.2.5) are not necessary to define full mutual independence. The minimum set requires $2^n - n - 1$ expressions, and the remainder can be derived from these. This will be demonstrated later. One essential set of four for $n = 3$ is

$$
\begin{aligned}
P(A_1 \mid A_2) &= P(A_1) \\
P(A_1 \mid A_3) &= P(A_1) \\
P(A_2 \mid A_3) &= P(A_2) \\
P(A_1 \mid A_2 A_3) &= P(A_1)
\end{aligned}
\qquad (2.2.6)
$$

2.2.3 Setting Up the Mathematical Model

The setting up of the correct events or mathematical model is vital to the success of probability analysis. The following rules can be used as a guide in identifying events.

1.   The events must be as simple as possible. Distinguishable events should not be lumped together. We may be interested primarily in

combined events, such as the event of failure in the last example due to *any* cause; however, we must begin the analysis with simple events.
2.   We must determine whether or not events are independent.

2.2.4 Population

The concept of a *population* is important when we have frequency data available or a recurrence of an event. A population is a real-world collection of things that are to be tested or measured in some way. The results of the tests correspond to the elementary events of our abstract model. Definition of the population is arbitrary, but must be carefully done so as to include only legitimate members, and not omit valid members. For example, depending on the purposes of our analysis, we could define a population of ball bearings as all bearings of a given configuration and any size, or we could limit the population to all bearings of a given configuration and size. The population may include hypothetical members not yet in existence, and may be finite or infinite.

2.3 THEOREMS OF PROBABILITY

We are concerned with predicting the probabilities of complex events from the known probabilities of simple events or other complex events. We wish to develop predictive equations for this purpose. They will all be developed from the mathematical model represented by Venn diagrams. The area of events in the Venn diagram may represent the number of elementary events in a given event, so that the probabilities, based on frequencies, are proportional to the areas. If there is no set of elementary events, we can still let the areas represent probabilities. The theorems that we shall establish could individually be shown to correspond to reality, but if we accept the mathematical model on which they are based as an approximation of reality, this is not necessary. A very large number of theorems can be written down; we shall derive the most useful ones.

1.   Let us begin with some simple relations. It is clear that

$$P(A) + P(\overline{A}) = 1 \qquad\qquad (2.3.1)$$

and from (2.2.1), (2.2.2), and (2.2.3),

$$P(A + A) = P(A) \qquad\qquad (2.3.2)$$

$$P(AA) = P(A) \qquad\qquad (2.3.3)$$

$$P(BC) = 1 - P(\overline{B} + \overline{C}) \qquad\qquad (2.3.4)$$

2. If $A_1, A_2, \ldots, A_n$ are part of a set of mutually exclusive and collectively exhaustive events in a sample space, it is clear from the model that

$$P\left(\sum_{i=1}^{n} A_i\right) = \sum_{i=1}^{n} P(A_i) \qquad (2.3.5)$$

We are using summation notation to represent the union of events. Thus for $n = 2$, (2.3.5) becomes

$$P(A_1 + A_2) = P(A_1) + P(A_2)$$

In words, this says that the probability of an elementary event occurring in A_1 or A_2 or both* is the sum of the probabilities of the event occurring in A_1 and A_2 separately, if A_1 and A_2 are events in a sample space. If n includes all events in a sample space, then from our definition of probability measure,

$$P\left(\sum_{i=1}^{n} A_i\right) = \sum_{i=1}^{n} P(A_i) = 1 \qquad (2.3.6)$$

The summation theorem may be extended to events not mutually exclusive. Consider the Venn diagram of Fig. 2.5 with A and B in general intersecting. To determine the probability of an elementary event occurring in A or B or both, we use the combined area of A and B. Since the area corresponding to AB occurs twice if we add A and B, it must be subtracted once, giving

$$P(A + B) = P(A) + P(B) - P(AB) \qquad (2.3.7)$$

This can be extended to more than two events, but the result becomes somewhat messy.

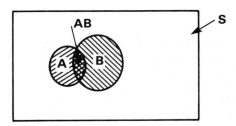

FIG. 2.5 Union of intersecting events.

---

*We shall sometimes leave out the "or both," but it is always implied in an "or"-type combination.

3. Let us now consider another combination theorem where two events intersect. We assume in general that they are not independent. We wish to define $P(A|B)$. Recall that we considered B a sample space for this probability. And the area of B containing elementary events in A is AB. So the relative area, which defines the probability in a sample space, is

$$P(A|B) = \frac{P(AB)}{P(B)}$$

This gives us the product rule for combined probabilities.

$$P(AB) = P(A|B)P(B) \tag{2.3.8}$$

It can be easily generalized for n events, without restriction on their nature. They need not be independent, mutually exclusive, or collectively exhaustive.

$$P(A_1 A_2 \cdots A_n) = P(A_1)P(A_2|A_1)P(A_3|A_1 A_2) \cdots P(A_n|A_1 A_2 \cdots A_{n-1})$$
$$\tag{2.3.9}$$

If the events are independent, the theorem has the simple form

$$P(A_1 A_2 \cdots A_n) = P(A_1)P(A_2) \cdots P(A_n) \tag{2.3.10}$$

If the events are mutually exclusive, the theorem becomes

$$P(A_1 A_2 \cdots A_n) = 0 \tag{2.3.11}$$

This must be true since there is no intersection. Equation (2.3.11) can be written more generally for any subset of mutually exclusive events.

$$P(A_j A_k \cdots A_m) = 0 \tag{2.3.12}$$

It is important to realize that all the relationships that we are identifying from the Venn diagram model are really only different ways of defining the model, and are not independent. It can be shown that all other expressions can be derived from (2.3.8).

4. A simple but important result derives from (2.3.3). Consider er the certain event $(A + \overline{A})(B + \overline{B})$, which gives

$$P[(A + \overline{A})(B + \overline{B})] = 1$$

Expanding the combined event gives

$$P(AB + A\overline{B} + \overline{A}B + \overline{A}\overline{B}) = 1$$

The four joint probabilities must constitute a mutually exclusive and collectively exhaustive set, so (2.3.5) applies and we have

$$P(AB) + P(A\overline{B}) + P(\overline{A}B) + P(\overline{A}\overline{B}) = 1 \tag{2.3.13}$$

We can apply (2.3.5) to any two of these.

$$P(AB) + P(A\overline{B}) = P(AB + A\overline{B}) = P[A(B + \overline{B})]$$

We next use (2.3.6).

$$P(AB) + P (A\overline{B}) = P(A)P[(B + \overline{B})|A]$$

But

$$P[(B + \overline{B})|A] = 1$$

So the final result is

$$P(AB) + P(A\overline{B}) = P(A) \tag{2.3.14}$$

5. A somewhat similar theorem follows from taking (2.3.8) and adding the conditioning event B to it.

$$P(AB|B) = P(A|BB)P(B|B)$$

We note that BB = B and $P(B|B) = 1$, so that

$$P(AB|B) = P(A|B) \tag{2.3.15}$$

6. Let $A_1 A_2 \cdots A_n$ be an exclusive and exhaustive set of events and B any other event. Applying (2.3.8) gives

$$P(A_i B) = P(B|A_i)P(A_i)$$

We now add all equations of the set above.

$$\sum_{i=1}^{n} P(A_i B) = \sum_{i=1}^{n} P(B|A_i)P(A_i)$$

Since the A_i's are exclusive, the $A_i B$'s are exclusive and (2.3.5) applies.

$$\sum_{i=1}^{n} P(A_i B) = P\left(\sum_{i=1}^{n} A_i B\right) = P\left(B \sum_{i=1}^{n} A_i\right) = P(B)$$

Substituting into the preceding equation gives

$$P(B) = \sum_{i=1}^{n} P(B|A_i)P(A_i) \tag{2.3.16}$$

7.   Let $A_1 A_2 \cdots A_j$ be an exclusive but not exhaustive set of events and B any other event. We are interested in an expression for $P(B|\sum A_j)$. Applying (2.3.8) gives

$$P\left(B \left| \sum_{i=1}^{n} A_i \right.\right) = \frac{P(B \sum_{i=1}^{n} A_i)}{P(\sum_{i=1}^{n} A_i)} = \frac{P(\sum_{i=1}^{n} BA_i)}{P(\sum_{i=1}^{n} A_i)}$$

We now use (2.3.5).

$$P\left(B \left| \sum_{i=1}^{n} A_i \right.\right) = \frac{\sum_{i=1}^{n} P(BA_i)}{\sum_{i=1}^{n} P(A_i)}$$

Finally, we apply (2.3.8) again to the numerator.

$$P\left(B \left| \sum_{i=1}^{n} A_i \right.\right) = \frac{\sum_{i=1}^{n} P(B|A_i)P(A_i)}{\sum_{i=1}^{n} P(A_i)} \tag{2.3.17}$$

Equations (2.3.16) and (2.3.17) are rather specialized and typical of the large number of probability relations that can be worked out.

2.3.1 Conditions for Independence

Recall (2.2.5), repeated below, which defines full mutual independence for n simple probabilities.

$$P(A_i|A_j A_k \cdots A_m) = A_i \tag{2.3.18}$$

This represents a very large number of relations, and we shall demonstrate, without rigorous proof, that only $2^n - n - 1$ are functionally independent. Consider the case for $n = 3$. We gave $2^n - n - 1$ = 4 expressions representing mutual independence in (2.2.6), repeated below.

$$P(A_1|A_2) = P(A_1)$$

$$P(A_1|A_3) = P(A_1)$$

$$P(A_2|A_3) = P(A_2) \qquad (2.3.19)$$

$$P(A_1|A_2A_3) = P(A_1)$$

If A_1 is independent of A_2, then A_2 must be independent of A_1. From this it follows that

$$P(A_2|A_1) = P(A_2)$$

$$P(A_3|A_1) = P(A_3) \qquad (2.3.20)$$

$$P(A_3|A_2) = P(A_3)$$

The remaining expressions are

$$P(A_2|A_1A_3) = P(A_2)$$
$$\qquad (2.3.21)$$
$$P(A_3|A_1A_2) = P(A_3)$$

To illustrate that these can be derived from the first four, we first take (2.3.8) for events A and B and invert it, to give

$$P(A|B)P(B) = P(B|A)P(A) \qquad (2.3.22)$$

Applying this to the second of the expressions in (2.3.21) gives

$$P(A_3|A_1A_2)P(A_1A_2) = P(A_1A_2|A_3)P(A_3)$$

Using (2.3.8) on the right-hand side gives

$$P(A_3|A_1A_2)P(A_1A_2) = P(A_1|A_2A_3)P(A_2|A_3)P(A_3)$$

From independence expressions already defined this becomes

$$P(A_3|A_1A_2)P(A_1A_2) = P(A_1)P(A_2)P(A_3)$$

But by (2.3.10),

$$P(A_1A_2) = P(A_1)P(A_2)$$

So it is clear that

$$P(A_3 | A_1 A_2) = P(A_3)$$

We can demonstrate the remaining expression similarly.

An alternative way of defining full mutual independence is the use of (2.3.10). It must be written for all possible combinations of all possible numbers of the n simple events. Thus for n = 3, we have

$$P(A_1 A_2 A_3) = P(A_1)P(A_2)P(A_3)$$

$$P(A_1 A_2) = P(A_1)P(A_2)$$

$$P(A_1 A_3) = P(A_1)P(A_3) \qquad\qquad (2.3.23)$$

$$P(A_2 A_3) = P(A_2)P(A_3)$$

again giving $2^n - n - 1$ expressions.

2.4 GENERALIZATION OF PROBABILITY LAWS

We shall begin by defining rather carefully some different kinds of probability. A *simple probability* is the probability of a single event occurring without condition (except the sample space or the general conditions of the problem). $P(A_i)$ is a simple probability. A *joint probability* is the probability of the intersection of two or more events. $P(A_i A_j A_k)$ is a joint probability. A *combined probability* is the probability of the union of two or more events, such as $P(A_i + A_j + A_k)$. A *compound probability* is any probability involving more than one event. A *basic probability* is a particularly significant type of joint probability where *all* simple events in the sample space, or their complements, are joined. An example is $P(A_1 \bar{A}_2 A_3 A_4)$, where n = 4.

If we know all the simple probabilities, and they are all mutually independent, the prediction of almost any compound probability is quite straightforward, using the theorems developed in Sec. 2.3. If they are not all mutually independent, or if some compound probabilities are known rather than all the simple probabilities, the problem may become much more difficult if we have more than two events. There has been a tendency in the past to treat such problems as an exercise in ingenuity, in selecting the proper theorems from the maze available. However, a direct, systematic approach can be used.

First we must establish how many probabilities must be known to solve a problem completely. To do this we must work with basic probabilities. They have two important and valuable properties that we shall demonstrate below. They are:

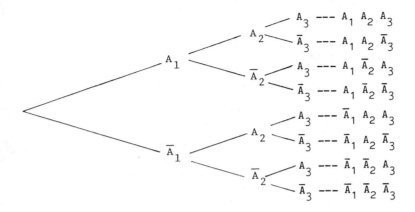

FIG. 2.6 Event network.

1. Any compound or simple probability can be expressed as a linear sum of the basic probabilities.
2. The basic probabilities are a mutually exclusive and collectively exhaustive set, and thus sum to 1.

The set of basic probabilities may be generated in two ways. The first is using an event network, shown in Fig. 2.6. Each basic event is generated by following a unique path. Thus each basic event is unique; and any elementary event that occurs in a basic event cannot occur jointly in any other basic event. Consider, for example, event $A_1A_2A_3$. All other basic events contain either $\overline{A}_1$, $\overline{A}_2$ or $\overline{A}_3$, and therefore any elementary event lying in $A_1A_2A_3$ cannot be in any other basic event. Thus basic events are mutually exclusive. Also, since it is clear from the way they are generated by the network that they are exhaustive, their probabilities must sum to 1.

The second way that they are generated has been suggested by Tribus (1969). The compound event

$$(A_1 + \overline{A}_1)(A_2 + \overline{A}_2) \cdots (A_n + \overline{A}_n)$$

is certain. Therefore, the probability of this event occurring is 1. We expand this with $n = 3$.

$$P[(A_1 + \overline{A}_1)(A_2 + \overline{A}_{+2})(A_3 + \overline{A}_3)] = 1 \qquad (2.4.1)$$

$$P[A_1A_2A_3 + A_1A_2\bar{A}_3 + A_1\bar{A}_2A_3 + A_1\bar{A}_2\bar{A}_3$$
$$+ \bar{A}_1A_2A_3 + \bar{A}_1A_2\bar{A}_3 + \bar{A}_1\bar{A}_2A_3 + \bar{A}_1\bar{A}_2\bar{A}_3] = 1 \qquad (2.4.2)$$

Since the basic events are exclusive, they may be written

$$P(A_1A_2A_3) + P(A_1A_2\bar{A}_3) + P(A_1\bar{A}_2A_3) + P(A_1\bar{A}_2\bar{A}_3)$$
$$+ P(\bar{A}_1A_2A_3) + P(\bar{A}_1A_2\bar{A}_3) + P(\bar{A}_1\bar{A}_2A_3) + P(\bar{A}_1\bar{A}_2\bar{A}_3) = 1$$

$$(2.4.3)$$

It is clear that 2^n basic probabilities are generated.

It is fairly obvious how any compound probability may be expressed in terms of the basic probabilities. Consider first $P(A_1)$. For A_1 to occur, we must have the occurrence of $A_1A_2A_3$ or $A_1A_2\bar{A}_3$ or $A_1\bar{A}_2A_3$ or $A_1\bar{A}_2\bar{A}_3$. Writing this logic in equation form gives

$$P(A_1) = P(A_1A_2A_3 + A_1A_2\bar{A}_3 + A_1\bar{A}_2A_3 + A_1\bar{A}_2\bar{A}_3) \qquad (2.4.4)$$

By (2.3.5) this is

$$P(A_1) = P(A_1A_2A_3) + P(A_1A_2\bar{A}_3) + P(A_1\bar{A}_2A_3) + P(A_1\bar{A}_2\bar{A}_3)$$

$$(2.4.5)$$

This is a generalization of (2.3.14). *We can generate an expression for any joint probability in a similar way, by scanning through the set of basic events and adding to the expression any that contain the given joint event.* For example,

$$P(A_2\bar{A}_3) = P(A_1A_2\bar{A}_3) + P(\bar{A}_1A_2\bar{A}_3) \qquad (2.4.6)$$

The logic for combined probabilities is equally clear. Consider the event $A_1 + A_2$. For this to occur we must have the occurrence of $A_1A_2A_3$ or $A_1A_2\bar{A}_3$ or $A_1\bar{A}_2A_3$ or $A_1\bar{A}_2\bar{A}_3$ or $\bar{A}_1A_2A_3$ or $\bar{A}_1A_2\bar{A}_3$, giving the expression

$$P(A_1 + A_2) = P(A_1A_2A_3) + P(A_1A_2\bar{A}_3) + P(A_1\bar{A}_2A_3)$$
$$+ P(A_1\bar{A}_2\bar{A}_3) + P(\bar{A}_1A_2A_3) + P(\bar{A}_1A_2\bar{A}_3) \qquad (2.4.7)$$

In this case as long as either A_1 or A_2 occur in a basic event it qualifies for the expression. *We can generate an expression for any combined probability by scanning through the set of basic events and adding to the expression any that contains any one of the simple events in the given combined event.*

This is sufficient to convert any compound probability into an expression for a sum of basic events. Conditional probabilities are first treated by (2.3.8). If we have $P(A_1\bar{A}_2|A_3)$, this becomes

$$P(A_1\bar{A}_2|A_3) = \frac{P(A_1\bar{A}_2A_3)}{P(A_3)} \qquad (2.4.8)$$

If the value of $P(A_1\bar{A}_2|A_3)$ is known to be X, the expression becomes

$$P(A_1\bar{A}_2A_3) - XP(A_3) = 0 \qquad (2.4.9)$$

or

$$-XP(A_1A_2A_3) + (1 - X)P(A_1\bar{A}_2A_3) - XP(\bar{A}_1A_2A_3) - XP(\bar{A}_1\bar{A}_2A_3) = 0$$

$$(2.4.10)$$

Since any unknown probability can be written in terms of basic probabilities, if we know all the basic probabilities, we have completely solved the problem. Since there are 2^n basic probabilities, this represents in general the number of unknowns to be determined, unless we happen to know by the terms of the problem one or more of the basic probabilities. To determine these unknowns we have one equation representing the fact that the basic probabilities must sum to 1. This is the single expression required to define the mathematical model. We must know then $2^n - 1$ additional quantities or relationships which, when written in terms of basic probabilities, provide enough equations to make the problem determinate. Thus if we have three simple probabilities in a problem, or n = 3, we must know $2^n - 1$ or seven probabilities or relationships of some kind related to the problem. If n = 2, we must know three measures of probability related to the problem.

Before examining specific examples, let us look at some common quantities that we may know. An important case is if all simple probabilities are independent. Recall that full mutual independence can be expressed by $2^n - n - 1$ equations of the type

$$P(A_i A_j \cdots A_m) = P(A_i)P(A_j) \cdots P(A_m) \qquad (2.4.11)$$

If we know all $P(A_1), P(A_2), \ldots, P(A_n)$, then we have $2^n - 1$ expressions, enough to make the problem determinate. In fact, in this case, any compound probability can be easily and directly obtained from the theorems, without recourse to solving for the set of basic probabilities. However, if all simple probabilities are now known, and some of the known measures are compound probabilities, then a direct solution is not possible. Even worse, equations represented by

(2.4.11) will be nonlinear in terms of basic probabilities, making the problem in general rather difficult to solve. However, ingenious use of the theorems may avoid the use of basic probabilities and the solution of a set of simultaneous nonlinear algebraic equations.

We may not necessarily have all simple events mutually independent. We might, for example, have only one independence relationship, such as

$$P(A_i A_j) = P(A_i)P(A_j) \tag{2.4.12}$$

Sufficient additional probability measures of any kind must be known to make up a total of $2^n - 1$.

It might be noted that (2.4.11) makes no mention of complementary events. The independence or otherwise of $\overline{A}_i$ is implied by the independence of A_i, since they have the fixed relationship

$$P(A_i + \overline{A}_i) = 1 \tag{2.4.13}$$

Therefore, no additional relationships are required, such as

$$P(\overline{A}_i \overline{A}_j) = P(\overline{A}_i)P(\overline{A}_j) \tag{2.4.14}$$

or

$$P(A_i \overline{A}_j) = P(A_i)P(\overline{A}_j) \tag{2.4.15}$$

However, there is a situation where the complementary event must be included. Suppose that we know from physical arguments that A_i is independent of A_j, *for a given value of* A_k. This yields *two* independence relationships that are functionally independent.

$$P(A_i A_j | A_k) = P(A_i | A_k)P(A_j | A_k)$$
$$P(A_i A_j | \overline{A}_k) = P(A_i | \overline{A}_k)P(A_j | \overline{A}_k) \tag{2.4.16}$$

It can be argued that these must be functionally independent since $\overline{A}_k$ is a different situation than A_k.

Other relationships are commonly derived from a situation where events are mutually exclusive. If the simple events are also collectively exhaustive, there will be $2^n - n - 1$ relationships based on (2.3.12).

$$P(A_i A_j \cdots A_m) = 0 \tag{2.4.17}$$

To set these up, we must take all possible combinations of n events taken n at a time, n events taken $n - 1$ at a time, and so on, down to n events taken two at a time. The number of these is determined from the theory of permutations and combinations in Appendix B. For example, if n = 4 we have

$$P(A_1A_2A_3A_4) = 0$$

$$P(A_1A_2A_3) = 0 \qquad P(A_1A_2) = 0 \qquad P(A_2A_4) = 0$$

$$P(A_1A_2A_4) = 0 \qquad P(A_1A_3) = 0 \qquad P(A_3A_4) = 0$$

$$P(A_1A_3A_4) = 0 \qquad P(A_1A_4) = 0$$

$$P(A_2A_3A_4) = 0 \qquad P(A_2A_3) = 0$$

We must know n additional probability measures or relationships for the problem to be determinate. If, for example, we know all simple probabilities, any compoind probability can be directly derived using the theorems.

It should be noted that it is not always necessary to know $2^n - 1$ probability relationships or measures to solve for a particular probability, although the whole problem may not be determinate. For example, let n = 6 and all simple events be independent. If we are interested only in estimating $P(A_1A_3A_5)$, we need only know $P(A_1)$, $P(A_3)$, and $P(A_5)$, since by (2.3.9),

$$P(A_1A_3A_5) = P(A_1)P(A_3)P(A_5)$$

It is clear that for high-order problems we must resort to the digital computer, and even its memory capacity will soon be strained. If n = 10, we must solve 984 simultaneous equations. However, it may be possible to reduce the order by decomposition or by separation.

In the computer solution it is convenient to have available a subroutine to generate the basic probabilities; a subroutine to express any joint probability in terms of basic probabilities; a subroutine to express any combined probability in terms of basic probabilities where each term is a simple probability; a subroutine for expressing all possible combinations of n events taken n at a time, n − 1 at a time, and so on; and a subroutine to generate factorials. These are given in Appendix A. The main program must be written to suit a given type of problem. If a compound probability is a combined type with each term a joint probability, each term may be separately transformed to basic probabilities and the results simply added.

The reader may at this point be worrying about how we would in practice determine the "known" probability measures in a problem. We have implied how this may be done in our discussion of the meaning of probability in Sec. 2.1. We may use frequency counts from observed data, or estimate probabilities purely by judgment, or some

combination of both. This is discussed in more detail in the following chapters.

The use of the theory discussed above is illustrated in the following examples. The first three examples do not use basic probabilities.

EXAMPLE 2.4

We begin with an abstract example to illustrate concepts. We
have a population of five discs marked 1, 3, 7, 8, and 13. We
select two discs without replacement. The events that we are
interested in are as follows:

A = both numbers are greater than 5

B = both numbers are odd

C = the sum of the numbers exceeds 12

We wish to determine $P(A)$, $P(B)$, $P(C)$, $P(B + C)$, $P(A|B)$,
$P(B|C)$, and $P(C|A)$.

The following elementary events may occur:

1,3   3,7   7,8   8,13

1,7   3,8   7,13

1,8   3,13

1,13

These are *not* the simple events. We may define the simple
events A, B, and C to suit our problem. This choice of simple
events may be quite a difficult one. Often, however, the elementary events are the simple events.

We scan these 10 elementary events for the occurrence of
A, B, and C, giving the following:

A: 7,8   7,13   8,13
B: 1,3   1,7   1,13   3,7   3,13   7,13
C: 1,13   3,13   7,8   7,13   8,13

Using a frequency count, we obtain

$P(A) = 0.3$, $P(B) = 0.6$, $P(C) = 0.5$

The compound events are

```
   AB:   7,13
B + C:   7,8   7,13   8,13   1,3   1,7   1,13   3,7   3,13
 A|B:   7,13
 B|C:   1,13   3,13   7,13
 C|A:   7,8   7,13   8,13
```

Again using a frequency count, we get

$$P(AB) = 0.1$$

$$P(B + C) = 0.8$$

$$P(A|B) = 1/6$$

$$P(B|C) = 3/5$$

$$P(C|A) = 3/3$$

EXAMPLE 2.5 Sheet Metal Tolerances

A component is made up of two pieces of sheet metal of the same nominal thickness, spot-welded together. The event of any piece of sheet metal being received from the steel mill within tolerance is A. We know from specification that

$$P(A_1) = P(A_2) = P(A) = 0.992$$

where A_1 and A_2 correspond to the particular two pieces used in the part. Because there is a high likelihood that both pieces will be made from the same stock sheet, A_1 and A_2 are not independent events. Thus we may also know that

$$P(A_1|A_2) = P(A_2|A_1) = 0.999$$

We would expect that if piece 1 is good, there would be a higher probability of piece 2 being good.

We wish to estimate the probability of the part being good, or $P(A_1A_2)$. We can apply (2.3.8):

$$P(A_1A_2) = P(A_1|A_2)P(A_2)$$

$$= 0.999 \times 0.992 = 0.9910$$

EXAMPLE 2.6 Car Wheel Bolts

There are five wheel bolts on a car. The probability of losing any one bolt is 0.00001. If more than one bolt is lost, the wheel will come off. We wish to estimate the risk of this catastrophic event. The elementary events are A_1, A_2, A_3, A_4, and A_5— the loss of each bolt.

$$P(\text{lost wheel}) = (A_1A_2 + A_1A_3 + A_1A_4 + A_1A_5 + A_2A_3$$

$$+ A_2A_4 + A_2A_5 + A_3A_4 + A_3A_5 + A_4A_5)$$

We now assume that the elementary events are independent, and recognize that A_1A_2, A_1A_3, and so on, are mutually exclusive events. We therefore use (2.3.10) and (2.3.5) to get

$$P(\text{lost wheel}) = P(A_1)P(A_2) + P(A_1)P(A_3) + \cdots + P(A_4)P(A_5)$$

$$= 10 \times 0.0001^2 = 10^{-9} \text{ in 1 mile}$$

A practical interpretation would be that we would expect, on the average, one lost wheel in 10^9 vehicle miles.

EXAMPLE 2.7 Device Failure

We wish to predict the probability that a device will be operated correctly. Successful operation is certain if the operator remembers to actuate two switches. It requires skill, and success is uncertain if he forgets to actuate one or both switches. The following events are defined.

A_1 = switch 1 is not actuated

A_2 = switch 2 is not actuated

A_3 = operation of device fails

Since n = 3, there must be seven known probability relations. The following probabilities are determined by judgment and trials or be definition.

$$P(A_1|\overline{A}_2) = 0.35$$

$$P(A_2|\overline{A}_1) = 0.35$$

$$P(A_1|A_2) = 0.73$$

$$P(A_3|A_1\overline{A}_2) = 0.35$$

$$P(A_3|A_2\overline{A}_1) = 0.35$$

$$P(A_3|A_1A_2) = 0.66$$

$$P(A_3|\overline{A}_1\overline{A}_2) = 0$$

We require $P(\overline{A}_3)$.

In this example we shall apply the general method using basic probabilities. We first convert all conditional probabilities.

$$P(A_1|\overline{A}_2) = \frac{P(A_1\overline{A}_2)}{P(\overline{A}_2)} = 0.35$$

or

$$P(A_1\overline{A}_2) - 0.35P(\overline{A}_2) = 0$$

and similarly

$$P(A_2\overline{A}_1) - 0.35P(\overline{A}_1) = 0$$

$$P(A_1A_2) - 0.73P(A_2) = 0$$

$$P(A_3A_1\overline{A}_2) - 0.35P(A_1\overline{A}_2) = 0$$

$$P(A_3A_2\overline{A}_1) - 0.35P(A_2\overline{A}_1) = 0$$

$$P(A_3A_1A_2) - 0.66P(A_1A_2) = 0$$

$$P(A_3\overline{A}_1\overline{A}_2) = 0$$

Converting to basic probabilities gives

$$0.65P(A_1\overline{A}_2A_3) + 0.65P(A_1\overline{A}_2\overline{A}_3) - 0.35P(\overline{A}_1\overline{A}_2A_3)$$
$$- 0.35P(\overline{A}_1\overline{A}_2\overline{A}_3) = 0$$

$$0.65P(\overline{A}_1A_2A_3) + 0.65P(\overline{A}_1A_2\overline{A}_3) - 0.35P(\overline{A}_1\overline{A}_2A_3)$$
$$- 0.35P(\overline{A}_1\overline{A}_2\overline{A}_3) = 0$$

$$0.27P(A_1A_2A_3) + 0.27P(A_1A_2\overline{A}_3) - 0.73P(\overline{A}_1A_2A_3)$$
$$- 0.73P(\overline{A}_1A_2\overline{A}_3) = 0$$

$$0.65P(A_1\overline{A}_2A_3) - 0.35P(A_1\overline{A}_2\overline{A}_3) = 0$$

$$0.65P(\overline{A}_1A_2A_3) - 0.35P(\overline{A}_1A_2\overline{A}_3) = 0$$

$$0.33P(A_1A_2A_3) - 0.66P(A_1A_2\overline{A}_3) = 0$$

$$P(\overline{A}_1\overline{A}_2A_3) = 0$$

$$P(A_1A_2A_3) + P(A_1A_2\overline{A}_3) + P(A_1\overline{A}_2A_3) + P(A_1\overline{A}_2\overline{A}_3) + P(\overline{A}_1A_2A_3)$$
$$+ P(\overline{A}_1A_2\overline{A}_3) + P(\overline{A}_1\overline{A}_2A_3) + P(\overline{A}_1\overline{A}_2\overline{A}_3) = 1$$

We can put these expressions in matrix form.

$$\begin{bmatrix} 0 & 0 & 0.65 & 0.65 & 0 & 0 & -0.35 & -0.35 \\ 0 & 0 & 0 & 0 & 0.65 & 0.65 & -0.35 & -0.35 \\ 0.27 & 0.27 & 0 & 0 & -0.73 & -0.73 & 0 & 0 \\ 0 & 0 & 0.65 & -0.35 & 0 & 0 & 0 & 0 \\ 0 & 0 & 0 & 0 & 0.65 & -0.35 & 0 & 0 \\ 0.33 & -0.66 & 0 & 0 & 0 & 0 & 0 & 0 \\ 0 & 0 & 0 & 0 & 0 & 0 & 1 & 0 \\ 1 & 1 & 1 & 1 & 1 & 1 & 1 & 1 \end{bmatrix} \begin{bmatrix} A_2A_2A_3 \\ A_1A_2\overline{A}_3 \\ A_1\overline{A}_2A_3 \\ A_1\overline{A}_2\overline{A}_3 \\ \overline{A}_1A_2A_3 \\ \overline{A}_1A_2\overline{A}_3 \\ \overline{A}_1\overline{A}_2A_3 \\ \overline{A}_1\overline{A}_2\overline{A}_3 \end{bmatrix}$$

$$= [0\ 0\ 0\ 0\ 0\ 0\ 0\ 1]^T$$

Using a standard numerical technique, we obtain the following solution for the basic probabilities.

$$P(A_1A_2A_3) = 0.2720 \qquad P(\overline{A}_1A_2A_3) = 0.05335$$

$$P(A_1A_2\overline{A}_3) = 0.1401 \qquad P(\overline{A}_1A_2\overline{A}_3) = 0.09907$$

$$P(A_1\overline{A}_2A_3) = 0.05335 \qquad P(\overline{A}_1\overline{A}_2A_3) = 0.0000$$

$$P(A_1\overline{A}_2\overline{A}_3) = 0.09907 \qquad P(\overline{A}_1\overline{A}_2\overline{A}_3) = 0.2831$$

The desired probability is

$$P(\overline{A}_3) = P(A_1A_2\overline{A}_3) + P(A_1\overline{A}_2\overline{A}_3) + P(\overline{A}_1A_2\overline{A}_3) + P(\overline{A}_1\overline{A}_2\overline{A}_3)$$

$$= 0.6213$$

2.5 BAYES' THEOREM

Bayes' theorem is a special probability relationship derived as follows. For two events A and B, (2.3.8) gives

$$P(A|B) = \frac{P(AB)}{P(B)}$$

and

$$P(B|A) = \frac{P(AB)}{P(A)}$$

Eliminating P(AB) between these gives

$$P(A|B) = \frac{P(A)P(B|A)}{P(B)} \tag{2.5.1}$$

This is the simplest version of Bayes' theorem; and it is usually amplified by a further definition of the events. P(A) is defined as a prior probability for A, which is its probability of occurring if we do not know B, or have not observed B. P(A|B) is called the *posterior probability* of A occurring, which is its probability of occurring after we have observed that B has occurred. There is really nothing special about Bayes' theorem. It is a convenient formulation of a probability law; but it solves nothing that cannot be solved by the general method of the previous sections.

If one accepts the concept of subjective probability, it is possible to conceive a situation where P(A) is defined subjectively without support of frequency data. Event B is defined as new data that become available which provide evidence about the occurrence of A. We wish to determine the updated or posterior probability P(A|B). Bayes' theorem can be interpreted as applying to this situation and conveniently used to calculate P(A|B). For this reason statisticians who accept the concept of subjective probability have historically been called Bayesian statisticians. The term is somewhat misleading because there is still nothing inherently special about Bayes' theorem as a probability law, even in this context. It is simply a convenient rule for solving a particular type of problem in which certain probabilities are known. The fact that one of them is subjective makes no difference. Once again, the general method could be used rather than Baye's theorem.

The formal Bayesian method used in statistical inference is a further development which is open to considerable question, even for a person wholly committed to the use of subjective probability. This is discussed further in Sec. 3.6.

The most common usage of the theorem is in a situation where A is one of a set of mutually exclusive events, A_i. Equation (2.5.1) now has the form

$$P(A_i|B) = \frac{P(A_i)P(B|A_i)}{P(B)} \tag{2.5.2}$$

P(B) can often be more easily evaluated if it is put in the following form, using (2.3.16).

$$P(A_i|B) = \frac{P(A_i)P(B|A_i)}{\sum_{j=1}^{n} P(B|A_j)P(A_j)} \qquad (2.5.3)$$

EXAMPLE 2.8 Part Testing

The measurement problem is a common example in which the prior probability is known. A manufactured part has a known acceptance rate of 0.95 if the testing method is certain. However, a go-nogo measuring device is being used which is only 99% reliable. That is, there is a 1% chance that it will accept a bad part, and a 1% chance that it will reject a good part. What is the chance of a good part being accepted?

We define our events as

A = a part is good

B = a part is accepted as good

In order to use (2.5.3) we further define A as

A_1 = a part is good

A_2 = a part is bad

The known probabilities are

$P(B|A_1) = 0.99$

$P(B|A_2) = 0.01$

$P(A_1) = 0.95$

$P(A_2) = 0.05$

Now using (2.5.3) we get the required probability.

$$P(A_1|B) = \frac{0.95 \times 0.99}{0.99 \times 0.95 + 0.01 \times 0.05} = 0.99944$$

It is important to note again that in the use of Bayes' theorem above, it is simply another probability law. The same result could be achieved by using the general method of Sec. 2.4. This example is also useful in making clear the distinction between the events $A_1|B$ and $A_1 B$. The former is the event that a part is good, knowing that it has been measured as good. The second event, $A_1 B$, is the event that a part is good *and* it has been measured as good. The distinction is a subtle and difficult one. We can use (2.3.8) to obtain $P(A_1 B)$:

$$P(A_1 B) = P(BA_1) = P(B|A_1)P(A_1) = 0.99 \times 0.95 = 0.9405$$

and also to get $P(B)$:

$$P(B) = \frac{P(A_1 B)}{P(A_1|B)} = \frac{0.0405}{0.99944} = 0.94103$$

It will perhaps help to understand why $P(A_1|B)$ and $P(A_1 B)$ are different if we think in terms of observed frequencies. $P(A_1 B)$ tells us that out of 10,000 parts examined, we would expect on the average 9405 to be both good and measured as good. Whereas $P(A_1|B)$ tells us that out of 9410 measured as good, we would expect 0.99944×9405 or 9399 actually to be good.

We should also note that the probability of a part being good, $P(A)$, could have been subjective, and nothing would be changed except that Bayes' theorem would now be interpreted as an update of $P(A)$ given one observation B.

A confusing aspect of Bayes' theorem is that it is restricted to a rather narrow class of problems, whereas authors commonly imply that it is a powerful and general method of updating probability estimates. One would like to use it in the more general situation where we have a subjective probability $P(A)$, and B is the event of observing a sample of events corresponding to A. Thus B equals the set $a_1 a_2 a_3 \cdots a_m$, where a_i is an observed occurrence or nonoccurrence of A. Using Bayes' rule gives

$$P(A|B) = \frac{P(A)P(B|A)}{P(B|A)P(A) + P(B|\overline{A})P(\overline{A})}$$

What can we do with $P(B|A)$? It could be written

$$P(B|A) = P(a_1 a_2 a_3 \cdots a_m|A) = P(a_1|A)P(a_2|A) \cdots P(a_m|A)$$

We would describe $a_i|A$ as the event of observing A has occurred, if it has occurred. But $P(a_i|A)$ is 0 if a_i is a nonoccurrence of A, and 1 if it is an occurrence of A. Bayes' theorem is clearly meaningless in this context. It does become meaningful only if there is some uncertainty in the observation, so that $P(a_i|A)$ is not 0 or 1.

PROBLEMS

Problem 2.1 We wish to estimate the probability of any individual having a flat tire and being able to change it. Assume that the event of a flat tire occurring in 50,000 miles is A, and $P(A)$ is known. We also assume that we know $P(B\,|\,C)$ and $P(B\,|\,\overline{C})$, where B is the event that the driver can change it and C is the event that the driver is a man. Give the expression for the required probability.

Problem 2.2 In Example 2.2, what is the value of $P(A_1|\overline{A}_2)$?

Problem 2.3 A company wishes to estimate the probability that a given device will fail during the warranty period. The following events are defined.

F = failure of a device

C = failure claimed by customer

R = device returned to manufacturer by customer

I = returned device inspected by manufacturer

The manufacturer does not require the customer to return all devices that the customer perceives as failed because of the cost involved. The manufacture also does not inspect all returned parts that are presumed failed by the customer, again for reasons of cost. The customer may be in error in perceiving the device as failed; it may simply require adjustment, or proper operation, or an adjoining device may actually be the one at fault. Also, the manufacturer's inspection may be in error and incorrectly judge a failed device as good, or vice versa. In these terms the required probability is $P(F)$, which can be thought of as a corrected value for $P(C)$.

The manufacturer has frequency records as follows:

S   = total sales volume

N_R = number actually returned

N_C = failures claimed by customer

N_I = number inspected

N_{IF} = number inspected that are judged failed

The manufacturer is also able to estimate the probability that its inspection will pass as nonfailed, a device that really has not failed, and also the probability that the inspection will pass as okay a device that really has failed. (This problem is adapted from a paper by Yun and Kalivoda.*)

Problem 2.4 A designer is trying to decide on the best design for a particular job. Two different designs are available, A and B. They are being compared on the basis of three important variables; efficiency, capacity, and cost. The designer is able to define the following probabilities.

$P(A_1)$ = probability that A has greater efficiency than B

$P(A_2)$ = probability that A has greater capacity than B

$P(A_3)$ = probability that A costs less than B

He also has good reason to believe that efficiency is independent of capacity for a given cost, but no other independent relationship can be assumed. He is able to estimate the following probability measures.

$P(A_1|A_3) = 0.45$

$P(A_1|\overline{A}_3) = 0.65$

$P(A_2|A_3) = 0.80$

$P(A_2|\overline{A}_3) = 0.60$

$P(A_3) = 0.45$

Which design should he choose?

Problem 2.5 Two types of pile-driving machines are available, A and B. There are four soil types in which they may operate. The probabilities of the pile drivers operating successfully in each type of soil has been established by experience as shown in the following table. The probability of any particular soil being encountered is also shown in the table.

*K. W. Yun and F. E. Kalivoda (1977), A Model for Estimation of the Product Warranty Return Rate, *Proc. 1977 Annual Reliability and Maintainability Symp.*, pp. 31-37.

	Soil 1	Soil 2	Soil 3	Soil 4
Design A	0.910	0.623	0.715	0.316
Design B	0.370	0.500	0.765	0.875
Probability of soil occurring	0.317	0.186	0.132	0.365

What are the probabilities that design A and design B will work in any situation where conditions are unknown?

REFERENCES

Ang, A. H. S., and W. H. Tang (1975), *Probability Concepts in Engineering Planning and Design*, Vol. 1: *Basic Principles*, Wiley, New York.

Churchman, C. W. (1961), *Prediction and Optimal Decision: Philosophical Issues of a Science of Values*, Prentice-Hall, Englewood Cliffs, N.J.

Good, I. J. (1965), *The Estimation of Probabilities*, MIT Press, Cambridge, Mass.

Lindley, D. V. (1965), *Introduction to Probability and Statistics from a Bayesian Viewpoint*, Part 1: *Probability*, Cambridge University Press, Cambridge.

Savage, L. J. (1954), *The Foundation of Statistics*, Wiley, New York.

Schlaifer, R. (1961), *Introduction to Statistics for Business Decisions*, McGraw-Hill, New York.

Siddall, J. N. (1982), *Optimal Engineering Design: Principles and Applications*, Marcel Dekker, New York.

Tribus, M. (1969), *Rational Descriptions, Decisions, and Designs*, Pergamon, New York.

3

Probability Distributions

The concept of a random variable and of probability functions in the context of design • Characteristic measures in distributions • Commonly used analytical functions for discrete and continuous random variables • Multiple random variables • The concept of an infinite-simal event to represent the event of a specific value of a continuous random variable

A discussion of Bayes' theorem as applied to density functions

3.1 THE CONCEPT OF A RANDOM VARIABLE

The concept of a random variable should be examined by first referring back to Sec. 1.1 and Fig. 1.1, in which the hierarchy of design variables was illustrated. We saw that the designer specified directly the *independent design variables*. For example, he might designate a shaft diameter to be 3.250 in. However, he recognizes that he must accept some uncertainty in this size, so he also specifies bounds on the uncertainty in the form of tolerances. Or he might specify a particular steel for a part, and again recognize the inevitability of uncertainty in the yield stress value by specifying a guaranteed minimum yield stress.

Thus we must distinguish between design variation, in which the designer varies the nominal value of a design variable to achieve an optimum design, and stochastic variation, which is a variation about the nominal value which the designer can control to some extent but not predict. Returning to our illustration, the designer can require that the shaft diameter can vary only within the bounds 3.249 to 3.255, but he cannot predict what measured diameter a specific shaft manufactured next month will have.

In our hierarchy of design variables, we also have *design char-acteristics*, which are design variables dependent on the primary or independent quantities. The designer predicts the values of these characteristics by using engineering modeling. Since they are a func-tion of variables that can be random, they also can be random. The prediction of their random nature is the topic of *probabilistic analysis*, discussed in Chapter 6. An example is the amount of steam condensed per hour in a condenser. The temperature of the cooling water, the quantity of the steam, the flow rate of the cooling water, and the roughness of the fluid passages are all random independent variables that affect the randomness of the condensation rate.

In addition to random variables embodied physically in a device or system, the engineer is also concerned with random quantities used to measure the environment in which a device operates. An example is the loading on a structure, which may be applied at random inter-vals in random amounts, such as wind loading on a building, or the force on a bulldozer blade. Other examples are ice loading on trans-mission lines, corrosive conditions in which a device operates, and wave heights impacting on a drilling tower.

These state of nature random variables are quite different in type from design random variables, and correspond to the traditional type of random variable with which scientists work in observing na-ture. They either are, or can be derived from, an existing population that can be observed. The amount of data that is observed depends to some extent on accessibility, but also on the resources available to make the measurements.

Design random variables are quite different in nature; since they represent random quantities that may not even exist. Their random nature may be inferred by comparison with existing similar parts, or by tests of prototype parts, or purely by judgment.

Random variables can be discrete or continuous. Discrete ran-dom design variables are rather rare in the design of components or devices, but are not uncommon in systems design. Environmental ran-dom discrete variables also occur. Examples are the number of defec-tive parts in 1 day's production, the number of loads carried by a truck in a week, or the number of harvester combines out of a pro-duction run of 10,000 that had a field breakdown in the first season. Examples of continuous random variables are innumerable: yield stress for a steel alloy, diameter of a shaft, resistance of a resistor, maximum horsepower of an engine, and so on.

In engineering our concern is with the probability of occurrence of events, and we must relate random variables to events. If the variable is discrete, the event is the occurrence of any given value, such as 612 combines breaking down in the first season. The set of events defined by the random variable is predetermined in size and mutually exclusive and collectively exhaustive. On the other hand, for continuous variables, we cannot enumerate a finite number of events having a specific value, and therefore an event is defined as the occurrence of a value within a specified interval. A set of events in this case need not be defined. We may only be interested, for example, in the event that a continuous variable have a value greater than some specification.

3.2 PROBABILITY FUNCTIONS

3.2.1 The Concept of a Probability Density Function

Because of the fundamental importance of the probability density function in probabilistic design it is extremely important that its concept be thoroughly understood. It is the basic tool for codifying and communicating uncertainty about the value of a continuously varying variable.

There are several ways of developing the concept of a density function. First of all, it can be related to a histogram. If a sample of values of a random variable is analyzed to give frequencies in specified intervals, we can plot the results as a histogram, as shown in Fig. 3.1. The intervals must span the range of the sample, and the number of observations or values of x occurring in each interval is counted. The number of intervals is arbitrary.

If the ordinates are divided by the sample size, we obtain relative frequencies, as shown in Fig. 3.2. This histogram can be used to estimate probabilities of x having a value in a *specified interval*. Thus, in the i-th interval, R_i is an estimate of the fraction of times that x can be expected to have a value in the i-th interval, or the probability that x will have such a value. Similarly, x will have an estimated probability of occurring within the first three intervals of

$$P(x \text{ within first three intervals}) = R_1 + R_2 + R_3 \qquad (3.2.1)$$

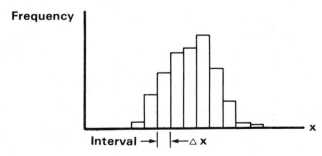

FIG. 3.1 Frequency histogram.

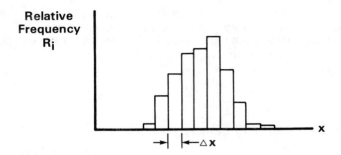

FIG. 3.2 Relative frequency histogram.

We will improve the accuracy of the estimate if we draw a smooth
curve through the midpoints of the bars, as shown in Fig. 3.3 (see
subroutine SMOOTH in Appendix A). This also has the advantage that
we can work with intermediate points such as x_s. With this in view we
rewrite (3.2.1) as

P(x having values within first three intervals)

$$= \sum_{i=1}^{3} \frac{R_i \, \Delta x}{\Delta x} = \frac{\text{area of first three bars}}{\Delta x} \qquad (3.2.2)$$

Following this we could write a similar expression for any point x_s,
and use the smoothed curve.

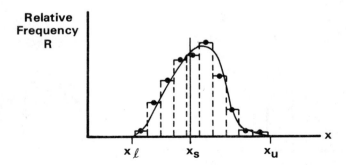

FIG. 3.3 Smoothed relative frequency curve.

$$P(x < x_s) = \int_{x_\ell}^{x_s} \frac{R \ dx}{\Delta x}$$
(3.2.3)

It would clearly be desirable to generalize the curve, and make it more convenient to use, if we change the vertical scale of Fig. 3.3 to a new quantity $f(x)$ related to R by

$$f(x) = \frac{R}{\Delta x}$$
(3.2.4)

Now (3.2.3) becomes

$$P(x < x_s) = \int_{x_\ell}^{x_s} f(x) \ dx$$
(3.2.5)

The total area under the histogram is

$$\sum_{i=1}^{m} R_i \ \Delta x = \Delta x$$
(3.2.6)

and the total area under the density function curve is the area under the histogram divided by Δx.

$$\int_{-\infty}^{\infty} f(x) \ dx = 1$$
(3.2.7)

The function $f(x)$ can be used in a similar way to obtain the probability of x having values in any interval, say $a \leqslant x < b$.

$$P(a \leqslant x < b) = \int_{a}^{b} f(x) \ dx$$
(3.2.8)

It now becomes clear that the probability density function is an elegant way of codifying the random nature of x—if there is a high likelihood of x having values in a certain region, the curve will have high values over this region.

Many authors in probability texts axiomatically define the density function as a function with the following properties:

$$\int_{-\infty}^{\infty} f(x) \ dx = 1$$
(3.2.9)

where $f(x)$ is single valued and nonnegative. They go on to say that if x is thought of as a quantity (variable) that can randomly have any value (possibly within bounds), then the event of this quantity being in the interval $a \leqslant x < b$ is said to have the probability

$$P(a \leqslant x \leqslant b) = \int_{a}^{b} f(x) \, dx \qquad (3.2.10)$$

Others begin with a definition of the cumulative distribution function.

$$F_{x}(x_{i}) = P(x \leqslant x_{i}) \qquad (3.2.11)$$

The probability density function $f(x)$ is then defined as

$$f(x) = \frac{dF(x)}{dx} \qquad (3.2.12)$$

and

$$P(x < x_{i}) = F(x_{i}) = \int_{-\infty}^{x_{i}} f(x) \, dx \qquad (3.2.13)$$

This relationship is illustrated in Fig. 3.4.

It is sometimes necessary when working with density function notation to show an argument other than the random variable. In this

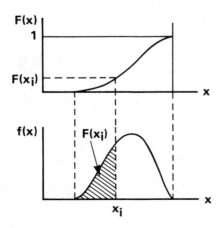

FIG. 3.4 Relationship between the cumulative distribution function and the probability density function.

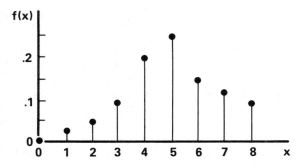

FIG. 3.5 Discrete distribution.

context the random variable will be shown as a subscript to avoid ambiguity. For example, $f_{x_1}(x_2/a)$ represents the probability density function for x_1 when x_1 equals x_2/a.

3.2.2 The Probability Function for a Discrete Variable

If the random variable is discrete, the probability distribution is represented directly by a probability mass function, as illustrated in Fig. 3.5. The function $f(x)$ now represents the actual probability of the value x occurring.

3.2.3 Characteristic Measures of a Random Variable

The random nature of a variable is commonly represented in a limited way by a central measure and a "scatter" measure. The central measure may be the mean, median, or mode.

The *mean* value is a weighted average, in which the weighting factors are the probabilities associated with each value. The mean is commonly designated μ, and for a discrete variable is defined by

$$\mu = \sum_{i=1}^{n} x_i f(x_i) \qquad (3.2.14)$$

where

        n = set size

        x_i = i-th discrete value

        $f(x_i)$ = probability mass function

For the continuous variable, it is perhaps easier to first think of it as having n equal discrete intervals over its range. Now the mean is approximately

$$\mu = \sum_{i=1}^{n} x_i f(x_i) \, \Delta x \tag{3.2.15}$$

where x_i is now the midinterval value of x. This expression can be made exact by converting to an integral.

$$\mu = \int_R xf(x) \, dx \tag{3.2.16}$$

The mean can also be called the expected value of x, designated $E(x)$. We shall see later that is a special case of the more general concept of an expected value.

The *mode* is the most likely value of the random variable, and corresponds to the maximum value of the probability density function, or probability mass function for a discrete variable. A density function can be multimodal, as shown in Fig. 3.6. This usually results from combining two different populations.

The *median* is that value of the random variable for which any other sampled value is equally likely to be above or below. Mathematically it is defined by

$$0.5 = \int_{-\infty}^{\check{x}} f(x) \, dx = F(\tilde{x}) \tag{3.2.17}$$

where $\check{x}$ is the median. This can be generalized to give fractiles or percentiles.

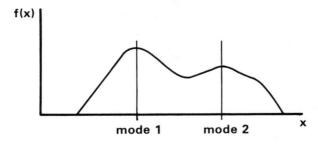

FIG. 3.6 Bimodal density function.

$$\xi = \int_{-\infty}^{\check{x}_\xi} f(x) \ dx = F(\tilde{x}_\xi) \tag{3.2.18}$$

Thus $x_{0.1}$ is the value for x corresponding to a cumulative distribution function value of 0.10, or 10% probability that a sampled x value will be less than $x_{0.1}$; and it is called the one-tenth factile, or 10 percentile.

The *standard deviation*, σ, is the commonly used characteristic measure of dispersion, or "width" of the density function. It is defined by

$$\sigma = \left[\int_R (x - \mu)^2 f(x) \ dx \right]^{1/2} \tag{3.2.19}$$

The square of the deviation is the variance or second central moment. The concept of moments is discussed in Chapter 4. It is sometimes more meaningful to use a relative deviation, usually called the *coefficient of variation*.

$$\Delta = \frac{\sigma}{\mu} \tag{3.2.20}$$

The *upper and lower bounds* should also be included in the characteristic measures, defined in Fig. 3.7.

3.3 THEORETICAL DISCRETE DISTRIBUTIONS

3.3.1 Occurrence or Nonoccurrence of an Event

Discrete distributions are generated by examining probabilities associated with the occurrence or nonoccurrence of a given specified event,

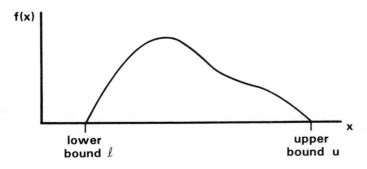

FIG. 3.7 Bounds of a random variable.

or a sequence of events. Certain assumptions are conveniently made about the occurrence of these events, which often model closely real-life events, and which lead to several simple analytical distributions.
    The assumptions are the following:

1. Each event has only two outcomes, occurrence and nonoccurrence.
2. If there are a sequence of events, or each event can occur at different times or in different places, the probability of each occurring is the same.
3. Different events of a given type are statistically independent.

    Mathematicians sometimes call the events "trials," and if they satisfy the assumptions listed above, they are called *Bernoulli trials*.
    These particular kinds of events may be used to model engineering situations such as:

1. A batch of 10,000 components are being manufactured, each with a known probability of p of being incorrectly manufactured. What is the probability that 90% of them will be satisfactory?
2. A company sells 2,000 new combine harvesters, and has estimated, using reliability theory, that the probability of no failure for the first season for each is 0.90. The marketing group believes that the company's reputation for reliability will remain intact if no more than 25% have a breakdown in the first season. What is the probability of this being exceeded?
3. Extreme value theory can be used to estimate the probability that a device will be subjected to a load exceeding a specified load in any single year. What is the probability that 10% of the devices will be exposed to a greater load over a 5-year period?

3.3.2 The Binomial Distribution

The models described above are generalized by the binomial distribution, which gives the probability of occurrences when there are n possible events or trials, and the known probability of occurrence in each individual trial is p. The random variable is x, the discrete number of occurrences.
    The number of sequences in which the event occurs x times is developed from combinatorial theory in Appendix B, and equals

$$\binom{n}{x} = \frac{n!}{x! \, (n - x)!} \tag{3.3.1}$$

The probability of getting x occurrences in any one sequence is based on the product rule for combined independent events. If the event occurs in the first x trials,

P(x events occurring)

= P(occurring in trial 1) · P(occurring in trial 2) ··· P(occurring in trial x) · P(not occuring in trial x + 1) ··· P(not occurring in trial n)

$$= p^x(1 - p)^{n-x} \qquad (3.3.2)$$

We can now use the summation rule for mutually exclusive events to combine (3.3.1) and (3.3.2) to obtain the *binomial distribution*

$$f(x; p, n) = \binom{n}{x} p^x (1 - p)^{n-x} \qquad (3.3.3)$$

where the notation f(x; p, n) means the density function of a random variable x with distribution parameters p and n. It can be shown that this distribution has the following mean and variance:

$$\mu = np, \quad \sigma^2 = np(1 - p) \qquad (3.3.4)$$

EXAMPLE 3.1 Device Failure

A device contains n "identical" components, or in statistical terms components from the same population. For a given life, the survival of the device requires no failure of at least r of the n components. We model this situation using the binomial distribution for the components, letting k be the general number of successes in n trials, where a trial is the use of a component in the device. If we require that at least r out of n components in the device survive for nonfailure of the device, the successful event is

no failure of 4 components

or no failure of 4 + 1 components
: :
or no failure of n components

This can be represented by the cumulative probability distribution for the binomial.

$$P(\text{no device failure}) = \sum_{k=r}^{n} \binom{n}{k} p^k (1 - p)^{n-k} \qquad (3.3.5)$$

where p is the probability of no failure for each component.

3.3.3 The Multinomial Distribution

The concept of Bernoulli trials can be extended to more than two outcomes to a trial. We now have k mutually exclusive and collectively

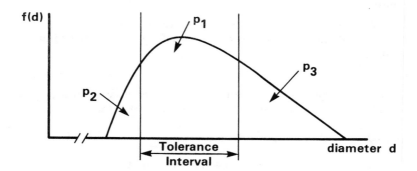

FIG. 3.8 Probability density function for shaft diameters.

exhaustive outcomes with known probabilities $p_1, p_2, \ldots, p_k$. The trials are still independent. The probability mass function is multivariable.

$$f(x_1, x_2, \ldots, x_k; n, p_1, p_2, \ldots, p_n)$$

$$= \frac{n!}{x_1! x_2! \cdots x_k!} \, p_1^{x_1} p_2^{x_2} \cdots p_n^{x_n} \qquad (3.3.6)$$

where x_i is the number of occurrences of the i-th outcome. A derivation of this expression can be found in Hahn and Shapiro (1967).

EXAMPLE 3.2   Shaft Diameter

A shaft is inspected after machining, and a critical diameter can be satisfactory, too large or too small. The distinction is important since oversize parts can be reworked, whereas undersize parts cannot. We know the probability distribution for the finished diameter, shown in Fig. 3.8. From this we can obtain the probabilities for the three events p_1, p_2, and p_3. The number made per day is n.

The probability mass function for the number occurring of each type in a day's output is then

$$f(x_1, x_2, x_3; n, p_1, p_2, p_3) = \frac{n!}{x_1! x_2! x_3!} \, p_1^{x_1} p_2^{x_2} p_3^{x_3}$$

$$\qquad (3.3.7)$$

Multivariate distributions are discussed more generally in Sec. 3.5.

3.3.4 The Hypergeometric Distribution

The binomial distribution requires that the trials be independent so that the probability of a success must not be dependent on the previous trials. For this to be satisfied the population must be considerably larger than the sample size n—at least 10 times.

Suppose, for example, that 200 parts are made in a day and 100 are inspected. After 50 are checked the population being drawn from is down to 150, and the probability of failure will not remain constant.

The population size, N, must now be a parameter of the distribution, which is [see Derman et al. (1973)]

$$f(x; N, n, k) = \frac{\binom{k}{x}\binom{N-k}{n-x}}{\binom{N}{n}} \tag{3.3.8}$$

where x is the number of successful items in a sample of n, when it is known that there are k successful items in the population of N.

3.3.5 The Geometric and Negative Binomial Distributions

If we have a sequence of Bernoulli trials we may be interested in the random variable representing the number of trials between successive occurrences of the event.

EXAMPLE 3.3 Railway Ties

A railway track is being inspected for faulty ties. It is considered acceptable to have at least k ties in a row between defective ones. If we want a probability of better than 0.99 that this will occur, what must be the proportion of bad ties in the section being inspected?

We designate p as the desired quantity, and also consider it to be the probability that any given tie will be bad. We also assume that this probability is independent for each tie. Then, based on the product rule for combined events, the probability mass function for the number of ties between successive bad ones is

$$f(x; p) = p(1 - p)^{x-1} \tag{3.3.9}$$

This is the *geometric distribution*. Its mean value is $1/p$.

However, our specified probability of 0.99 is the complement of the event that $x \leqslant k - 1$. This is the cumulative distribution function corresponding to x equal to $k - 1$, and is

$$F(k - 1; p) = \sum_{x=1}^{k-1} p(1 - p)^{x-1} \qquad (3.3.10)$$

We can solve this expression for p.

The *negative binomial* distribution (sometimes called Pascal) is a generalization of the geometric. We now define x as the number of Bernoulli trials until the k-th occurrence of an event. It has the probability mass function [see Ang and Tang (1975)]

$$f(x; k, p) = \binom{x - 1}{k - 1} p^{k}(1 - p)^{x-k} \qquad (3.3.11)$$

The probability must be zero for x < k.

3.3.6 The Poisson Process

Some engineering events can be closely modeled by a Poisson process. This simply assumes that the number of events occurring in a unit interval is constant on the average, and that the events are all independent. The interval may be in time or space.

We let

λ = mean occurrence rate, events/unit interval

x = random variable, number of occurrences in interval t

Then the probability mass function is symbolized as $f(x; \lambda, t)$ and represents the probability that there will be x occurrences in interval t. The required function is derived as follows. We define an infinitesimal interval dt, which can have one occurrence; but more than one would have a negligible probability. The probability of such an occurrence is λ dt, by virtue of the definition of λ. An expression for the probability of x occurrences in t + dt is derived by noting that the event can occur in two mutually exclusive ways. We can have x occurrences in t *and* none in dt, *or* x − 1 occurrences in t *and* 1 in dt. In expression form this is

$$f(x; \lambda, t + dt) = f(x; \lambda, t)(1 - \lambda\,dt) + f(x - 1, \lambda, t)\lambda\,dt$$

$$(3.3.12)$$

using the simple probability laws. Rearranging to define a derivative, we get

$$\frac{f(x; \lambda, t + dt) - f(x; \lambda, t)}{dt} = \lambda[f(x - 1, \lambda, t) - f(x; \lambda, t)]$$

$$(3.3.13)$$

As it approaches zero,

$$\frac{d[f(x; \lambda, t)]}{dt} = \lambda[f(x - 1, \lambda, t) - f(x; \lambda, t)] \qquad (3.3.14)$$

with boundary condition

$$f(x; \lambda, 0) = 0 \qquad (3.3.15)$$

This has the solution

$$f(x; \lambda, t) = \frac{(\lambda t)^x e^{-\lambda t}}{x!} \qquad (3.3.16)$$

Subprogram POISS in Appendix A can be used to evaluate this function.

EXAMPLE 3.4 Quarry Machine

A designer is concerned with designing a system for handling boulders in a quarry. The proposed equipment does not have the capacity to handle more than 10 boulders of weight W or greater in 1 day. The mean rate of occurrence of such boulders is 5 per day. What is the probability that it will encounter more than 10?

Using Poisson's distribution, and assuming that the rate of occurrence is constant, the probability for 11 is

$$f(11; 5, 1) = \frac{5^{11} e^{-5}}{11!} = 0.008242$$

This is quite a low probability, so we generate a distribution for x in the following table.

x	0	1	2	3	4	5	6
f(x; 5, 1)	0.00674	0.03369	0.08422	0.14037	0.17547	0.17547	0.14622

x	7	8	9	10	11	12	13
f(x; 5, 1)	0.10444	0.06528	0.03627	0.01813	0.00824	0.00343	0.00132

x	14	15	16	17	18
f(x; 5, 1)	0.00047	0.00016	0.00005	0.00001	0.00000

This is illustrated in Fig. 3.9.

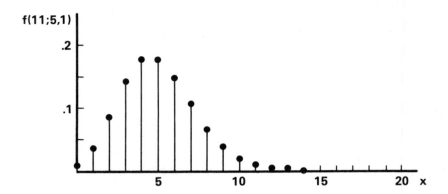

FIG. 3.9 Example of a Poisson distribution.

The addition of the significant probabilities for x > 10 gives us the estimated probability of exceeding 10 per day—a total of 0.01368. Thus we would expect only about 14 days in 1000 to have any problem.

The designer then gets the idea of getting less expensive and lower capacity equipment, and also to store any unprocessed boulders in one day and handle them on days when there is surplus capacity. So she models alternate equipment that will handle 6 per day, and a 5-day operation. A similar table could be generated for the function f(x; 5, 5). The equipment can handle 30 boulders in 5 days, so P(x > 30) is required. We shall use function POISS to calculate the sum of significant values of P(x) for x > 30. The result is 0.13669. This turns out to be too large, so the designer adjusts the parameters to achieve a reasonable value.

A practical example such as this brings to our attention the problem of estimating λ, and also the problem of verifying that it is really constant. These questions are discussed in Sec. 5.3.

It is important to keep in mind that the Poisson process is a physical model, and the engineer must use judgment to decide if the model is adequate for a given application. If, for example, successive events are not independent, a different model must be used and the engineer might consider Markov chains, discussed in Chapter 7.

3.3.7 Other Discrete Distributions

Discrete distributions having analytical form are all based on physical modeling, whereas this is not true of all continuous distributions. They cannot be applied unless the engineer is satisfied that the modeling is appropriate. Other types are discussed in Kendall and Stewart

(1977), Derman et al. (1973), Patil (1965), Johnson and Kotz (1969), and Hastings and Peacock (1975).

3.4 THEORETICAL CONTINUOUS DISTRIBUTIONS

3.4.1 Introduction

The relationship between the continous distributions and physical modeling is much more tenuous than for the discrete distributions. Physical arguments can only be used to indicate a likelihood that a given variable will have a certain theoretical distribution, but the primary basis for choosing a distribution is experience with similar physical quantities. The exponential and gamma distributions are perhaps an exception, since their modeling is based on the Poisson process. The normal function can only in a special case be justified strongly by physical arguments. The extreme value distribution can be derived from physical modeling, but it is a special type that is covered in Chapter 8.

Theoretical distributions tend to be used because of their convenience, despite rather limited physical justification. There is little need for this in design applications, and purely numerical representation of distributions can be quite adequate when the computer is used.

However, if there is evidence that a theoretical distribution can be fitted well to a random variable, this is useful information, and more confidence can be placed in small-sample information used to define the distribution parameters.

3.4.2 The Gamma Function

It may be recalled that for a Poisson process, with a mean rate of occurrence per unit interval of λ, the mass probability function is

$$f(x; \lambda, t) = \frac{(\lambda t)^x e^{-\lambda t}}{x!} \qquad (3.4.1)$$

and it gives the probability that x events will occur in a given interval t, where *x is the random variable*. We now envisage a situation, still with a Poisson process, when we wish to consider t as the random variable rather than x.

To achieve this we let x be k, now a constant. We also must note that $f(x; \lambda, t)$ is a probability, and cannot be converted directly to a density function by simply redefining the variable. We make the transformation by working with the cumulative distribution function for the random variable t. By definition

$$F(t) = P(k \text{ occurrences in interval } t \text{ or less}) \qquad (3.4.2)$$

The complement of this is

$1 - F(t) = P(not$ having k occurrences in t or less)

$\quad\quad\quad = P(0$ occurrences in t or 1 occurrence in t

$\quad\quad\quad \cdots or$ k $-$ 1 occurrences in t)

$\quad\quad\quad = P(0$ occurrences in t)

$\quad\quad\quad + P(1$ occurrence in t)

$\quad\quad\quad + \cdots$

$\quad\quad\quad + P(k - 1$ occurrences in t)   (3.4.3)

These probabilities are given by the Poisson distribution. Thus

$$F(t) = 1 - \sum_{x=0}^{k-1} \frac{(\lambda t)^x e^{-\lambda t}}{x!} \tag{3.4.4}$$

The derivative of this gives the gamma density function.

$$f(t) = \frac{\lambda(\lambda t)^{k-1} e^{-\lambda t}}{(k-1)!} \tag{3.4.5}$$

The mean and variance are

$$\mu = \frac{k}{\lambda} \tag{3.4.6}$$

$$\sigma^2 = \frac{k}{\lambda^2} \tag{3.4.7}$$

The gamma density function can also be generalized for noninteger k. It can be shown that it has the form

$$f(t) = \frac{\lambda(\lambda t)^{k-1} e^{-\lambda t}}{\Gamma(k)} \tag{3.4.8}$$

where $\Gamma(k)$ is the standard gamma function

$$\Gamma(k) = \int_0^\infty s^{k-1} e^{-s} \, ds \tag{3.4.9}$$

In this application a Poisson process would not be modeled, and k and λ are treated as parameters that are adjusted to fit the data.

EXAMPLE 3.5 Quarry Machine (Continued)

A careful distinction must be made between the Poisson and gamma distributions. Returning to Example 3.4, we used the Poisson distribution when our concern was with a random number of boulders occurring. We would use the gamma distribution if we were concerned with a random number of days to encounter a given number of boulders. We could, for example, use the density function to estimate the following probability, calling on FGAMAI:

P(an accumulation of 20 boulders encountered between 6 and 8 days)

$$= \int_6^8 f(t) \, dt \qquad (3.4.10)$$

This is illustrated in Fig. 3.10.

```
      DIMENSION FT(51)
      DO 1 I=1,51
      T=6.+.02*FLOAT(I-1)
1     FT(I)=FGAMAI(5.,20,T)
      PROB=FSIMP(FT,2.,51)
      WRITE(6,2) PROB
2     FORMAT(//,65H PROBABILITY OF ACCUMULATING 20 BOULDERS BETWEEN 6 AN
     1D 8 DAYS IS ,F7.5)
      STOP
      END
```

PROBABILITY OF ACCUMULATING 20 BOULDERS BETWEEN 6 AND 8 DAYS IS   .03910

In every case the event must be carefully defined.

3.4.3 The Exponential Function

The exponential density function is a special case of the gamma, when k equals 1; and the variate is the interval up to first occurrence. The function thus is

$$f(t) = \lambda e^{-\lambda t} \qquad (3.4.11)$$

with

$$\mu = \frac{1}{\lambda} \qquad (3.4.12)$$

$$\sigma^2 = \frac{1}{\lambda^2} \qquad (3.4.13)$$

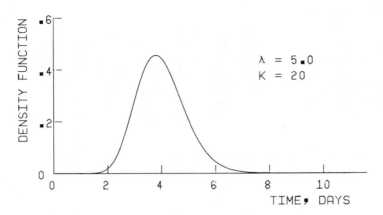

FIG. 3.10 Example of a gamma distribution.

An obvious application is the random variable time to failure of a component or device, when the mean failure rate λ is constant. The cumulative distribution function has the convenient form

$$F(t) = 1 - e^{-\lambda t} \qquad\qquad (3.4.14)$$

so that the probability of *survival* up to time t, or the probability of having a life of t or better, is

$$P(\text{life} \geq t) = 1 - F(t) \qquad\qquad (3.4.15)$$

This is an example of the *reliability function*, discussed in detail in Chapter 12.

Another way to approach the exponential function is to imagine that failure is due to encountering an obstacle, which is equally likely to occur at any time, such as a nail or pothole causing failure of a tire. Let us assume that there is an unknown n number of obstacles encountered per unit of time, each having an equal and finite probability p of causing failure. The average frequency of failure, or failure rate, will be

$$\lambda = np \qquad\qquad (3.4.16)$$

The probability of encountering the i-th obstacle that causes a failure in time t is given by the Poisson distribution.

$$f(i; \ \lambda t) = \frac{(\lambda t)^{i} e^{-\lambda t}}{i!}$$

If i is zero, then we have

$$P(\text{zero failures}) = e^{-\lambda t} \tag{3.4.17}$$

which corresponds to (3.4.15). It is important to note that the total number of obstacles encountered, nt, must be rather large.

It is interesting to see what happens if nt is relatively small. We can now use the binomial distribution

$$f(i; p, nt) = \frac{(nt)!}{i!(nt - i)!} \, p^i(1 - p)^{nt-i} \tag{3.4.18}$$

Substituting for $p = \lambda t$, we have

$$f(i; \lambda, nt) = \frac{(nt)!}{i!(nt - t)!} \, (\lambda t)^i(1 - \lambda t)^{nt-i} \tag{3.4.19}$$

Again we let i be zero.

$$f(0; \lambda, nt) = (1 - \lambda t)^{nt} \tag{3.4.20}$$

For nt large it can be shown that

$$(1 - \lambda t)^{nt} \underset{\sim}{\sim} e^{-\lambda t} \tag{3.4.21}$$

3.4.4 The Normal Function

The normal or Gaussian function is historically the dominant theoretical probability function in the theory of statistics, and has a central role in the theory of statistical inference. Although it has been widely used in scientific work to represent populations arising from natural phenomena, and in error theory, it use in engineering work is much more limited. Its main disadvantages are that it must be symmetrical, and the tails go to infinity at both ends.

The probability density function has the form

$$f(x) = \frac{1}{\sigma\sqrt{2\pi}} \, \exp\left(-\frac{(x - \mu)^2}{2\sigma^2}\right), \quad -\infty < x < \infty \tag{3.4.22}$$

where the parameters are conveniently

μ = mean value

σ = standard deviation

The cumulative distribution function does not have an analytical form.

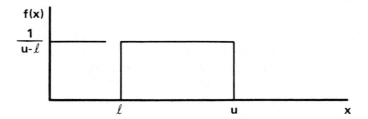

FIG. 3.11 Uniform distribution.

The normal distribution may be theoretically justified in certain rather limited circumstances by the *Central Limit Theorem*. If the random variable is the sum of a large number of independent elements, each of which has a small effect on the sum, the distribution of the variable tends to be normal. This is true no matter what distribution the elements have. However, cases where this has practical application in engineering are rare.

The *log normal distribution* is the normal with the variable transformation

$$y = \ln x \qquad\qquad (3.4.23)$$

So that if y is normally distributed, we can use the theory of Sec. 6.3 to show that

$$f(x) = \frac{1}{\sqrt{2\pi}(\ln x)\delta^2}\exp\left[-\frac{(\ln x - \check{x})^2}{2\delta^2}\right], \quad x \geqslant 0 \qquad (3.4.24)$$

$$= 0, \quad x < 0$$

with parameters δ and $\check{x}$. It turns out that $\check{x}$ is the median of the distribution.

3.4.5 The Uniform Distribution

The very simple uniform distribution is illustrated in Fig. 3.11. It expresses the simple fact that x is equally likely to have any value between ℓ and u. Its analytical form is

$$f(x) = \frac{1}{u - \ell}, \quad \ell \leqslant x \leqslant u$$

$$= 0, \quad x < \ell \text{ or } x > u \qquad\qquad (3.4.25)$$

The mean and variance are

$$\mu = \frac{\ell + u}{2} \tag{3.4.26}$$

$$\sigma^2 = \frac{(u - \ell)^2}{12} \tag{3.4.27}$$

3.4.6 The Beta Distribution

The beta distribution has the advantage of finite bounds on both ends, and good flexibility in shape. The density function is

$$f(x) = \frac{1}{B(q, r)} \frac{(x - \ell)^{q-1}(u - x)^{r-1}}{(u - \ell)^{q+r-1}} , \quad \ell \leqslant x \leqslant u$$

$$= 0, \quad \text{elsewhere} \tag{3.4.28}$$

where q and r are parameters, ℓ and u are the bounds, and $B(q, r)$ is the beta function.

$$B(q, r) = \int_0^1 t^{q-1}(1 - t)^{r-1} \, dt \tag{3.4.29}$$

where $t = (x - \ell)/(u - \ell)$. In terms of the gamma function it is

$$B(q, r) = \frac{\Gamma(q)\Gamma(r)}{\Gamma(q + r)} \tag{3.4.30}$$

The mean, variance, and mode are

$$\mu = \ell + \frac{q}{q + r} (u - \ell) \tag{3.4.31}$$

$$\sigma^2 = \frac{qr}{(q + r)^2(q + r + 1)} (u - \ell)^2 \tag{3.4.32}$$

$$\tilde{x} = \ell + \frac{1 - q}{2 - q - r} (u - \ell) \tag{3.4.33}$$

Its versatility is illustrated in Fig. 3.12.

3.4.7 The Weibull Distribution

The Weibull distribution is very widely used in engineering applications. Its possible shapes are somewhat similar to the beta density function, except that the right tail is infinite. It has the advantage

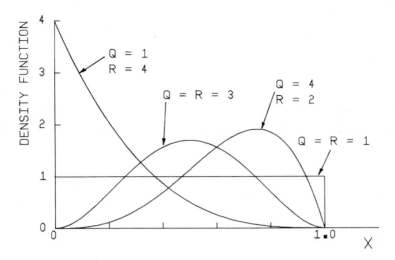

FIG. 3.12 Beta function for different parameter values.

of convenient analytical forms for the cumulative distribution function and the hazard* function. Also, it is convenient to obtain a linear plot for the cumulative distribution function by variable transformation. These plots are important in data fitting and are discussed in Sec. 5.3.

The density function is

$$f(x) = \frac{\beta}{\eta - \ell} \left(\frac{x - \ell}{\eta - \ell}\right)^{\beta-1} \exp\left[-\left(\frac{x - \ell}{\eta - \ell}\right)^{\beta}\right], \quad x \geq \ell$$
$$= 0, \quad x < \ell \tag{3.4.34}$$

where β and η are parameters and ℓ is the lower bound. The cumulative distribution function has the form

$$F(x) = 1 - \exp\left[-\left(\frac{x - \ell}{\eta - \ell}\right)^{\beta}\right] \tag{3.4.35}$$

and the hazard function is

$$h(x) = \frac{\beta}{\eta - \ell} \left(\frac{x - \ell}{\eta - \ell}\right)^{\beta-1} \tag{3.4.36}$$

It is also possible to derive the following statistical parameters:

---

*The hazard function is discussed in Chapter 11.

$$\mu = (\eta - \ell)\Gamma\left(1 + \frac{1}{\beta}\right) \qquad (3.4.37)$$

$$\sigma^2 = (\eta - \ell)\left[\Gamma\left(1 + \frac{2}{\beta}\right) - \Gamma^2\left(1 + \frac{1}{\beta}\right)\right] \qquad (3.4.38)$$

$$\check{x} = (\eta - \ell)(\ln 2)^{1/\beta} \qquad (3.4.39)$$

$$\check{x}_\xi = (\eta - \ell)\left(\ln \frac{1}{1 - \xi}\right)^{1/\beta} \qquad (3.4.40)$$

where

μ = mean

σ^2 = variance

$\check{x}$ = median

ξ = fractile = $F(x_\xi)$

$\check{x}_\xi$ = value for x corresponding to fractile

$\Gamma(\cdot)$ = gamma function, defined in Sec. 3.4.2

The Weibull function is extensively used in reliability work, where it appears to represent most mechanical components quite well.

3.4.8 Other Distributions

Many other distributions have been used in statistical work. Useful catalogs of them can be found in Hastings and Peacock (1975), Patil (1965), and Johnson and Kotz (1970).

3.5 MULTIPLE RANDOM VARIABLES AND MULTIVARIATE DISTRIBUTIONS

3.5.1 Multiple Random Variables

Very commonly, in engineering work, we are concerned with events related to more than one random variable. Consider, for example, the I-beam shown in Fig. 3.13. We assume that the only possible modes of failure are yielding of the flange in direct stress, and yielding of the web in shear. The probability of no failure is

$$P(\text{no failure}) = P(\sigma_f < S_t \text{ and } \tau_w < S_s) \qquad (3.5.1)$$

where

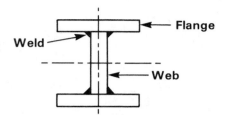

FIG. 3.13 I-beam.

σ_f = stress in flange, assumed deterministic

S_t = yield stress of flange material, a random variable

τ_w = shear stress in web, assumed deterministic

S_s = yield stress in shear of web material, a random variable

If the web and flanges are made of different thickness of stock material, and from different batches of steel, the two events in (3.5.1) will be stochastically independent and we can use the product rule to get

$$P(\text{no failure}) = P(\sigma_f < S_t)P(\tau_w < S_s) = \int_{\sigma_f}^{\infty} f(S_t) \, dS_t \int_{\tau_w}^{\infty} f(S_s) \, dS_s$$

$$(3.5.2)$$

However, if the flanges and web have the same thickness and are cut from the stock plate, or even if they are from the same batch of steel, they will not be stochastically independent events, and (3.5.2) cannot be used. We shall see in the following section how this problem is handled.

A second illustration of multiple random variables is the performance of an internal combustion engine. We assume that it is being designed and there is uncertainty in the prediction of the maximum horsepower p and the fuel consumption rate c. The probability of satisfactory engine performance is based on specifications as follows:

$$P(\text{satisfactory performance}) = P(p > 227 \text{ and } c < 0.185) \quad (3.5.3)$$

The random variables p and c are *functionally dependent* on a set of independent variables such as dimensions, fuel characteristics, and so on, that we shall designate $x_1, x_2, \ldots, x_n$. Thus we can write the general functional relationships

$$p = g_1(x_1, x_2, \ldots, x_n) \quad\quad (3.5.4)$$

$$c = g_2(x_1, x_2, \ldots, x_n) \tag{3.5.5}$$

We shall examine in Chapter 6 techniques for determining the random nature of dependent variables such as p and c, knowing the random nature of the independent variables x_1, x_2, $\ldots$, x_n. Assuming for the moment that we can do this and have f(p) and f(c), it is clear that p and c are functionally dependent because they are functions of the same variables. If this is true, then separate events associated with them, such as p > 228 and c < 185, must be *stochastically dependent*.

Although this is generally true, the inverse is not and stochastically dependent variables are not necessarily functionally dependent.

3.5.2 Multivariate Distributions and Their Relationships

The problem of determining probabilities of combined events associated with more than one random variable was discussed in Sec. 3.5.1. To develop solution techniques it is necessary to define the concept of a multivariate distribution.

The extension to a two-variable probability density function is again perhaps most easily grasped if we begin with a two-dimensional frequency histogram. We can think of each histogram bar as a *cell*, and any values of x_1 *and* x_2 falling within a cell contribute to the frequency count for that cell. The cell is identified by the midinterval values of x_1 and x_2. This is illustrated in Fig. 3.14. The procedure for converting this to a *joint probability density function*, f(x_1,

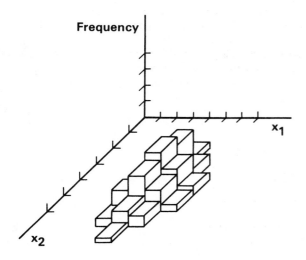

FIG. 3.14 Two-dimensional frequency histogram.

x_2), is similar to that used in Sec. 3.2.1 for a single variable. This density function can be used to obtain the probability of joint events associated with x_1 and x_2.

Returning to our I-beam illustration in the preceding section, we had the two random variables S_t and S_s. If they are not stochastically independent, their combined random nature is defined by their joint density function $f(S_t, S_s)$; and the event expressed in (3.5.1) could be calculated from the joint density function as

$$P(\text{no failure}) = P(\sigma_f < S_t \text{ and } \tau_w < S_s)$$

$$= \int_{-\infty}^{\sigma_f} \int_{-\infty}^{\tau_w} f(S_t, S_s) \, dS_t \, dS_s \tag{3.5.6}$$

and similarly for the expression (3.5.3):

$$P(\text{satisfactory performance}) = P(p > 227 \text{ and } c < 0.185)$$

$$= \int_{227}^{\infty} \int_{-\infty}^{0.185} f(p,c) \, dp \, dc \tag{3.5.7}$$

In these expressions one would actually use the upper bound for ∞ and the lower bound for $-\infty$.

The concept of an *infinitesimal event* is a useful tool in developing relationships between probability functions. It is the event of a random variable having a value between x and x + dx, as shown in Fig. 3.15. The probability of this event is f(x) dx. It could be thought of in a sense as the probability of the random variable having a given value x.

We can use this concept to develop a product rule for joint density functions. The joint infinitesimal event for random variables x_1 and x_2 has the probability $f(x_1, x_2) \, dx_1 \, dx_2$. However, if the ran-

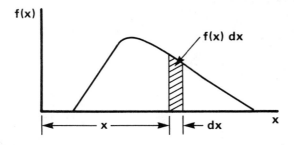

FIG. 3.15 Probability of an infinitesimal event.

dom variables are independent, their infinitesimal events will be independent. Using (2.3.10), we get

$$P(x_1 \text{ occurring in } dx_1 \text{ and } x_2 \text{ occurring in } dx_2)$$

$$= P(x_1 \text{ occuring in } dx_1)P(x_2 \text{ occurring in } dx_2)$$

$$= f(x_1) \ dx_1 f(x_2) \ dx_2 \tag{3.5.8}$$

It follows that for stochastically independent random variables

$$f(x_1, \ x_2) = f(x_1)f(x_2) \tag{3.5.9}$$

The concept of a joint density function can be extended to more than two variables, and expression (3.5.9) can be corresponding extended.

When variables are dependent it is meaningful to have a conditional density function, $f(x_1|x_2)$. Again using infinitesimal events, we can write, by use of (2.3.8),

$$P(x_1 \text{ occurring in } dx_1|x_2 \text{ occurs in } dx_2) = f(x_1|x_2) \ dx_1$$

$$= \frac{f(x_1,x_2) \ dx_1 \ dx_2}{f(x_2) \ dx_2}$$

$$\tag{3.5.10}$$

or

$$f(x_1|x_2) = \frac{f(x_1, \ x_2)}{f(x_2)} \tag{3.5.11}$$

Thus it is meaningful to have single density functions $f(x_1)$ and $f(x_2)$ even when x_1 and x_2 are dependent. These are called the *marginal density function*, and using the infinitesimal event approach, we can show that

$$f(x_1) = \int_{-\infty}^{\infty} f(x_1|x_2)f(x_2) \ dx_2$$

$$= \int_{-\infty}^{\infty} f(x_1, \ x_2) \ dx_2 \tag{3.5.12}$$

The joint cumulative distribution function is also analogous to that for a single variable, and is defined as

$$F(x_1, x_2) = \int_{-\infty}^{x_1} \int_{-\infty}^{x_2} f(x_1, x_2) \, dx_1 \, dx_2 \tag{3.5.13}$$

Conversely,

$$f(x_1, x_2) = \frac{\partial^2 F(x_1, x_2)}{\partial x_1 \, \partial x_2} \tag{3.5.14}$$

3.5.3 Characteristic Measures for Multiple Random Variables

Characteristic measures are obtained by extensions of the definitions for single random variables in Sec. 3.2.3. Thus the *mean* of x_1 is

$$\mu_1 = \int_{R_1} \int_{R_2} x_1 f(x_1, x_2) \, dx_1 \, dx_2 \tag{3.5.15}$$

The *variance* of x_1 is

$$\sigma_1^2 = \int_{R_1} \int_{R_2} (x_1 - \mu_x)^2 f(x_1, x_2) \, dx_1 \, dx_2 \tag{3.5.16}$$

We can also have a *covariance*

$$\text{cov}(x_1, x_2) = \int_{R_1} \int_{R_2} (x_1 - \mu_1)(x_2 - \mu_2) f(x_1, x_2) \, dx_1 \, dx_2 \tag{3.5.17}$$

If the variables are independent, we have

$$\text{cov}(x_1, x_2) = \int_{R_1} (x_1 - \mu_1) f(x_1) \, dx_1 \int_{R_2} (x_2 - \mu_2) f(x_2) \, dx_2 \tag{3.5.18}$$

If this is expanded, it immediately follows that the covariance is zero.

Books on statistics define a dimensionless coefficient of correlation,

$$\rho = \frac{\text{cov}(x_1, x_2)}{\sigma_1 \sigma_2} \tag{3.5.19}$$

which is used as a measure of the linear relationship between x_1 and x_2.

3.6 BAYES' THEOREM FOR DENSITY FUNCTIONS

It would be desirable to be able to update a subjective estimate of a density function by using a sample of observations; no matter how small or large the sample is. Bayes' theorem has been used for this by an extension of the procedure used in Sec. 2.5.

The density function must be defined by an assumed analytical distribution, and the parameters of this function are themselves considered as random variables with subjective density functions. And it is these parameter distributions that are actually updated by the Bayesian process.

A general expression can easily be derived in the case of a single parameter distribution. We let the observed sample of values be x_1, x_2, ..., x_n; and let the parameter be λ. Its prior distribution, estimated subjectively, is $f(\lambda)$; and its posterior distribution is $f(\lambda|x_1, x_2, ..., x_n)$. The reader is reminded again that this is *not* the distribution for the primary random variable x, but rather for the parameter λ in the analytical density function for x.

The basic form of Bayes' theorem is, from (2.5.3),

$$P(A_i|B) = \frac{P(B|A_i)P(A_i)}{\sum_j P(B|A_j)P(A_j)} \qquad (3.6.1)$$

The infinitesimal event $d\lambda$ is now used, and this event corresponds to A_i. The event of observing the specific set of values x_1, x_2, ..., x_n corresponds to B. Equation (3.6.1) converts to

$$f(\lambda|x_1 x_2 \cdots x_n) = \frac{P(x_1 x_2 \cdots x_n|\lambda)f(\lambda)}{\int_R P(x_1 x_2 \cdots x_n|\lambda)f(\lambda)\ d\lambda} \qquad (3.6.2)$$

When the posterior distribution of λ is obtained, its mean is then used with the primary distribution, $f(x)$. To use Bayes' rule in the foregoing form, we must be able to interpret the data sample in a way that makes $P(x_1 x_2 \cdots x_n|\lambda)$ meaningful. To illustrate, we let the density function for x be

$$f(x) = \lambda e^{-\lambda x} \qquad (3.6.3)$$

and let x be the time to failure of a component. If the data sample is the times from testing n items to failure, the probability in (3.6.2) has no meaning, since x_i is an instantaneous event and has zero probability of occurring. Continuous variables can only have a probability of

occurring within a given interval. In this situation, a different form
of Bayes' theorem must be used, given below. However, if the sample
is instead the number of components failed out of a large population in
a specified time T, the Poisson distribution can be used, (3.3.16).

$$P(x_1 x_2 \cdots x_n | \lambda) = \frac{(\lambda T)^n e^{-\lambda T}}{n!} \qquad (3.6.4)$$

Now (3.6.2) can be evaluated for a specified prior $f(\lambda)$.

An alternative form of Bayes' theorem can be derived from
(3.5.11), using any two random variables x_1 and x_2. Two similar
expressions can be written

$$f(x_1 | x_2) = \frac{f(x_1, x_2)}{f(x_2)} \qquad (3.6.5)$$

$$f(x_2 | x_1) = \frac{f(x_1, x_2)}{f(x_1)} \qquad (3.6.6)$$

Eliminating $f(x_1, x_2)$ between them gives

$$f(x_1 | x_2) = \frac{f(x_1 | x_2) f(x_1)}{f(x_2)} \qquad (3.6.7)$$

Using (3.5.12), this can be given the alternative form

$$f(x_1 | x_2) = \frac{f(x_2 | x_1) f(x_1)}{\int_{R_1} f(x_2 | x_1) f(x_1) \, dx_1} \qquad (3.6.8)$$

This is a general relationship between random variables and requires
no assumptions of subjective distributions or updating of a prior distribution.

However, it can be used in this way to solve our illustration of
updating the subjective prior distribution of λ, a single parameter of
an assumed analytical distribution for x. Equation (3.6.2) is now replaced by a version of (3.6.8).

$$f(\lambda | x) = \frac{f(x | \lambda) f(\lambda)}{\int_R f(x | \lambda) f(\lambda) \, d\lambda} \qquad (3.6.9)$$

We can now let the data sample be the more usual one where it is the
result of testing n parts to failure. However, out of the set x_1, x_2,

..., x_n we can only use one value at a time. Beginning with x_1, we substitute it for x in (3.6.9).

$$f(\lambda|x_1) = \frac{f(x_1|\lambda)f(\lambda)}{\int_R f(x_1|\lambda)f(\lambda) \, d\lambda} \tag{3.6.10}$$

This can be evaluated for a given assumed prior distribution $f(\lambda)$ and an assumed analytical distribution $f(x; \lambda)$. A second application of (3.6.9) can be used, substituting x_2 for x, and letting the prior distribution be $f(\lambda|x_1)$.

$$f(\lambda|x_2) = \frac{f(x_2|\lambda)f(\lambda|x_1)}{\int_R f(x_2|\lambda)f(\lambda|x_1) \, d\lambda} \tag{3.6.11}$$

This updating recurrence can be repeated for all n sample values.

There are other situations where Bayes' theorem is applied in different ways (Ang and Tang, 1975; Breipohl, 1970; Grohowski et al., 1976; Kapur and Lamberson, 1977; Olsson, 1968; Stewart, 1979; Weir, 1968); however, in each case the subjective prior distribution is for a distribution parameter or a probability, not for a real quantity.

Although Bayesian statistans lay claim to the use of subjective probability, they do not really do so in the full sense in this application. They are still mired in the statistician's preoccupation with the estimation of parameters for an analytical distribution, and the determination of confidence limits associated with the parameters. They do not really treat the primary distribution (the distribution of the actual random variable) as subjective. Rather, as indicated above, they consider the distribution of a parameter as subjective, and update it with Bayes' theorem. With sufficient updating, the variance of this secondary distribution approaches zero, and the statistician is satisfied that he or she has removed all uncertainty about the true value of this parameter. The statistician also accepts a third level of risk estimation, or compounding of uncertainty, by using the parameter distribution to establish confidence limits for its estimation. Five percent confidence limits are invariably used, and invariably treated as though they were absolute bounds.

This procedure provides an illusion of rigor which is comforting, but it has very dubious validity on the following counts.

1. No attempt is included to assess the inherent suitability of the analytical function being used.
2. The modeling of a distribution parameter as a random variable is an abstract concept, and the engineer cannot relate it to reality by judgment. The subjective choice of an a priori distribution is thus quite arbitrary, and not meaningful in the true subjective

sense. Mathematicians tend to use standard functions, such as the gamma, which are mathematically convenient, but without other justification.
3.   The risk level represented by confidence limits is too abstract to be useful to an engineer in assessing risk. As noted above, the result is that confidence limits are treated as bounds.

Philosophically, one must question how we can expect the Bayesian approach to be meaningful. If we are to accept that probability distributions of engineering quantities are subjective, the random variable itself must be treated directly as subjective. The concept of a random parameter of a distribution has no basis in reality, and therefore cannot be an acceptable grounds for exercising engineering judgment, which must be related to observed realities. This is the only way that the engineer can relate risk to reality by judgment; the result is a codification of the engineer's best judgment and experience, and *should be treated as exact* until he or she has new experience. If this is accepted, both the concepts of random parameters of a distribution, and confidence limits, are meaningless.

The question remains: Given a priori f(x) for the primary random variable, and new data, how do we update f(x) in a rigorous manner? In principle, we should be able to apply Bayes' theorem directly to the prior f(x) to get a posterior f(x). However, Bayes' theorem simply does not work in this direct manner, and we have to "know" the general form of the primary distribution. Other approaches are possible and are discussed in Sec. 5.5.

REFERENCES

Ang, A. H.-S., and W. H. Tang (1975), *Probability Concepts in Engineering Planning and Design*, Vol. 1: *Basic Principles*, Wiley, New York.

Breipohl, A. M. (1970), *Probabilistic Systems Analysis: An Introduction to Probabilistic Models, Decisions, and Applications of Random Processes*, Wiley, New York.

Derman, C., L. J. Gleser, and I. Olkin (1973), *A Guide to Probability Theory and Application*, Holt, Rinehart and Winston, New York.

Grohowski, G., W. C. Hausman, and L. R. Lamberson (1976), A Bayesian Statistical Inference Approach to Automotive Reliability Testing, *J. Quality Technol.*, Vol. 8, No. 4, pp. 197-208.

Hahn, G. J., and S. S. Shapiro (1967), *Statistical Models in Engineering*, Wiley, New York.

Hastings, N. A. J., and J. B. Peacock (1975), *Statistical Distribution*, Halsted Press, New York.

Johnson, N. L., and S. Kotz (1969), *Distributions in Statistics: Discrete Distributions*, Houghton Mifflin, Boston.

Johnson, N. L., and S. Kotz (1970), *Continuous Univariate Distributions*, Vols. 1 & 2, Houghton Mifflin, Boston.

Kapur, K. C., and L. R. Lamberson (1977), *Reliability in Engineering Design*, Wiley, New York.

Kendall, M. G., and A. Stuart (1977), *The Advanced Theory of Statistics*, Vol. 1: *Distribution Theory*, 4th ed., Macmillan, New York.

Olsson, J. E. (1968), Implementation of a Bayesian Reliability Measurement Program, *Ann. Assurance Sci.*, Vol. 1, No. 1, pp. 372-379.

Patil, G. P., ed. (1965), *Classical and Continuous Distributions*, Statistical Publishing Society, Calcutta.

Stewart, L. (1979), Multiparameter Univariate Bayesian Analysis, *J. Am. Stat. Assoc., Theory and Methods*, Vol. 74, No. 367, pp. 684-693.

Weir, W. T. (1968), Bayesian Reliability Evaluation, *Ann. Assurance Sci.*, Vol. 1, No. 1, pp. 344-346.

Moments of a Distribution

Definitions of moments · Estimation of moments · Algebra of expected values · Transformation of moments

4.1 DEFINITION OF MOMENTS

The idea of moments arises from the more general concept of expected value. If we let $g(x)$ be any function of a continuous random variable x, its expected value is defined as

$$E[g(x)] = \int_R g(x)f(x)\ dx \qquad (4.1.1)$$

Similarly, for a discrete random variable

$$E[g(x)] = \sum_i g(x_i)f(x_i) \qquad (4.1.2)$$

We shall limit the following discussion to continuous variables, with obvious estension to discrete variables.

We have seen earlier that the random nature of a variable can be expressed by probability functions. It can also be defined by a series of moments, which are a special kind of expected value.

$$m_i = E(x^i) = \int_R x^i f(x)\ dx \qquad (4.1.3)$$

for $i = 1$ to ∞. The quantity m_i is the *ith moment about the origin*. The first moment is the mean value of x, usually designated μ.

More commonly, moments are defined with respect to the mean, and are called the *central moments*.

$$c_i = E[(x - \mu)^i] = \int_R (x - \mu)^i f(x) \, dx \qquad (4.1.4)$$

We shall see in Chapter 5 that it is possible to generate an approximate density function, knowing the lower moments.

Traditionally, only the first four moments have been commonly used. The second central moment is the variance. The third is known to relate to the symmetry of the density function and is incorporated in the dimensionless *coefficient of skewness*.

$$\alpha_1 = \frac{c_3}{\sigma^3} \qquad (4.1.5)$$

The fourth is related to the flatness in the *coefficient of kurtosis*.

$$\alpha_2 = \frac{c_4}{\sigma^4}$$

The concepts of expectation and moments can be extended to jointly distributed random variables. The general expected value of a function of two random variables is

$$E[g(x_1, x_2)] = \int_R \int_R g(x_1, x_2) f(x_1, x_2) \, dx_1 \, dx_2 \qquad (4.1.6)$$

By definition, the joint moments of a joint distribution are

$$m_{rs} = E[x_1^r x_2^s] = \int_{R_1} \int_{R_2} x_1^r x_2^s f(x_1, x_2) \, dx_1 \, dx_2 \qquad (4.1.7)$$

The central moments are similarly defined.

$$c_{rs} = E[(x_1 - \mu_1)^r (x_2 - \mu_2)^s] \qquad (4.1.8)$$

where μ_1 and μ_2 correspond to m_{10} and m_{01}.

$$\mu_1 = \int_{R_1} \int_{R_2} x_1 f(x_1, x_2) \; dx_1 \; dx_2 \tag{4.1.9}$$

The moment c_{11} is the covariance.

4.2 ESTIMATION OF MOMENTS

Moments can be estimated directly from a sample of observed values of a random variable, x_1, x_2, ..., x_n. The mean is given by

$$\mu = \frac{1}{n} \sum_{i=1}^{n} x_i \tag{4.2.1}$$

The second central moment, or variance, is estimated by

$$\sigma^2 = \frac{1}{n-1} \sum_{i=1}^{n} (x_i - \mu)^2 \tag{4.2.2}$$

An explanation for the $(n-1)$ in the denominator can be found in standard texts on statistics. The higher central moments are estimated by

$$c_j = \frac{1}{n} \sum_{i=1}^{n} (x_i - \mu)^j \tag{4.2.3}$$

Joint moments are estimated in similar fashion. We let x_{ki} be the i-th observed value of the k-th random variable. It should be noted that the observations occur in sets. Thus the i-th observed value of all variables related to a device all arise from the i-th device observed. The moments are

$$\mu_k = \frac{1}{n} \sum_{i=1}^{n} x_{ki} \tag{4.2.4}$$

$$c_{rs} = \frac{1}{n} \sum_{i=1}^{n} (x_{1i} - \mu_1)^r (x_{2i} - \mu_2)^s \tag{4.2.5}$$

Expressions for joint moments for more than two variables could be written, but the notation becomes rather messy. Justification for these expressions can be found in texts on statistics.

Sometimes the information is only available in grouped form. The range has been divided into k intervals, and frequencies have been counted for values occurring in each interval. The following quantities are used:

q_i = frequency in i-th interval

Δx = interval size

x_i^* = mean value of x in the i-th interval

n = sample size

The moments are estimated by the following expressions (Sveshnikov, 1978):

$$\mu = \sum_{i=1}^{k} x_i^* \frac{q_i}{n} \qquad (4.2.6)$$

$$c_j = \sum_{i=1}^{k} (x_i^* - \mu)^j \frac{q_i}{n}, \quad j = 2, \ldots \qquad (4.2.7)$$

If x_i^* is unknown, it may be approximated by the midpoint of the interval.

4.3 ALGEBRA OF EXPECTED VALUES

A few expressions in the algebra of expected values are sometimes useful, and can easily be verified by using the basic definition. If the quantity c is a constant, then

$$E(c) = c \qquad (4.3.1)$$

$$E(cx) = cE(x) \qquad (4.3.2)$$

If x_i is the i-th of n random variables and c_i is the i-th of a corresponding set of constants, then

$$E\left(\sum_{i=1}^{n} c_i x_i\right) = \sum_{i=1}^{n} c_i E(x_i) \qquad (4.3.3)$$

If $g_1(x)$ and $g_2(x)$ are any two functions of x, then

$$E[g_1(x) + g_2(x)] = E[g_1(x)] + E[g_2(x)] \tag{4.3.4}$$

4.4 TRANSFORMATION FROM MOMENTS ABOUT THE ORIGIN TO CENTRAL MOMENTS

By the binomial theorem we can expand the expected value argument for the general central moment.

$$(x - \mu)^i = x^i + (-1)i\mu x^{i-1} + (-1)^2 \frac{i(i-1)}{2!}\mu^2 x^{i-1} + \cdots + (-1)^i \mu^i$$

$$= \sum_{j=0}^{i} (-1)^j \frac{i}{j! \, (i-j)!}\mu^i x^{i-j} \tag{4.4.1}$$

Taking the expected value of both sides gives

$$E[(x - \mu)^i] = \sum_{j=0}^{i} (-1)^j \frac{i!}{j!(i-j)!}\mu^j E(x^{i-j}) \tag{4.4.2}$$

Using our previous notation for moments yields

$$c_i = \sum_{j=0}^{i} (-1)^j \frac{i!}{j!(i-j)!}\mu^j m_{i-j} \tag{4.4.3}$$

Subroutine CONVERT in Appendix A can be used for this conversion.

REFERENCE

Sveshnikov, A. A. (1978), *Problems in Probability Theory, Mathematical Statistics and Theory of Random Functions*, Dover, New York.

5

Generation of Probability
Density Functions

Most of the writers on inference ignore totally the problem of
choice of probability model, so for instance we are supposed to
know that x_1, x_2, ..., x_n are a random sample from $N(\mu, \sigma^2)$,
but we know nothing about μ and σ^2. Really, I say to you,
just how artificial can one be? [Kempthorne, 1971]

*A survey of various methods of generating density functions for real-
life random variables, both when sample data are available and when
nothing but the engineer's experience and judgment are available ·
Comparative tests are presented for different candidate analytical
distributions*

5.1 INTRODUCTION

A key problem in probabilistic design is the generation of density
functions from available information about the random nature of the
variable. Traditionally, the information is in the form of a sample of
observed values of the variable. These may arise from tests, field
measurements, meteorological information, and so on. However, the
information may be less explicit than this—just the general experience
of the engineer with similar devices.

Means must be found to codify this information in a density
function, no matter in what form the information exists. This can
be done in general in four ways: by analyzing the data to yield a
nonparametric or numerically defined distribution, by fitting one of
the standard analytical distributions such as the Weibull, by deter-

mining the maximum entropy distribution, or by subjective codification of judgment.

The methods will be limited to single-variable density functions. In some cases the extension to multivariable distributions is possible and relatively straightforward.

5.2 NUMERICALLY DEFINED DISTRIBUTIONS

The numerically defined distribution is obtained by the following classical algorithm, which also gives some insight into the meaning of a probability density function.

1. Determine the range of the data and divide it into equally spaced intervals. The number of intervals is essentially based on judgment.
2. Count the frequency of observations in each interval.
3. Establish a set of points at the midpoint of each interval corresponding to the frequencies. This is illustrated in Fig. 5.1. The bounds will usually lie outside the sample data bounds and must be set by judgment.
4. Define a smooth curve through the points using subroutine SMOOTH from Appendix A.
5. Obtain the area under the curve by a computer numerical integration subroutine and divide each smoothed station value by the area to make the final area equal to 1. Subroutine FNORM can be used.

This method requires a relatively large number of values to give meaningful results— 30 or 40 at least. The algorithm is embodied in HIST.

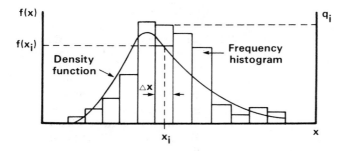

FIG. 5.1 Obtaining a density function from a histogram.

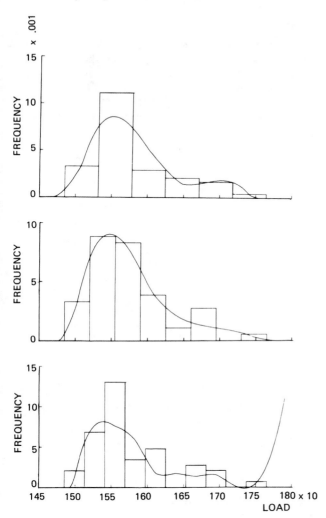

FIG. 5.2 Histogram and numerically defined density function from a data sample of 52.

The number of intervals must be selected by judgment. If there are too many, the frequency in any one bar will be low, and there will be considerable scatter from a fitted curve. If there are too few intervals, the frequency curve will not be meaningfully defined, although the bars will appear to give a smoother curve.

EXAMPLE 5.1   Tensile Tests of SAE 1008 Steel

The data associated with the following sample result from tensile
tests on laboratory specimens made from SAE 1008 sheet steel.
The values are the load in pounds on the specimens when lower
yielding begins. The example illustrates in Fig. 5.2 the generation of a histogram and a nonparametric numerically defined
density function for a sample size of 52. Subroutine HIST is
used.

```
DIMENSION X(52),XFREQ(11),FREQ(10),XDENS(101),FDENS(101),W1(10),
1W2(10)
COMMON/PLOTPL/IFLAG
IFLAG=2
READ(5,1)X
FORMAT(10F7.0)
NSAMP=52
BL=1460.
BU=1790.
IPLOT=1
MINT=101
NORM=1
N=4
LABEL=4HLOAD
NBARS=10
CALL HIST(X,NSAMP,NBARS,BU,BL,NORM,IPLOT,N,LABEL,MINT,XFREQ,FREQ,
1XDENS,FDENS,W1,W2)
CALL PLOTPL(1,XDENS,FDENS,MINT,LABEL,N,XDENS(MINT),XDENS(1))
CALL PLOTPL(0,XDENS,FDENS,MINT,LABEL,N,XDENS(MINT),XDENS(1))
STOP
END
```

1550.	1550.	1560.	1525.	1560.	1535.	1520.	1565.	1510.	1555.
1550.	1695.	1625.	1545.	1565.	1655.	1685.	1560.	1525.	1545.
1565.	1675.	1610.	1665.	1535.	1545.	1565.	1510.	1540.	1570.
1535.	1565.	1600.	1585.	1765.	1520.	1570.	1600.	1515.	1485.
1560.	1550.	1575.	1550.	1560.	1535.	1610.	1625.	1585.	1600.
1685.	1660.								

5.3 FITTING ANALYTICAL DISTRIBUTIONS
BY PARAMETER ESTIMATION

There are a large number of standard analytical distributions available
(see Chapter 3), one or more of which may suitably represent the
population. They are in functional form and contain one to four parameters which must be estimated from the sample data. The designer
may know from previous experience with similar random variables
which distribution is most likely to satisfactorily fit the data. *Note
that this is subjective use of information.* Lacking previous experience, the designer may try fitting several distributions and pick the
best.

Statistical techniques are available for comparing methods, such as the chi-square method, but because of our basic commitment to subjective probabilities, no absolute measure of satisfactory fit is possible. It must be left to judgment. A designer can accept curves that would shock a statistician. A plot of the proposed analytical curve superimposed on a numerically defined distribution is one possible eyeball way of using judgment.

The subject of *statistical inference* provides the mathematical basis for the estimation of distribution parameters. We shall only discuss the elements of it here; the interested reader is referred to one of the many standard texts on the subject.

The sample used to represent the population of values of the random variable must be randomly selected; each sample value must be independent of all others; and the population must remain unchanged in nature from one sample value to another.

Three commonly used alternative methods are available to estimate distribution parameters using the sample information: the method of moments, the method of maximum likelihood, and the Bayesian method. Arguments for not using the third method are given in Sec. 3.6, and it will not be discussed further here. In some distributions the first two methods give the same result, but in others they do not. The choice of the best method is not clear cut, but it can be argued that the method of moments is preferable in small samples. Many other statistical techniques also exist (Kendall and Stuart, 1979).

The *method of moments* uses an obvious and straightforward approach in which the sample moments are used as estimators of the population moments. Expressions for calculating sample moments were given in Sec. 4.2. In some distributions the first one or two moments (the mean and the variance) are the actual parameters. In most other cases there is an analytical relationship between the moments and the parameters, so that the latter can be calculated in terms of the moments. It is intuitively clear that accuracy of estimation is proportional to the sample size.

The *method of maximum likelihood* follows the approach of calculating the parameter values which are most likely to yield the available sample data. With a given sample x_1, x_2, ..., x_n, the probability of obtaining these particular values is based on the infinitesimal event.

$$P(x_1 x_2 \cdots x_n) = f(x_1; \theta_1, \theta_2, \ldots, \theta_m) \, dx$$

$$f(x_2; \theta_1, \theta_2, \ldots, \theta_m) \, dx \cdots f(x_n; \theta_1, \theta_2, \ldots, \theta_m) \, dx$$

$$(5.3.1)$$

where $f(x_i; \theta_1, \theta_2, \ldots, \theta_m)$ is the specified analytical density function with parameters $\theta_1, \theta_2, \ldots, \theta_m$. If the differentials are deleted, the expression is called the *likelihood function*.

$$L = f(x_1; \theta_1, \theta_2, \ldots, \theta_m) f(x_2; \theta_1, \theta_2, \ldots, \theta_m)$$
$$\cdots f(x_n; \theta_1, \theta_2, \ldots, \theta_m) \tag{5.3.2}$$

The values of the θ's most likely to have yielded the sample values are those that maximize L. The method of the calculus is used, giving

$$\frac{\partial L}{\partial \theta_j} = 0, \quad j = 1, m \tag{5.3.3}$$

Sometimes it is more convenient to use

$$\frac{\partial(\ln L)}{\partial \theta_j} = 0, \quad j = 1, m \tag{5.3.4}$$

The maximum location will be the same because of the monotonic nature of ln L.

The method will be made clearer by a simple illustration, using the exponential distribution. The density function is

$$f(x) = \lambda e^{-\lambda x} \tag{5.3.5}$$

where λ is the parameter to be determined. The likelihood function is

$$L = \lambda^n \exp\left(-\lambda \sum_{i=1}^{n} x_i\right) \tag{5.3.6}$$

It is convenient in this case to find the maximum of ln L, rather than L.

$$\ln L = n \ln \lambda - \lambda \sum_{i=1}^{n} x_i \tag{5.3.7}$$

The derivative is set equal to zero:

$$\frac{d}{d\lambda}(\ln L) = \frac{n}{\lambda} - \sum_{i=1}^{n} x_i = 0 \tag{5.3.8}$$

and solving for λ gives

$$\lambda = \frac{n}{\sum_{i=1}^{n} x_i} \tag{5.3.9}$$

Thus in this case the method gives the same result as the method of moments, since for the exponential distribution

$$\lambda = \frac{1}{\mu}$$

A major topic in the classical theory of statistical inference is *interval estimation of parameters*, by which a statistician draws conclusions like the following. "The 95% confidence limits on the estimation of the mean of x are 756 and 925." This statement means that if different samples of the same size were analyzed, on the average 95% of them will contain the true population mean in the confidence limits. It is more loosely interpreted as meaning that the true mean has a value between the confidence limits with a probability of 95%.

The topic of interval estimation of confidence limits is thus an attempt to provide a theoretical means of estimating the accuracy of the parameter estimates. Confidence limits are not used in this text on the grounds that the technique does not provide a meaningful result in the engineering sense—it does not assist the engineer to relate judgment to reality. The argument is pursued in detail in Appendix C.

5.4 FITTING ANALYTICAL DISTRIBUTIONS USING RANKS

Possibly a more efficient way of using all the information and judgment available in the estimation of parameters in analytical distributions is by use of order statistics and ranks (Gumbel, 1958; Guttman et al., 1965; Hahn and Shapiro, 1967; Lipson and Sheth, 1973), in which a linearized cumulative distribution plot is made from a ranked sample ordered from low to high. The ith member of a ranked sample can be thought of as a random variable having different values for every sample of the same size obtained. It is called the *ith order statistic*; and we shall designate it o_i. The corresponding value of the cumulative distribution function of the primary variable x is called the *ith rank*, and is defined by

$$r_i = F_x(o_i) = \int_{-\infty}^{o_i} f_x(x) \, dx \tag{5.4.1}$$

where $f_x(x)$ represents the density function, which we do not, of course, know. The ith rank must also be a random variable, since it is a function of o_i. This is illustrated in Fig. 5.3.

For a given observed value of o_i, which we can call x_i, from a specific sample, we would like to estimate a corresponding value of

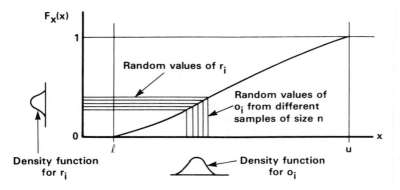

FIG. 5.3 Illustration of order statistics and ranks as random variables.

$F(x_i)$, in order to make our plot. The best that we can do is to use the mean value of r_i, called the mean rank, and if this plotted against the observed value x_i, a rough representation of the cumulative distribution function is obtained.

It would seem intuitively clear that the ratio i/n, where n is the sample size, is a rough measure of $F_X(x_i)$. We shall see in Sec. 8.2 that the distribution of r_i is independent of the primary distribution $f_X(x)$, and it has a mean value given by

$$\bar{r}_i = \frac{i}{(n+1)} \tag{5.4.2}$$

It is difficult to estimate whether or not the fit made by a curve passing through $\bar{r}_i$ versus x_i is satisfactory unless the function can be linearized. Consider, for example, the Weibull function with ℓ equal to zero.

$$F(x) = 1 - \exp\left[\left(-\frac{x}{\eta}\right)^{\beta}\right] \tag{5.4.3}$$

where η and β are the parameters. If we take the double log of both sides, we get

$$\ln \ln \frac{1}{1 - F(x)} = \beta \ln x - \beta \ln \eta \tag{5.4.4}$$

If we transform variables so that

$$y = \ln \ln \frac{1}{1 - F(x)} \quad \text{and} \quad z = \ln x \tag{5.4.5}$$

then we have the linear relationship

$$y = \beta z - \beta \ln \eta \qquad (5.4.6)$$

The value of y corresponding to the sample value of x is estimated by substituting the mean rank for $F(x)$.

$$y_i = \ln \ln \frac{1}{1 - [i/(n + 1)]} \qquad (5.4.7)$$

If the values of y and z are plotted corresponding to sample values of $i/(n + 1)$ and x_i, they should approximate a straight line if the distribution can be adequately represented by a straight line. There will always be some scatter because the observed ranks are random variables, but the trend should be straight.

Parameter values can be determined from the plot, by noting that

$$\begin{aligned} \text{slope} &= \beta \\ \text{intercept} &= - \beta \ln \eta \end{aligned} \qquad (5.4.8)$$

Subroutine MRWEIB generates a least-squared "error" fit of y versus z and plots the straight line for a Weibull distribution.

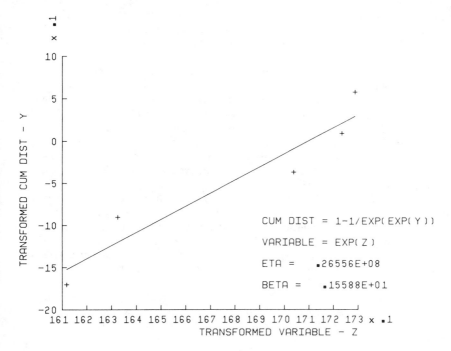

FIG. 5.4 Mean-rank Weibull plot for life to failure.

EXAMPLE 5.2   Life to Failure

Five components are tested to failure and have the following lives: 25.1, 12.3, 32.2, 10.0, and 30.5×10^6 hours. Previous experience has established that the population is well represented by a Weibull distribution with zero lower bound. The following program illustrates how MRWEIB is called. The straight-line fit is shown in Fig. 5.4. Although the scatter seems large, it is not unusual for such a small sample.

```
DIMENSION X(5),W(10)
DATA X/25.1E6,12.3E6,32.2E6,10.0E6,30.5E6/
NSAMP=5
IPLOT=1
CALL MRWEIB(X,NSAMP,IPLOT,ETA,BETA,W)
STOP
END
```

Obtaining the fit is not quite as easy if the lower bound is not zero. In this event the variable is transformed to $(x - \ell)$, and the value of ℓ is found by trial to give the minimum least-squared "error." An example is shown in Fig. 5.5 for the random variable "percent of grain damaged in a grain handling device," with a sample size of 11. Plots were made assuming different values of ℓ until the best-fit straight line was achieved. Figure 5.5 also illustrates the use of Weibull probability paper, with the scales transformed to give a linear plot directly.

MRWEIB can be used very conveniently to try different lower bounds. The lower bound is subtracted from the values in a loop, and these values are used in the call to MRWEIB. No other change is necessary. It must be kept in mind, however, that η will correspond to the transformed variable $(x - \ell)$, and not the original variable x. The parameter β is unaffected.

This method gives meaningful results with very small samples, as low as four or five; however, this is acceptable only if it is known from previous experience that a given distribution function does adequately represent similar populations. Practical aspects of the method are developed in considerable detail by Hahn and Shapiro (1967), and they illustrate that a sample size of at least 50 is required to confirm that an adequate straight-line fit really has been achieved with an unknown population. However, this is not always available, and the engineer must make the best estimate with the information in hand, combined with judgment.

A least-squares fit can be obtained numerically on the computer, rather than a purely graphical solution, but a plot is important to judge the adequacy of the straight line. Both are provided by MRWEIB.

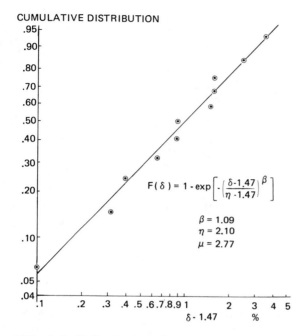

CUMULATIVE DISTRIBUTION

$$F(\delta) = 1 - \exp\left[-\left(\frac{\delta - 1.47}{\eta - 1.47}\right)^{\beta}\right]$$

$\beta = 1.09$
$\eta = 2.10$
$\mu = 2.77$

$\delta - 1.47$ %

FIG. 5.5 Weibull plot of grain damage, where δ is the percent of grain damaged, and the lower bound has been found to be 1.47.

The exponential distribution can also be given a straight-line plot by means of a variable transformation, but some functions require a different approach, illustrated here using the normal function. The vertical scale is set up by first constructing an auxiliary scale for the standard variate.

$$s = \frac{x - \mu}{\sigma} \qquad (5.4.9)$$

Corresponding values of the cumulative distribution function are laid out on the actual scale adjacent to the variate scale, as shown in Fig. 5.6. The horizontal axis is for the variable x, using a linear or arithmetic scale. The theoretical line for F(x) will be straight because of the linear relation between x and s. Normal probability paper is commercially available in which the scale for F(x) is rationalized in convenient intervals. General-purpose probability paper cannot be constructed for all theoretical distributions, nor can simple transformations be used to linearize all distributions.

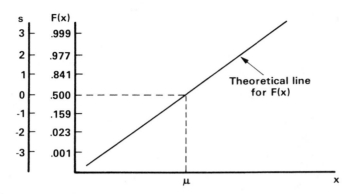

FIG. 5.6 Setup for normal probability paper.

The use of the least-squares principle in establishing the straight-line fit is an arbitrary hypothesis for defining "best fit" and should not be confused with its application in regression theory, where it is based on the maximum likelihood theory [see, e.g., Guest (1961)]. There is a problem in assigning weights to the deviations. Fitting the line using deviations in y_i implies different weights than if we used, for example, deviations in $F(x_i)$ directly.

5.5 THE MAXIMUM ENTROPY METHOD

5.5.1 Jaynes' Principle

The maximum entropy method is based on Jaynes' principle: *The minimally prejudiced probability distribution is that which maximizes the entropy subject to constraints supplied by the given information.*

Jaynes (1957) argues that: "This is the only unbiased assumption that we can make; to use any other would amount to an arbitrary assumption of information which by hypothesis we do not have."

For a continuous random variable, entropy is defined as

$$S = - \int_R f(x) \ln [f(x)] \, dx \qquad (5.5.1)$$

and for a discrete random variable it is

$$S = - \sum_{i=1}^{n} f(x_i) \ln [f(x_i)] \qquad (5.5.2)$$

where f(x) is the probability density function and $f(x_i)$ is the probability mass function. This principle would seem to be a valuable approach to generating density or mass functions, in which the information incorporated in the constraints would be whatever data the engineer has available.

5.5.2 The Concept of Entropy

Before proceeding with the development of this approach it is desirable to attempt to understand why entropy is defined in the way it
is. Entropy is an obscure concept, and this obscurity is typical of
concepts that are essentially arbitrary abstract definitions, based on
intuition but with desirable and consistent characteristics. Since they
are arbitrary they cannot really be explained, and attempts to explain
them therefore usually become shrouded in obscurity and mystery.

Entropy is really quite a simple concept if one accepts that it is
arbitrary. It arises from an equally arbitrary concept called *information*, or sometimes *self-information*. These concepts are developed in
the subject of *information theory* primarily applied to communications,
but also applied to many other field. For more details the reader is
referred to Jones (1979) and Abramson (1963). The term "information" in this context is a restricted technical meaning useful in the
theory of information flow.

Information in this sense is defined as

$$I(x) = - \log f(x) \tag{5.5.3}$$

or for a discrete variable

$$I(x_i) = - \log f(x_i) \tag{5.5.4}$$

in which x is thought of as the event of x having a given value. $I(x)$
is thought of, in a restricted sense, as being a kind of measure of the
information associated with this event. The base used for the logarithm is not too important, since it only changes the value of the information by a constant. If base 2 is used, then $I(x)$ has units of
bits; and if base e is used, the units are called *nats*. This will be
illustrated below.

First, let us try to develop some feel for why information is so
defined. We again note that this is an intuitively arbitrary definition,
but with important desirable characteristics. The first question to be
answered is: Why is information a function of probability of the occurrence of a value of the random variable? This can be answered by
considering Fig. 5.7. We imagine that we are obtaining a stream of
sample data values, possibly from field measurements. The first four
are shown in the figure. We are more likely to get values such as x_1,
x_2, and x_3 than we are to get one such as x_4. Thus each value among

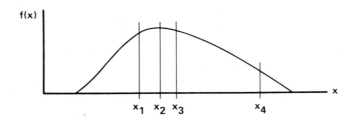

FIG. 5.7 Illustration of why information is associated with probability.

those in the most likely region gives us little additional information about the nature of the random variable. On the other hand, a rare value such as x_4 gives us useful information about the tails or low-probability regions. Thus it is meaningful to think of information content as being inversely proportional to the likelihood of the event occurring.

The second question about the definition is: Why is the logarithm used? The answer is simply that it is a fortuitous choice that gives consistent and desirable results. We can illustrate by a couple of examples, but first we note that if we use the logarithm, we must use a negative sign to achieve the desired inversion.

The first illustration is the case when we are using binary numbers with m digits to measure values of x. We also specify that any number is equally likely to occur. With m digits there are 2^m possible combinations, so the probability of the event of any one number occurring is $1/2^m$. And the information is

$$I(x) = - \log_2 \frac{1}{2^m} = m \text{ bits} \qquad (5.5.5)$$

which is consistent with the fact that each number has m bits of information in the customary sense.

The second illustration is the expression for the information of two joint but independent events. If the events are E_1 and E_2, and their probabilities are p_1 and p_2, then the information of the joint event is

$$I(E_1 E_2) = - \log(p_1 p_2) = -\log p_1 - \log p_2$$

$$= I(E_1) + I(E_2) \qquad (5.5.6)$$

Thus the information of joint independent events is additive, which seems intuitively correct and desirable.

We can also think of information as being simply another measure of probability, considering the definition as simply a transformation of

variable. This transformed quantity has certain desirable and convenient properties as a measure of probability.

These desirable characteristics become more evident when we
come to entropy. It is defined as the mean value of information, and
for a continuous variable it has the form [see (4.1.1)]

$$S[f(x)] = - \int_R f(x) \log [f(x)] \, dx \qquad (5.5.7)$$

The corresponding definition for a discrete variable is

$$S[f(x_i)] = - \sum_{i=1}^{n} f(x_i) \log [f(x_i)] \qquad (5.5.8)$$

If $f(x)$ is zero, the product $f(x) \log [f(x)]$ becomes indeterminate,
but approaches zero as $f(x)$ approaches zero.

Since information is a measure of uncertainty for individual values of x, then entropy is a measure of uncertainty for the whole range
of values of x. The greater the uncertainty, the larger will be the
entropy. We can confirm this feature by looking at the range of uncertainty possible. If there is no uncertainty, only one value of x can
occur. The corresponding value of entropy is easily obtained for the
discrete case. If x_j is the known value, then

$$f(x_i) \log [f(x_i)] = 0$$

for all i values. This is true for $i \neq j$ because of what was said above
if $f(x)$ equals zero. If $i = j$, then $f(x_i) = 1$, and $\log [f(x_i)]$ is zero.
Thus the entropy is zero in the event of certainty. For the continuous case we use the infinitesimal event shown in Fig. 5.8. If the value of x occurs in dx, the density function has the form shown with

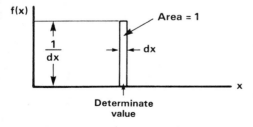

FIG. 5.8 Density function for a determinate value of x.

$$f(x) = \frac{1}{dx} \ , \quad \text{in the interval } dx$$

$$= 0, \quad \text{elsewhere}$$

The argument of the integral in (5.5.7) becomes

$$\frac{1}{dx} \ \log \left(\frac{1}{dx} \right) = 0$$

Again the entropy is zero if we can predict with certainty the value of x. At the other extreme we have complete uncertainty, codified in a uniform distribution. It can be demonstrated that in this case the entropy has its maximum possible value.

We begin this by establishing the following general inequality for y > 0.

$$\ln y \leqslant y - 1 \tag{5.5.9}$$

with equality only when y = 1. Since for $0 < t \leqslant 1$, $1/t \geqslant 1$, then

$$\int_y^1 \frac{dt}{t} \geqslant \int_y^1 dt, \quad 0 \leqslant y < 1 \tag{5.5.10}$$

Evaluating the integrals gives

$$- \ln y \geqslant 1 - y \tag{5.5.11}$$

or

$$\ln y \leqslant y - 1$$

and (5.5.9) is proven for $0 < y \leqslant 1$. For $t \geqslant 1$, $1/t \leqslant 1$ and

$$\int_1^y \frac{dt}{t} \leqslant \int_1^y dt, \quad y \geqslant 1 \tag{5.5.12}$$

Evaluating the integrals gives

$$\ln y \leqslant y - 1$$

and thus (5.5.9) applies for all $y \geqslant 0$.

We return to the value of entropy for a uniform distribution and note that for a continuous variable

$$f(x) = \frac{1}{u - \ell} \qquad (5.5.13)$$

where ℓ and u are the bounds. The entropy is (using natural logarithms)

$$S[f(x)] = \ln (u - \ell) \qquad (5.5.14)$$

For a discrete uniform distribution

$$f(x_i) = \frac{1}{n} \qquad (5.5.15)$$

where n is the number of discrete values. The entropy is

$$S[f(x_i)] = \ln (n) \qquad (5.5.16)$$

We shall first prove that this is the maximum possible entropy for a discrete distribution. We arbitrarily begin by using (5.5.9) in the following form, and require that all $f(x_i)$ be nonzero.

$$\sum_{i=1}^{n} f(x_i) \ln \frac{1}{nf(x_i)} \leqslant \sum_{i=1}^{n} f(x_i) \left[\frac{1}{nf(x_i)} - 1 \right] \qquad (5.5.17)$$

The right-hand side reduces to zero and the left-hand side can be converted as follows:

$$\sum_{i=1}^{n} f(x_i)\{\ln 1 - \ln [nf(x_i)]\} \leqslant 0$$

$$-\sum_{i=1}^{n} f(x_i) \ln [nf(x_i)] \leqslant 0$$

$$-\sum_{i=1}^{n} f(x_i) \ln [f(x_i)] \leqslant \ln n \quad \text{or} \quad S \leqslant \ln n \qquad (5.5.18)$$

For zero values of $f(x_i)$ the validity of (5.5.18) still applies since corresponding terms on the left-hand side will be zero. Recalling (5.5.16), we have proven that, for a discrete variable, the entropy is a maximum for a uniform distribution.

For the continuous variable we begin by an analogous form of (5.5.17)

$$\int_{\ell}^{u} f(x) \ln \left[\frac{1}{(u - \ell)f(x)} \right] dx \leqslant \int_{\ell}^{u} f(x) \left[\frac{1}{(u - \ell)f(x)} - 1 \right] dx$$

$$(5.5.19)$$

which reduces to

$$- \int_{\ell}^{u} f(x) \ln [f(x)] dx \leqslant \ln (u - \ell) \qquad (5.5.20)$$

Although there is an indeterminate form in the left-hand side of (5.5.19) when $f(x)$ is zero, the full argument of the integral becomes zero. Comparing this expression with (5.5.14), we see that for a continuous variable, entropy is again a maximum for a uniform distribution. Although we have used the natural logarithm above, the results are clearly valid for any logarithm. We have confirmed that the extreme values are consistent with our concept of entropy as a measure of uncertainty.

Jaynes' principle can now be considered intuitively likely because of the way entropy is defined, but it has not been formally proven. However, it does lead to consistent results in many applications; see, for example, Tribus (1969). Also, the following applications to the generation of density and mass functions show consistency.

5.5.3 Using Jaynes' Principle with a Uniform Distribution

If we have complete uncertainty except for the bounds, then we can use no constraints in conjunction with maximizing the entropy except the definition of a density function.

For the discrete distribution Jaynes' principle gives us

$$S = - \sum_{i=1}^{n} f(x_i) \ln [f(x_i)] = \text{maximum} \qquad (5.5.21)$$

$$\sum_{i=1}^{n} f(x_i) = 1 \qquad (5.5.22)$$

Using the classical method of the calculus with a Lagrangian multiplier $(\lambda + 1)$, we get

$$\frac{\partial}{\partial f(x_i)} \left\{ - \sum_{i=1}^{n} f(x_i) \ln [f(x_i)] + (\lambda + 1) \left[\sum_{i=1}^{n} f(x_i) - 1 \right] \right\} = 0$$

$$(5.5.23)$$

which reduces to

$$\lambda = \ln \ [f(x_i)] \tag{5.5.24}$$

or

$$f(x_i) = e^{\lambda} \tag{5.5.25}$$

The use of $(\lambda + 1)$ as a Lagrangian multiplier instead of λ was a trick to avoid the disappearance of λ. We must also satisfy (5.5.22), and substituting the last expression into this gives

$$ne^{\lambda} = 1$$

or

$$e^{\lambda} = \frac{1}{n} \tag{5.5.26}$$

But by (5.5.25)

$$f(x_i) = \frac{1}{n} \tag{5.5.27}$$

Thus we have demonstrated that if we have complete uncertainty, with no information to supplement Jaynes' principle, then it leads to a uniform distribution—as it should.

5.5.4 Using Jaynes' Principle with Moments

We are finally ready to see how Jaynes' principle can be used to generate a density function from sample data. We assume that we have no information to convince us that there are physical arguments that would suggest a particular analytical distribution; and therefore maximizing the entropy of the variable's distribution should lead to the least biased estimate of its form.

A convenient way to use the sample information is by evaluation of the sample moments, using the expressions given in Sec. 4.2 (Siddall and Diab 1975). It is more convenient mathematically to work with moments about the origin, and these can be obtained from central moments by using (4.4.3). The method will be illustrated for a continuous random variable; the corresponding development for a discrete variable is quite straightforward.

The basic formulation is as follows:

$$S = - \int_R f(x) \ \ln \ [f(x)] \ dx = \text{maximum} \tag{5.5.28}$$

$$\int_R f(x) \ dx = 1 \tag{5.5.29}$$

$$\int_R x^i f(x) \ dx = m_i, \quad i = 1, \ m \tag{5.5.30}$$

where m is the number of moments to be used and m_i is the ith moment about the origin, determined numerically from the sample.

We could attempt to solve this directly, using one of the techniques of numerical optimizations (Siddall, 1982); and this would lead to a discretized numerical definition of $f(x)$. However, it is possible to show that the density function inherently has a specific analytical form.

The classical method of the calculus is now applied to the expressions above (Siddall, 1982). The reader is reminded that we are adjusting $f(x)$ to achieve a maximum. We let $\bar{S}$ be the modified function, and use Lagrangian multipliers $\lambda_0, \lambda_1, \dots, \lambda_m$.

$$\bar{S} = S + (\lambda_0 + 1) \left[\int_R f(x) \ dx - 1 \right] + \sum_{i=1}^m \lambda_i \left[\int_R x^i f(x) \ dx - m_i \right]$$

$$\tag{5.5.31}$$

The multiplier $(\lambda_0 + 1)$ is actually used rather than λ_0 to give a more convenient result. Setting the derivative $d\bar{S}/df(x)$ equal to zero gives

$$- \int_R \{\ln \ [f(x \ ] + 1\} \ dx - (\lambda_0 + 1) \int_R dx - \sum_{i=1}^m \lambda_i \left(\int_R x^i \ dx \right) = 0$$

$$\tag{5.5.32}$$

Combining terms under the integral sign leads to

$$\int_R \left\{ - \ln \ [f(x)] - 1 + \lambda_0 + 1 + \sum_{i=1}^m \lambda_i x^i \right\} dx = 0 \tag{5.5.33}$$

For this expression to be zero, the argument must be zero.

$$\ln \ [f(x)] = \lambda_0 + \sum_{i=1}^m \lambda_i x^i \tag{5.5.34}$$

or

$$f(x) = \exp \left(\lambda_0 + \sum_{i=1}^{m} \lambda_i x^i \right) \qquad (5.5.35)$$

This is the analytical form for the *maximum entropy density function*; the problem remains to determine the values of the λ's.*
To achieve this we shall need two expressions, developed as follows. Equation (5.5.35) is substituted into (5.5.29).

$$\int_R \exp \left(\lambda_0 + \sum_{i=1}^{m} \lambda_i x^i \right) dx = 1 \qquad (5.5.36)$$

Multiplying by $e^{-\lambda_0}$ gives

$$e^{-\lambda_0} = \int_R \exp \left(\sum_{i=1}^{m} \lambda_i x^i \right) dx \qquad (5.5.37)$$

which leads to the first expression required.

$$\lambda_0 = -\ln \left[\int_R \exp \left(\sum_{i=1}^{m} \lambda_i x^i \right) dx \right] \qquad (5.5.38)$$

The second is obtained by differentiating (5.5.37) with respect to λ_i.

$$- e^{-\lambda_0} \frac{\partial \lambda_0}{\partial \lambda_i} = \int_R x^i \exp \left(\sum_{i=1}^{m} \lambda_i x^i \right) dx \qquad (5.5.39)$$

or

$$\frac{\partial \lambda_0}{\partial \lambda_i} = - \int_R x^i \exp \left(\lambda_0 + \sum_{i=1}^{m} \lambda_i x^i \right) dx \qquad (5.5.40)$$

Referring to (5.5.30) and (5.5.35), we see that this reduces to the second expression.

---

*To be rigorous, we should confirm that this solution is a global optimum. This is done in Appendix E.

$$\frac{\partial \lambda_0}{\partial \lambda_i} = -m_i \qquad (5.5.41)$$

In order to solve for the λ's we must set up a set of simultaneous equations. This is done by differentiating (5.5.38) with respect to λ_i.

$$\frac{\partial \lambda_0}{\partial \lambda_i} = -\frac{\int_R x^i \exp\left(\sum_{i=1}^{m} \lambda_i x^i\right) dx}{\int_R \exp\left(\sum_{i=1}^{m} \lambda_i x^i\right) dx} \qquad (5.5.42)$$

We can replace the left-hand side by $-m_i$, using (5.5.41).

$$m_i = \frac{\int_R x^i \exp\left(\sum_{i=1}^{m} \lambda_i x^i\right) dx}{\int_R \exp\left(\sum_{i=1}^{m} \lambda_i x^i\right) dx} \qquad (5.5.43)$$

This represents m simultaneous equations to be solved for λ_1, λ_2, ..., λ_m. Having these, λ_0 is obtained from (5.5.38). The equations above are put in a form more convenient for numerical solution as follows:

$$1 - \frac{\int_R x^i \exp\left(\sum_{i=1}^{m} \lambda_i x^i\right) dx}{m_i \int_R \exp\left(\sum_{i=1}^{m} \lambda_i x^i\right) dx} = R_i \qquad (5.5.44)$$

where the R_i's are the residuals that are reduced to near zero by a numerical technique. A solution can be obtained by using nonlinear programming to obtain the minimum of the sum of the squares of the residuals.

$$R = \sum_{i=1}^{m} R_i^2 = \text{minimum} \qquad (5.5.45)$$

Convergence is achieved when $R < \varepsilon$, or all $|R_i| < \varepsilon$, where ε is the specified acceptable error. Equation (5.5.38) is used to obtain λ_0. The integrals in (5.5.44) are evaluated numerically, and it is apparent that the bounds of the unknown density function must be known or assumed. The most successful nonlinear programming technique for this problem was found to be one proposed by Jacobson and Oksman (Siddall, 1982).

Because x^m must be evaluated in (5.5.44), there is risk of overflow in the computer. To circumvent this, the algorithm transforms the domain of x to between 0 and 1, using the transformation given in Appendix D.

The algorithm also requires an assumed starting point in order to begin the nonlinear programming solution. Experience with this algorithm has indicated that a good starting point is important for successful convergence, and therefore four alternative methods for determining a start are provided. The desired method can be preselected, or by default the algorithm will try them in the order given until one succeeds.

1. The distribution is assumed normal, in which case the parameters are estimated by

$$\lambda_1 = \frac{c_1}{c_2}, \quad \lambda_2 = -\frac{1}{2c_2}, \quad \lambda_3 = \lambda_4 = \cdots = \lambda_m = 0 \qquad (5.5.46)$$

2. The distribution is assumed to be uniform. This works better for J and U-shaped density functions

$$\lambda_1 = \lambda_2 = \cdots = \lambda_m = 0 \qquad (5.5.47)$$

3. A collocation method is used to calculate the x's. Numerical integration forms of (5.5.29) and (5.5.30) are written using m + 1 stations.

$$\sum_{j=1}^{m+1} a_j f(x_j) = 1$$

$$(5.5.48)$$

$$\sum_{j=1}^{m+1} a_j x_j^{i} f(x_j) = m_i, \quad i = 1, m$$

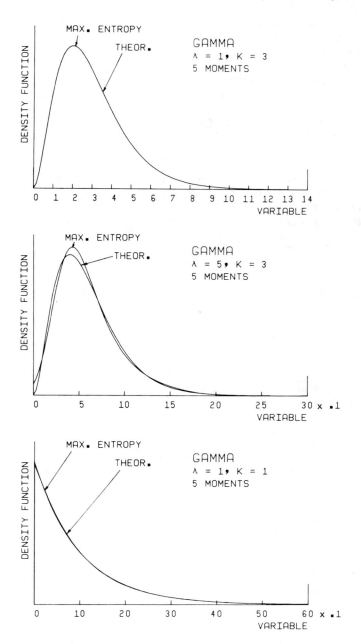

FIG. 5.9 Gamma and beta distributions.

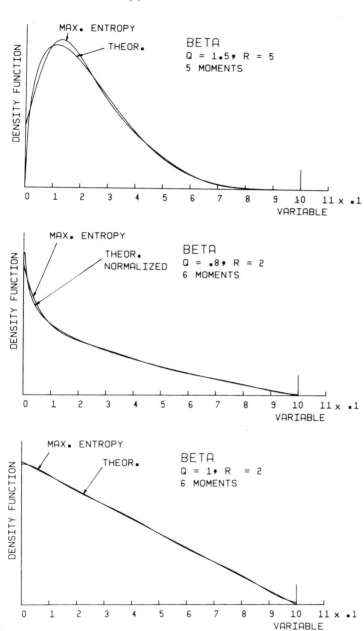

FIG. 5.9 (Continued)

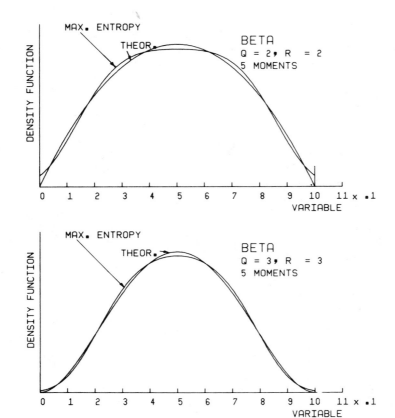

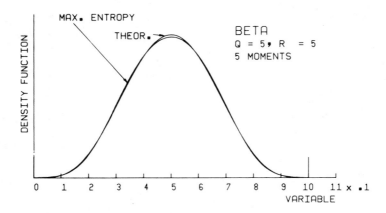

FIG. 5.10 Beta distributions.

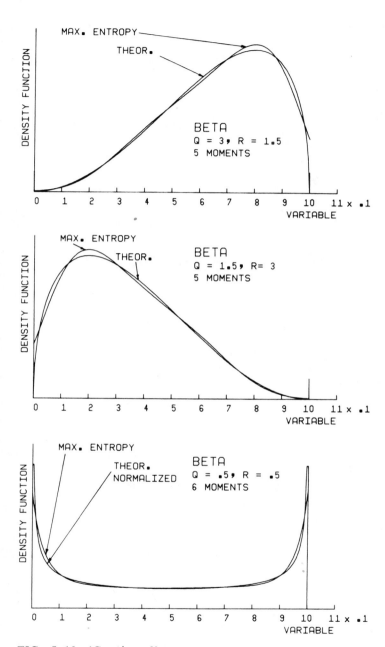

FIG. 5.10 (Continued)

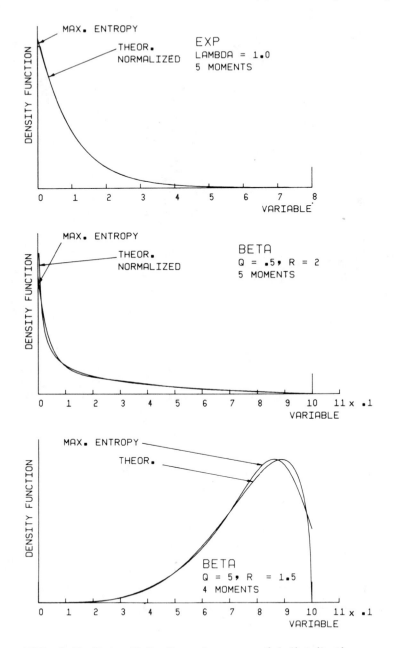

FIG. 5.11 Beta, Weibull, and exponential distributions.

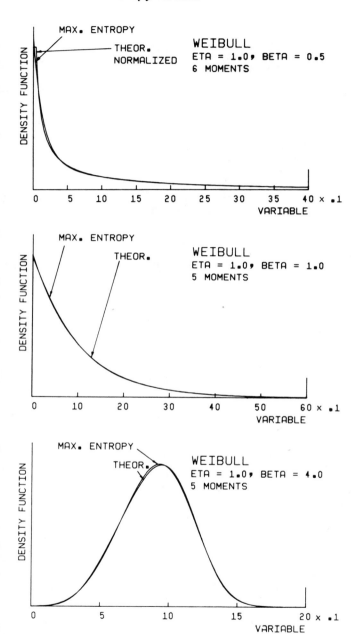

FIG. 5.11 (Continued)

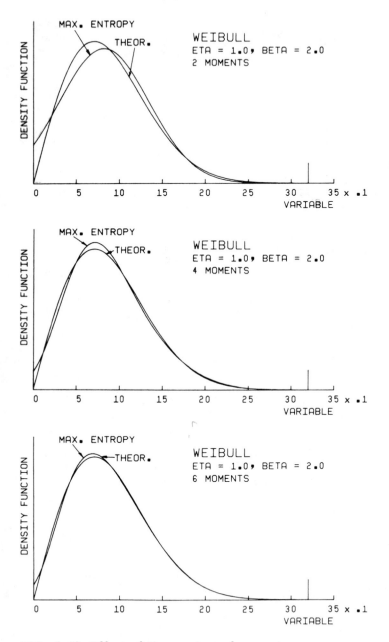

FIG. 5.12 Effect of the number of moments.

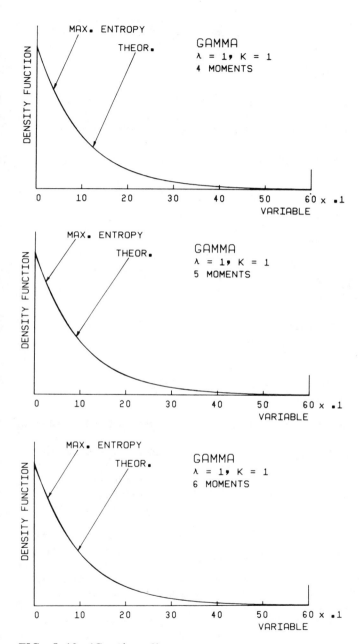

FIG. 5.12 (Continued)

where the a_j's are numerical integration multipliers. These $m + 1$ linear equations for the $m + 1$ unknown $f(x_j)$'s are solved and the and the results applied to (5.5.35).

$$f(x_j) = \exp (\lambda_0 + \lambda_1 x_j + \cdots + \lambda_m x_j^m), \quad j = 1, m + 1 \quad (5.5.49)$$

These can also be put in linear form and solved for the λ's.

4. A step-by-step start is used which begins using method 1 but with m set equal to 3. The algorithm then iterates with the values obtained for λ_0, λ_1, and λ_2, and with λ_3 equal to zero. This is repeated until m starting λ's are obtained.

The algorithm is summarized as follows.

1.   Input moments and bounds.
2.   Calculate moments about the origin.
3.   Transform the domain.
4.   Set up expressions for the residuals in (5.5.44).
5.   Select a starting method and calculate starting values.
6.   Call the optimization subroutine.
7.   Convergence achieved?
8.   *Yes*: Go to 12.
9.   *No*: All methods tried?
10.   *Yes*: Exit with failure.
11.   *No*: Select next method and go to 5.
12.   Calculate λ_0.
13.   Calculate λ's corresponding to original domain.
14.   Output results.

The algorithm also calculates a set of numerical values for the cumulative distribution function if desired.

The ability of the method to reproduce known distributions is illustrated in Figs. 5.9 to 5.11. These were generated by calculating moments from the analytical distribution and replotting the maximum entropy distribution on top of the original. We are thus, in effect, trying to reproduce distributions for which we have complete information.

The effect on the accuracy by increasing the number of moments used is indicated in Fig. 5.12 for the Weibull and gamma functions. The indications are that quite good density functions can be generated for most shapes using four or five moments. For some distributions, such as the normal, the exponential and the uniform, more than three or four moments makes no improvement.

The moments used in the illustrations are, of course, exact in the sense that the integral formulas were used. When sample moments are used, the accuracy would depend on the sample size. The method provides the best estimate of the distribution possible with the information available, but if the sample size becomes too small the estimates of the higher moments will tend to become meaningless.

A key problem is the choice of bounds. Sometimes at least one of them can be selected on physical grounds. Although one might be tempted simply to use a conservative estimate of the bounds based on the sample range, experience has indicated that badly selected bounds can give false rising tails on the maximum entropy distribution. Some iteration may be necessary to eliminate this. Thus the bounds are important items of input information, and more subjective than the data. Another possible iterative approach to selection of bounds is to use extreme value theory. This is discussed in Sec. 8.4.

In some applications the density function may be very sensitive to precision in the coefficients, and a large number of significant figures must be maintained in their values.

The basic algorithm is embodied in subroutine MEP 1, Appendix A.

EXAMPLE 5.3 Warehouse Floor Loads

This example illustrates the use of the maximum entropy method with a relatively large sample. The following table of data* represents observed live loads on a warehouse floor in lb/ft^2.

0.0	7.8	36.2	60.6	64.0	64.2	79.2	88.4	38.0	72.7
72.2	72.6	74.4	21.8	17.1	48.5	16.8	105.9	57.2	75.7
225.7	42.5	59.8	41.7	39.9	55.5	67.2	122.8	45.2	62.9
55.1	55.9	87.7	59.2	63.1	58.8	67.7	90.4	43.3	55.2
36.6	26.0	90.5	23.0	43.5	52.1	102.1	71.7	4.1	37.3
129.4	66.4	138.7	127.9	90.9	46.9	197.5	151.1	157.3	197.0
134.6	73.4	80.9	53.3	80.1	62.9	150.8	102.2	6.4	45.4
121.0	106.2	94.4	139.6	152.5	70.2	111.8	174.1	85.4	83.0
178.8	30.2	44.1	157.0	105.3	87.0	50.1	198.0	86.7	64.6
78.6	37.0	70.7	83.0	179.7	180.2	60.6	212.4	72.2	86.0
94.5	24.1	87.3	80.6	74.8	72.4	131.1	116.1	53.6	99.1
40.2	23.4	8.4	42.6	43.4	27.4	63.8	18.4	16.2	58.7
92.2	49.8	50.9	116.4	122.9	132.3	105.2	160.3	28.7	46.8
99.5	106.9	55.9	136.8	110.4	123.5	92.4	160.9	45.4	96.3
88.5	48.4	62.3	71.3	133.2	92.1	111.7	67.9	53.1	39.7
93.2	55.0	80.8	143.5	122.3	184.2	150.0	57.6	6.8	53.3
96.1	54.8	63.0	228.3	139.3	59.1	112.1	50.9	158.6	139.1
213.7	65.7	90.3	198.4	97.5	155.1	163.4	155.3	229.5	75.0
137.6	62.5	156.5	154.1	134.3	81.6	194.4	155.1	89.3	73.4
79.8	68.7	85.6	141.6	100.7	106.0	131.1	157.4	80.2	65.0
78.5	118.2	126.4	33.8	124.6	78.9	146.0	100.3	97.8	75.3
24.8	55.6	135.6	56.3	66.9	72.2	105.4	98.9	101.7	58.2

The maximum entropy distribution using five moments is called up in the following coding.

```
      DIMENSION X(220),SM(5),AL(7),XSPEC(1),PSPEC(1)
      READ(5,10)X
10    FORMAT(10F6.1)
```

---

*From Dunham et al. (1972).

```
        M=5
        NSAMP=220
        CALL  SMOM(X,M,NSAMP,SM)
        CALL  SORT(X,NSAMP)
C....  UPPER BOUND IS ESTIMATED FROM MAXIMUM X
        XMAX=X(NSAMP)
        XMIN=0.0
        XSPEC(1)=200.
        DO 4 I=1,3
        CALL  MEP1(M,SM,XMIN,XMAX,1,XSPEC,1,0,AL,PSPEC)
        WRITE(6,1)XMAX
1       FORMAT(////,20H  RESULTS FOR XMAX =,F7.1)
        WRITE(6,2)(AL(J),J=1,6)
2       FORMAT(51H  THE MAXIMUM ENTROPY DISTRIBUTION COEFFICIENTS ARE,//,(
        15E14.5))
        WRITE(6,3)PSPEC(1)
3       FORMAT(//,59H  THE PROBABILITY OF THE LOADING BEING LESS THAN 200
        1PSF IS,E14.5)
        XMAX=XMAX+10.
4       CONTINUE
        STOP
        END
```

```
RESULTS FOR XMAX =   229.5
THE MAXIMUM ENTROPY DISTRIBUTION COEFFICIENTS ARE

 -.69352E+01     .81214E-01    -.98970E-03     .49905E-05    -.14505E-07
 .21327E-10

THE PROBABILITY OF THE LOADING BEING LESS THAN 200 PSF IS    .96704E+00


RESULTS FOR XMAX =   239.5
THE MAXIMUM ENTROPY DISTRIBUTION COEFFICIENTS ARE

 -.67819E+01     .63374E-01    -.41700E-03    -.23696E-05     .25926E-07
 -.57575E-10

THE PROBABILITY OF THE LOADING BEING LESS THAN 200 PSF IS    .96995E+00


RESULTS FOR XMAX =   249.5
THE MAXIMUM ENTROPY DISTRIBUTION COEFFICIENTS ARE

 -.67212E+01     .56423E-01    -.19852E-03    -.51147E-05     .40660E-07
 -.85672E-10

THE PROBABILITY OF THE LOADING BEING LESS THAN 200 PSF IS    .97086E+00
```

The results show the extent of sensitivity to choice of the upper bound; the lower bound is established at zero. The corresponding curves are plotted in Fig. 5.13, superimposed on a normalized relative frequency histogram.

EXAMPLE 5.4 Fracture Toughness

The use of the maximum entropy method with moments is illustrated in this example for a relatively small sample. The following sorted data represent the fracture toughness in ksi$\sqrt{in}$. for

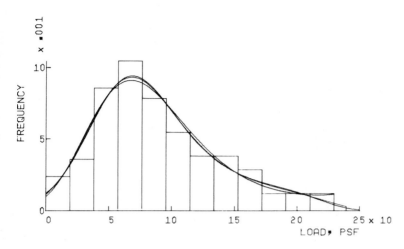

FIG. 5.13 Maximum entropy distributions for floor loadings, showing sensitivity to the upper bound.

nickel maraging steel*: 69.5, 71.9, 72.6, 73.1, 73.3., 73.5, 74.1, 74.2, 75.3, 75.5, 75.7, 75.8, 76.1, 76.2, 76.2, 76.9, 77.0, 77.9, 78.1, 79.6, 79.7, 79.9, 80.1, 82.2, 83.7, 93.7.

The coding for obtaining the maximum entropy distribution is given below.

```
      DIMENSION X(26),XCDF(1),AL(6),CDF(1),SM(5)
      DATA X/69.5,71.9,72.6,73.1,73.3,73.5,74.1,74.2,75.3,75.5,75.7,75.8
     1,76.1,76.2,76.2,76.9,77.0,77.9,78.1,79.6,79.7,79.9,80.1,82.2,83.7,
     293.7/
      M=4
      NSAMP=26
      CALL SMOM(X,M,NSAMP,SM)
      XMIN=X(1)
      XMAX=X(NSAMP)
      DO 3 I=1,6
      XMIN=X(1)-FLOAT(I-1)*.5
      CALL MEP1(M,SM,XMIN,XMAX,0,XCDF,1,0,AL,CDF)
      WRITE(6,1)XMIN,XMAX
1     FORMAT(15H  LOWER BOUND =,F7.2,16H    UPPER BOUND =,F7.2)
      WRITE(6,2)(AL(J),J=1,5)
2     FORMAT(51H  THE MAXIMUM ENTROPY DISTRIBUTION COEFFICIENTS ARE,//,
     15E14.5,///)
3     CONTINUE
      STOP
      END
```

*Data taken from Kies et al. (1965).

```
LOWER BOUND =  69.50    UPPER BOUND =  93.70
THE MAXIMUM ENTROPY DISTRIBUTION COEFFICIENTS ARE
  .10107E+05    -.52536E+03    .10187E+02    -.87324E-01    .27918E-03


LOWER BOUND =  69.00    UPPER BOUND =  93.70
THE MAXIMUM ENTROPY DISTRIBUTION COEFFICIENTS ARE
  .81682E+04    -.42974E+03    .84230E+01    -.72908E-01    .23511E-03


LOWER BOUND =  68.50    UPPER BOUND =  93.70
THE MAXIMUM ENTROPY DISTRIBUTION COEFFICIENTS ARE
  .68599E+04    -.36513E+03    .72298E+01    -.63142E-01    .20523E-03


LOWER BOUND =  68.00    UPPER BOUND =  93.70
THE MAXIMUM ENTROPY DISTRIBUTION COEFFICIENTS ARE
  .59635E+04    -.32081E+03    .64101E+01    -.56424E-01    .18464E-03


LOWER BOUND =  67.50    UPPER BOUND =  93.70
THE MAXIMUM ENTROPY DISTRIBUTION COEFFICIENTS ARE
  .53451E+04    -.29018E+03    .58430E+01    -.51771E-01    .17037E-03


LOWER BOUND =  67.00    UPPER BOUND =  93.70
THE MAXIMUM ENTROPY DISTRIBUTION COEFFICIENTS ARE
  .49072E+04    -.26847E+03    .54404E+01    -.48463E-01    .16020E-03
```

The bounds are first taken from the sample, and then iteration is used to adjust the bounds using the results from the previous solution by judgment from the plot. Successive plots are shown in Fig. 5.14. We see that lower bound adjustment gives a better curve. The upper bound tail is clearly due, in this case, to the outlier. The maximum entropy distribution faithfully reflects the rather large influence this outlier has on the moments. The engineer might wish to subjectively modify this tail if he or she believes that it is not truly representative of the population. The problem of outliers, and deciding whether or not to accept them, is discussed in Sec. 8.4.

This example illustrates the importance of the bounds as components of the designer's subjective judgment of the nature of the random variable. The fact that different bounds could lead to varying results for probability estimates should not be a matter of concern. It simply reflects the result of different judgmental inputs to risk estimation.

The curves in Fig. 5.14 faithfully reflect the designer's assumptions about the lower bound. If it is made "tight" or close to the smallest sample value, the initial value of the density function is relatively large, showing that there is a high probability of getting values near the low end. On the other hand, if the bound is stretched out, remote from the smallest observed value, it will pull the density function down into a tail, showing low probabilities of such low values occurring.

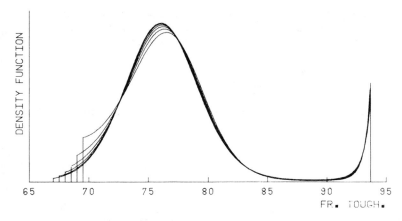

FIG. 5.14 Maximum entropy distribution for fracture toughness of steel, iterating to adjust the bounds.

The maximum entropy algorithm could be easily modified for use with a discrete variable. The integrals would simply be replaced by summations. It would also be possible, particularly if the number of discrete values is relatively small, to optimize the following directly.

$$S = - \sum_{j=1}^{n} f(x_j) \ln [f(x_j)] = \text{maximum} \qquad (5.5.50)$$

$$\sum_{j=1}^{n} f(x_j) = 1 \qquad (5.5.51)$$

$$\sum_{j=1}^{n} x_j^i f(x_j) = m_i, \quad i = 1, m \qquad (5.5.52)$$

See, for example, Griffeath (1972).

5.5.5 Using Jaynes' Principle with Ranks*

For small sample sizes the use of moments with Jaynes' principle has dubious validity. The higher moments become meaningless. In this

---

*The material in this section is an extension of an approach given by Siddall and Badawy (1979).

section a similar approach is used in which ranks are substituted for the moments. The concept of ranks is discussed in Sec. 5.4, and in more detail in Sec. 8.2. In the latter section the rank distribution is shown to be

$$f(r_i) = \frac{n!}{(i - 1)! \, (n - i)!} \, r_i^{i-1} (1 - r_i)^{n-1} \tag{5.5.53}$$

where r_i is the ith rank in an ordered sample of size n, and is defined by

$$r_i = \int_{-\infty}^{o_i} f_x(x) \, dx \tag{5.5.54}$$

The quantity o_i is the ith order statistic; and both o_i and r_i are random variables. The unknown value of the cumulative distribution function corresponding to a sample value of x is thus represented by the random variable r_i. This is illustrated in Fig. 5.15. In Sec. 5.4 we used the mean rank to estimate F(x) corresponding to a ranked observation. Now, however, we have to admit that all we know about r_i is its distribution and that its range is from 0 to 1. An encouraging aspect of the problem is that $f(r_i)$ is independent of the distribution of x, and depends only on i and n. Despite a constant range of 0 to 1, we would expect a more pinched distribution, with smaller variance, as n increases.

We shall take advantage of this pinching effect by discarding a small equal percentage of tail area on each end of $f(r_i)$, so that the retained area is q. We thus impose practical bounds on $F(o_i)$, expressed as

$$b_\ell \leqslant r_i \leqslant b_u \tag{5.5.55}$$

where the lower and upper bounds, b_ℓ and b_u, are obtained from

$$\frac{1 - q}{2} = \int_0^{b_\ell} f(r_i) \, dr_i \tag{5.5.56}$$

$$\frac{1 - q}{2} = \int_{b_u}^1 f(r_i) \, dr_i \tag{5.5.57}$$

Thus for a given q, we can use (5.5.53) to solve for b_ℓ and b_u. This information is used with Jayne's principle in the following expressions.

$$S = - \int_\ell^u f(x) \ln [f(x)] \, dx = \text{maximum} \tag{5.5.58}$$

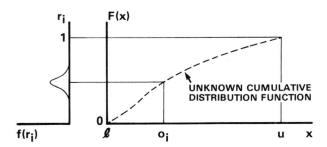

FIG. 5.15 Illustration of how the rank distribution represents the uncertainty about the value of F(x) corresponding to a ranked observation.

$$\int_{\ell}^{u} f(x) \, dx = 1 \tag{5.5.59}$$

$$b_{\ell} \leqslant \int_{\ell}^{d_i} f(x) \, dx \leqslant b_u, \quad i = 1, n \tag{5.5.60}$$

where

d_i = value of x for the ith member of a ranked sample

n = sample size

It is not possible to develop a theoretical distribution for f(x) in this case, but it would seem reasonable to fit an exponential function which is essentially the same as that used in the previous development. We shall designate the parameters γ.

$$f(x) = \exp\left(\gamma_0 + \sum_{j=1}^{m} \gamma_j x^j\right) \tag{5.5.61}$$

where m is an arbitrary integer of the order of 4 or 5. This approach corresponds to the Ritz method of solving variational problems.

Substituting (5.5.61) into (5.5.58), (5.5.59), and (5.5.60) now yields a discrete optimization problem with unknown γ's. For a given q, b_u and b_{ℓ} are not problem dependent, and it may be numerically more economical to select them from a precalculated table in each case.

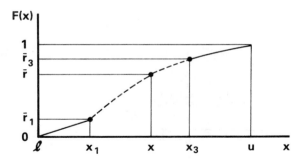

FIG. 5.16 Feasible estimate of the cumulative distribution function based on a ranked sample.

As before, the numerical search for an optimum requires a set of starting values for the γ's. The problem is more difficult in this case, that is, less likely to converge, because of sheer size. We could use the same starting techniques as before, but a feasible start may be essential. This can be achieved if we begin with a density function that we know satisfies (5.5.60). One that is associated with a cumulative distribution function drawn through the mean rank points should be feasible or close to it. This is illustrated in Fig. 5.16. In terms of the cumulative distribution function, the moments are

$$\mu = \int_{\ell}^{u} x \, \frac{dF}{dx} \, dx \qquad\qquad (5.5.62)$$

$$c_i = \int_{\ell}^{u} (x - \mu)^i \, \frac{dF}{dx} \, dx, \quad i = 2, m \qquad\qquad (5.5.63)$$

We can integrate piecewise as follows:

$$\mu = \int_{\ell}^{x_1} \frac{\bar{r}_1}{x_1 - \ell} \, x \, dx + \int_{x_1}^{x_2} \frac{\bar{r}_2 - \bar{r}_1}{x_2 - x_1} \, x \, dx$$

$$+ \cdots + \int_{x_k}^{x_{k+1}} \frac{r_{k+1} - r_i}{x_{k+1} - x_i} \, x \, dx + \cdots + \int_{x_n}^{u} \frac{1 - \bar{r}_n}{u - x_n} \, x \, dx$$

$$(5.5.64)$$

After substituting for the values of the mean ranks and integrating, this reduces to

$$\mu = \frac{\left(\ell + u + \sum\limits_{i=1}^{n} x_i\right)}{n + 1} \tag{5.5.65}$$

A similar treatment for the higher moments gives

$$c_j = \frac{1}{(n + 1)(j + 1)} \left[\frac{(x_1 - \mu)^{j+1} - (\ell - \mu)^{j+1}}{x_1 - \ell} \right.$$

$$+ \cdots + \frac{(x_{k+1} - \mu)^{j+1} - (x_k - \mu)^{j+1}}{x_{k+1} - x_k} \tag{5.5.66}$$

$$\left. + \cdots + \frac{(u - \mu)^{j+1} - (x_n - \mu)^{j+1}}{u - x_n} \right], \quad j = 2, m$$

The starting algorithm uses these moments to get a MEP1 solution, which provides the required starting γ's.

The method can be verified in a preliminary manner by generating theoretical samples (see Sec. 6.2) from analytical distributions and comparing the results from these samples with the originals. Figures 5.17 and 5.18 give the results of some trials. Plots based on "true samples" are also shown, obtained by inverting the theoretical cumulative distribution function in order to obtain variable values corresponding to mean ranks.

We would intuitively anticipate that the accuracy would improve as the sample size increases, as shown in Fig. 5.18, although as this occurs, numerical problems are more likely to be encountered with convergence of the algorithm. It would also be expected that accuracy would improve with the number of parameters used, as shown in Fig. 5.18, although round-off errors would tend to limit this.

The problem of selecting the initial bounds is also important in this algorithm, and iteration may be necessary.

EXAMPLE 5.5 Fracture Toughness

This is a repeat of Example 5.4, which generated the density function for fracture toughness of a steel using moments, and it will be of interest to see how they compare. A possible criterion

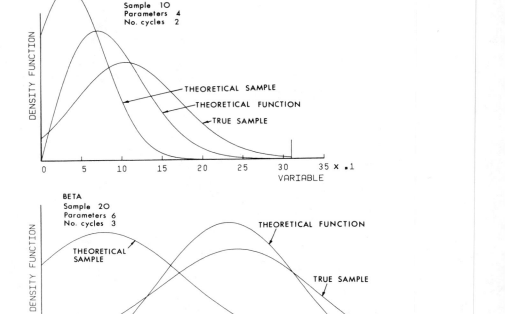

FIG. 5.17 Illustrations of the maximum entropy distribution using mean ranks.

is the Kolmogorov-Smirnov goodness-of-fit test given in Sec. 5.7. The results are generated by the following program.

```
      DIMENSION X(26),RP(6),CUM(1),PHI(52),XD(26),XP(1),W(76)
      DIMENSION SM(4),ARG(51),DENS(51)
      COMMON/MEP1/KPRINT,TOL,MAXFN
      DATA X/69.5,71.9,72.6,73.1,73.3,73.5,74.1,74.2,75.3,75.5,75.7,
     175.8,76.1,76.2,76.2,76.9,77.0,77.9,78.1,79.6,79.7,79.9,80.1,82.2,
     283.7,93.7/
C.... GENERATE AND PLOT MAX ENTROPY DIST BASED ON MEAN RANKS
      NL=5
      NSAMP=26
      QPROB=.99
      XMIN=67.0
      XMAX=93.7
      MINT=51
      NXP=0
```

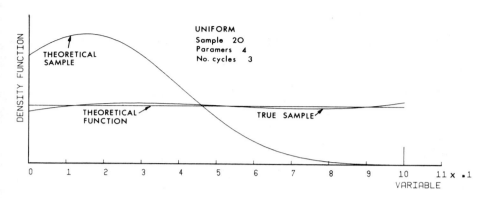

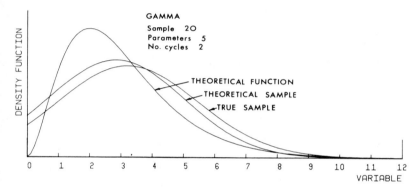

FIG. 5.17 (Continued)

```
        KDATA=1
        KSTART=1
        NCYCLE=2
        CALL MEP2(NL,NSAMP,X,XMIN,XMAX,QPROB,MINT,NXP,XP,KSTART,KDATA,
       1NCYCLE,RP,CUM,PHI,XD,W)
        CALL MEPOUT(RP,NL,0,XP,CUM)
        DO 1 I=1,MINT
        ARG(I)=FLOAT(I-1)/FLOAT(MINT-1)*(XMAX-XMIN)+XMIN
   1    DENS(I)=ENTRPF(RP,NL,ARG(I))
        CALL PLOTPL(1,ARG,DENS,MINT,8HFR TOUGH,8,XMAX,XMIN)
        CALL PLOTPL(1,ARG,DENS,MINT,8HFR TOUGH,8,XMAX,XMIN)
C.... GENERATE AND PLOT MAX ENTROPY DIST BASED ON MOMENTS
        MAXFN=200
        KPRINT=1
        CALL SMOM(X,4,NSAMP,SM)
        CALL MEP1(4,SM,XMIN,XMAX,0,XP,1,1,RP,CUM)
        CALL MEPOUT(RP,NL,0,XP,CUM)
        DO 2 I=1,MINT
   2    DENS(I)=ENTRPF(RP,NL,ARG(I))
        CALL PLOTPL(1,ARG,DENS,MINT,8HFR TOUGH,8,XMAX,XMIN)
        CALL PLOTPL(0,ARG,DENS,MINT,8HFR TOUGH,8,XMAX,XMIN)
        STOP
        END
```

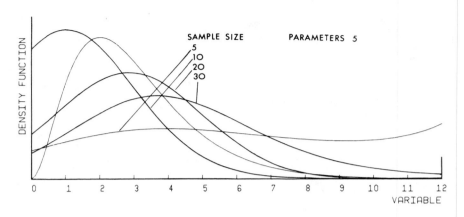

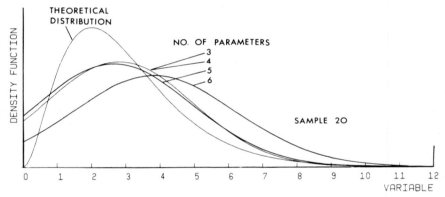

FIG. 5.18 Illustration of the effect on the maximum entropy distribution of changing the number of parameters and changing the sample size, using a theoretical sample from a gamma distribution: $\lambda = 1$, $k = 3$.

The two density functions are shown in Fig. 5.19.

EXAMPLE 5.6 Fatigue Strength

This example is taken from Lipson and Sheth (1973), and represents five values of fatigue strength of a member in ksi. In the reference a Weibull distribution was fitting having the form

$$f(x) = \frac{3.0}{60.3 - 50} \left(\frac{x - 50}{60.3 - 50} \right)^{2.0} \exp\left[-\left(\frac{x - 50}{60.3 - 50} \right)^{3.0} \right]$$

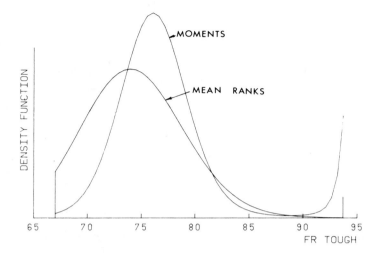

FIG. 5.19 Comparison of two kinds of maximum entropy distributions fitted to a data set.

It is interesting to compare this with a maximum entropy distribution based on ranks, generated by the following program.

```
        DIMENSION X(5),RP(6),CUM(1),PHI(10),XD(5),XP(1),W(29)
        DIMENSION ARG(51),DENS(51)
        DATA X/55.3,57.3,59.2,61.4,62.5/
C.... GENERATE AND PLOT WEIBULL
        XMIN=50.
        XMAX=70.
        MINT=51
        DO 1 I=1,MINT
        ARG(I)=XMIN+FLOAT(I-1)/FLOAT(MINT-1)*(XMAX-XMIN)
1       DENS(I)=FWEIB(ARG(I),XMIN,60.3,3.0)
        CALL PLOTPL(1,ARG,DENS,MINT,10HFAT STRENG,10,XMAX,XMIN)
C.... GENERATE AND PLOT MAX ENTROPY DIST
        NL=4
        NSAMP=5
        QPROB=.99
        NXP=0
        KDATA=1
        KSTART=1
        NCYCLE=1
        CALL MEP2(NL,NSAMP,X,XMIN,XMAX,QPROB,MINT,NXP,XP,KSTART,KDATA,
        1NCYCLE,RP,CUM,PHI,XD,W)
        CALL MEPOUT(RP,NL,NXP,XP,CUM)
        CALL ENTPL(XMIN,XMAX,NL,RP,10,10HFAT STRENG)
        CALL PLOTPL(0,ARG,DENS,MINT,10HFAT STRENG,10,XMAX,XMIN)
        STOP
        END
```

```
THE MAXIMUM ENTROPY DISTRIBUTION COEFFICIENTS ARE
-.23626E+03   .11647E+02 -.19284E+00   .10587E-02
```

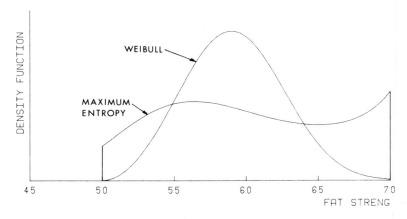

FIG. 5.20 Comparison of a Weibull and a maximum entropy density function fitted to a data set for fatigue strength having five values.

    The two curves are plotted in Fig. 5.20
    Although this application of the maximum entropy principle is of considerable theoretical interest, the software provided in PROBVAR (MEP2) should be used with caution. Full convergence to a maximum for the entropy cannot be guaranteed, and the software is not as robust as that used for the maximum entropy distribution using moments.
    It is also important to realize that a small sample alone provides inadequate information for defining a density function. It must be supplemented by engineering judgment, and this limits the usefulness of this method. Thus, in Example 5.6, we would be inclined to prefer the Weibull function—not because it may fit the data better, nor even because the Weibull function commonly gives a good fit for fatigue problems—but because our intuitive judgment tells us that in this instance, the Weibull *shape* better represents the random nature of fatigue life than does the maximum entropy curve shape.

5.6 SUBJECTIVE DISTRIBUTIONS

5.6.1 Subjective Sketching

If few or no data are available for building a density function for a random variable, the engineer can codify his or her judgment about its random nature by a subjectively sketched density function.
    The density function curve is the most direct expression of the random nature of a quantity, and is therefore the best vehicle for codifying that randomness. Familiarity with a wide variety of density function shapes would be useful to an engineer when sketching a new

one. Nevertheless, most engineers are not accustomed to codifying randomness in this way, and until the usage becomes more widespread and better documented, engineers will probably be more comfortable using various characteristic points, discussed below. Also, direct "eyeballing" may not be the best way to express judgment about the extreme tails of a density function, which sometimes are its most important zones.

When direct sketching of the density function curve is used, it is not necessary to have correct scaling of the function. The scale of the ordinate axis can be arbitrary, and the function later normalized to have an area equal to unity by obtaining the area of the sketched function using numerical integration, and then dividing each function value by this area. Subroutine FNORM in Appendix A provides this facility.

In the use of characteristic points, the engineer specifies points that are intuitively meaningful. The different possibilities are enumerated below, and related to the density function. In each case the subjectively estimated point is designated by an asterisk superscript.

Bounds

The upper and lower limits for the density function are specified, and shown in Fig. 5.21. In terms of the unknown density function, $f(x)$, they are expressed as

$$f(\ell^*) = f(u^*) = 0, \quad \text{or possibly a finite value} \tag{5.6.1}$$

If the engineer is not prepared to hazard estimates of an additional characteristic points, he or she simply sketches an arbitrary density function between them, possibly skewing it a bit one way or the other. The vertical scale is adjusted by numerical integration, using subroutine FNORM in Appendix A.

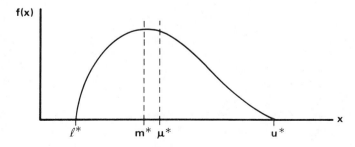

FIG. 5.21 Illustration of the bounds, the mean, and the mode.

Mode

The mode is the most likely value of the random variable, and corresponds to the maximum value of the density function, as indicated in Fig. 5.21.

$$\max\ [f(x)] = \tilde{x}^* \tag{5.6.2}$$

or

$$\frac{d[f(\tilde{x}^*)]}{dt} = 0 \tag{5.6.3}$$

An estimate of the mode can easily be combined with the bounds to give a somewhat more precisely defined density function.

Mean

The average value is a very familiar statistical parameter, shown in Fig. 5.21.

$$\int_R x\ f(x)\ dx = \mu^* \tag{5.6.4}$$

The mean can also be combined with the bounds as a basis for a sub-jective sketch of the density function. However, it is considerably more difficult to obtain a consistent curve. Trial and error must be used, adjusting a sequence of trial sketches until (5.6.4) is satisfied, and using numerical techniques embodied in subroutine CMOM.

Percentiles

A percentile is the variable value below which a given percentage of occurrences can be expected, illustrated in Fig. 5.22. The $\tilde{x}^*_{0.50}$

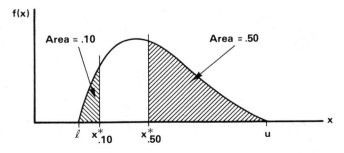

FIG. 5.22 Illustration of percentiles.

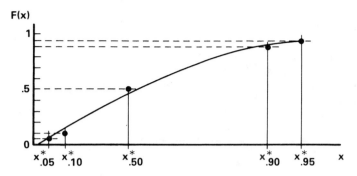

FIG. 5.23 Subjective cumulative distribution function.

percentile, for example, is the value at which one-half of the population values can be expected to occur. The commonest percentiles used are

$$\int_{\ell*}^{\check{x}_{0.10}^*} f(x) \, dx = 0.10 \qquad (5.6.5)$$

$$\int_{\ell*}^{\check{x}_{0.50}^*} f(x) \, dx = 0.50 \qquad (5.6.6)$$

$\check{x}_{0.10}^*$ and $\check{x}_{0.50}^*$.

The 10 percentile corresponds to a value of 0.10 for the cumulative distribution function and it is clear that estimating percentiles is really estimating the distribution function. A series of these will thus give a subjective distribution curve, as shown in Fig. 5.23. A smooth curve can then be fitted by eye or by using subroutines SMOOTH.

Selecting percentile (or fractile) points can be approached in different ways. For example, it may be more meaningful to an engineer to answer the following sequence of questions.

1. What value would have a 50-50 chance of being either exceeded or not exceeded? This is the $\check{x}_{0.50}^*$ percentile.
2. Given that the value is less than $\check{x}_{0.50}^*$, what value would have a 50-50 chance of being exceeded or not? This is the $\check{x}_{0.25}^*$ percentile.
3. Given that the value is greater than $\check{x}_{0.50}^*$, what value would have a 50-50 chance of being exceeded or not? This is the $\check{x}_{0.75}^*$ percentile.

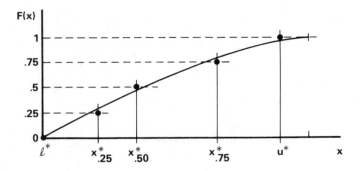

FIG. 5.24 Estimate of a cumulative distribution function using a sequence of percentiles.

The process could be iterated, in finer subdivisions, but would
soon begin to lose meaning. It might be helpful also to estimate the
bounds. The data would be plotted as a cumulative distribution function and smoothed by eyeball judgment. The result would look as
shown in Fig. 5.24.

If a cumulative distribution function is sketched rather than a
density function, the latter should always be obtained by using subroutine CUMTOF, and plotted. It is important to develop a catalog of
density functions as a basis of comparison.

5.6.2 The Evolutionary Method

The evolutionary method is based on the concept that if a component
is being examined for which no hard random data exist, it is probably
essentially similar, from a stochastic point of view, to some existing
component for which probability functions are known. The parts may
be physically very similar, with relatively minor modifications; or they
may be quite dissimilar. The important criteria are that the modeling
is essentially the same, and the environmental loadings are essentially
similar. If these criteria are satisfied, the hypothesis is that the *form*
of the probability function will be unchanged, and that only the *scale*
will change.

We would therefore have some unknown transformation of variable
parameter which changes the scale.

$$x_n - \ell_n = k_s (x_0 - \ell_0) \qquad (5.6.7)$$

where

x_n = variable for new component

x_0 = variable for old component

k_s = transformation parameter or *evolutionary* factor

ℓ_n = lower bound for new component

ℓ_0 = lower bound for old component

This is illustrated in Fig. 5.25 for $k_s > 1$.

The relationship between density functions is (see Sec. 6.3)

$$f_n(x_n) = \frac{1}{k_s} f_0(x_0) = \frac{1}{k_s} f_0 \left[\ell_0 + \frac{1}{k_s} (x_n - \ell_n) \right] \qquad (5.6.8)$$

If no data are available for the new component, the engineer would simply use judgment to decide on values of k_s and ℓ_n.

If small-sample information is available for the new component, we designate it $\overline{S} = [X_1, X_2, \ldots, X_n]$, in ranked form. This new information is used to update the old (a priori) density function, and the parameters k_s and ℓ_n are estimated by a least-mean-squared-error fit.

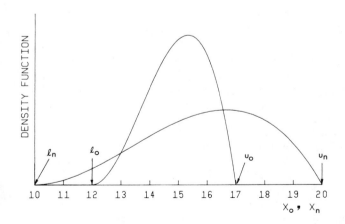

FIG. 5.25 Evolutionary change in the density function.

For a trial k_s and ln we determine

$$F_n(X_i) = \int_{ln}^{X_i} f_n(x) \, dx, \quad i = 1, n$$

$$= F_0 \left[\ell_0 + \frac{1}{k_s} (X_i - \ell_n) \right] \qquad (5.6.9)$$

The mean rank is calculated (see Sec. 5.4) as

$$\bar{r}_i = \frac{i}{n + 1}, \quad i = 1, n \qquad (5.6.10)$$

The "error" is

$$e_i = F_n(X_i) - \bar{r}_i, \quad i = 1, n \qquad (5.6.11)$$

The parameters k_s and ℓ_n are determined by minimizing the function

$$U = \sum_{i=1}^{n} e_i^2 = \text{minimum} \qquad (5.6.12)$$

Subroutine EVOL can be used to evaluate k_s and ℓ_n. It employs a standard optimizing technique from Siddall (1982).

5.6.3 The Consensus Method

It is important for an engineer to draw on all possible sources of information when setting up a subjective density function; and this may include immediate co-workers, plus people from other areas who are knowledgeable in the field. These areas could include sales, marketing, purchasing, production, and the like.

Some of these people will not be familiar with probabilistic formalities, and a simple informal discussion would be helpful. However, in some situations a formalized approach may be useful, based on the Delphi Method (Sackman, 1975). The essence of the method is to have a coordinator solicit judgments on the form of the density function (in this application) from a group of experts. The coordinator then returns the set to each of the experts in anonymous form and asks for a new judgment. This may be iterated until something near a consensus is achieved.

5.7 GOODNESS-OF-FIT TESTS

The two most common tests of how well data fits a proposed distribution are the Kolmogorov-Smirnov test and the chi-square test. The former is quite simple, and begins by ordering the sample from low to high. A stepwise cumulative frequency function is obtained using the expressions

$$Q(x) = 0, \quad x < x_1$$

$$= \frac{i}{n}, \quad x_i \leqslant x < x_{i+1} \tag{5.7.1}$$

$$= 1, \quad x \geqslant x_n$$

where the sample data are $x_1, x_2, \ldots, x_n$. We would expect from our mean rank plotting that this function would tend to follow the cumulative distribution function, as suggested in Fig. 5.26. The test quantity is the maximum different over the whole range between $Q(x)$ and $F(x)$.

$$T = \max_x |F(x) - Q(x)| \tag{5.7.2}$$

This test value can be used to compare different analytical functions in a measure of goodness of fit.

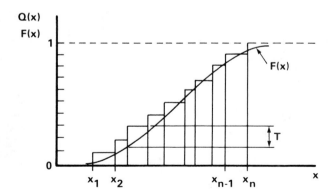

FIG. 5.26 Plot of a stepwise cumulative frequency function and the proposed theoretical cumulative distribution function for the Kolmogorov-Smirnov test.

The chi-square test requires that the range of the variable be divided into k equally spaced intervals, Δx wide. A relative frequency histogram is then obtained for the data, with the frequency in each interval being designated q_i. The theoretical frequency for each interval is then calculated from the proposed analytical density function $f(x)$.

$$e_i = \int_{\text{ith interval}} f(x) \, dx = f(x_i) \, \Delta x \qquad (5.7.3)$$

These are shown in Fig. 5.27. The test expression now is

$$T = \sum_{i=1}^{k} \frac{(q_i - e_i)^2}{e_i} \qquad (5.7.4)$$

The Kolmogorov-Smirnov test is preferable for relatively small samples, when a frequency histogram would become meaningless. The two methods are embodied in FITT1 and FITT2 in Appendix A.

In both methods it can be shown that T is a random variable, and the distributions can be found in books on statistics. These can be

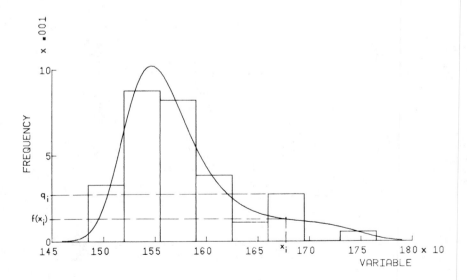

FIG. 5.27 Illustration for the chi-square test.

used to obtain significance levels to use in deciding whether to accept or reject a proposed theoretical distribution. However, this is not consistent with our rigorous definition of subjective probability, and the tests should be limited to comparing different candidate distributions.

5.8 CHOICE OF METHOD

The choice of method depends primarily on the kinds of information available. These can usefully be categorized as follows.

1. Fewer than about 5 observed values of the random variable are available and perhaps none.
2. Between 5 and about 50 observed values of the random variable are available.
   a. Experience with similar variables in similar devices is available.
   b. No such previous experience is available.
3. More than 50 observed values are available.

If an engineer is confronted with case 1 and has almost no hard data, one of the subjective techniques suggested in Sec. 5.6 must be used. If a relatively small sample of values is available, fewer than about 50, the engineer really cannot make a decision as to the best type of distribution to use, based on the data alone. In the event of case 2a, it may be acceptable to use a theoretical distribution that has been found to work well with similar variables in similar devices. However, this must be done with caution. It is not unusual to find that a theoretical distribution such as the Weibull or normal has been used in the past without adequate justification, and the true history should be carefully determined. Lacking this kind of relevant experience, an engineer can try several candidate distributions and compare them by means of the Kolmogorov-Smirnov goodness-of-fit test. However, it may be found that there is no significant difference between two or more types, and the choice must be arbitrary, or based on judgment. The conceptual basis for the maximum entropy distribution indicates that it should be used unless there is clear evidence that another distribution is better. The maximum entropy distribution based on ranks should be used if the sample size is relatively small, say less than 25.

In the event of case 3, with a relatively large sample, the goodness-of-fit test (chi-square) becomes more meaningful. With quite large samples, a histogram-based fit is probably the least biased choice, although it will be found that a maximum entropy distribution gives a close fit, and provides the convenience of a theoretical representation.

There is also the problem of choosing the best method of estimating parameters of a given theoretical distribution. Should one, for example, use mean-rank plotting, the method of moments, or the maximum likelihood method? Theoretical arguments have been used to favor one or the other, but the best guide is the goodness-of-fit test. The method of mean-rank plotting has the advantage that it also tests the suitability of the theoretical distribution being tried.

A strong consensus does not exist in the literature on criteria for choice. Some distributions, such as the normal and exponential, have been widely used on the basis primarily of mathematical convenience, rather than on the basis of an appropriate representation of the population. This is partly based on a property that mathematicians call *robustness* (Kendall and Stuart, 1979). Some estimation techniques are said to be robust if they give satisfactory results when the distribution is assumed normal when in reality the population is not normal. However, robustness is essentially limited to the estimation of the mean, and if we wish to use the distribution to estimate probabilities of events associated with a random variable, serious error can result. Nevertheless, the convenience and flexibility of using a single distribution is very attractive, and Haugen (1980), for example, has proposed a whole system of probabilistic design based only on the normal distribution.

Chapter 3 gives the most commonly used theoretical distributions and gives references for more in the literature. Chapter 8 gives distributions developed primarily for extreme value random variables, but there is no reason why they, too, could not be used as general-purpose candidates to represent random variables.

PROBLEMS

Problem 5.1 Write a user-oriented computer program to generate a smooth normalized probability density function, using a numerical definition of the function based on the frequency histogram.

Problem 5.2 Find from some source a set of sampled data, and use the method of plotting ranks to attempt to fit the data to each of the following analytical distributions using linear graph paper: exponential, extreme value, and Weibull. Do the same for the normal distribution using normal probability paper.

Problem 5.3 Using the same data as in Problem 5.2, determine the corresponding maximum entropy distribution. Plot curves for all four probability density functions and compare.

Problem 5.4 Use the chi-square method to compare the four distributions developed in Problems 5.2 and 5.3.

Problem 5.5 A type of diesel engine has had time to failure in hours recorded from operating records of 40 machines, as follows:*

0.13	121.58	2959.47	102.34
0.78	672.87	124.09	393.37
3.55	62.09	85.28	184.09
14.29	656.04	380.00	1646.01
54.85	735.89	298.58	412.03
216.40	895.80	678.13	813.00
1296.93	1057.57	861.93	239.10
952.65	470.97	1885.22	2633.98
8.82	151.44	862.93	658.38
29.75	163.95	1407.52	855.95

a. Do a mean-rank plot for an exponential distribution and obtain the parameter.
b. Do a mean-rank plot for a Weibull distribution and obtain the parameters.
c. Do a mean-rank plot for a normal distribution and obtain the parameters.
d. Generate a maximum entropy distribution using moments.
e. Generate a maximum entropy distribution using ranks.
f. Plot the results and compare.

Problem 5.6 An artesian well has had the following flows in acre-ft recorded over a period of years as follows:†

11,300	12,800	12,700	10,400	10,800	11,500	9,900	11,900
13,000	13,700	14,100	15,200	15,100	15,400	16,000	16,500
16,700							

Generate and plot a probability density function for this random variable. What is the mean flow?

Problem 5.7 Prove that the maximum entropy principle gives an exponential distribution if we know only the first moment and the bounds are from zero to infinity. *Hint:* Use equations (5.5.35) and (5.5.38).

Problem 5.8 Make a subjective estimate of some random variable with which you are familiar, but for which you have no sample data except possibly for one or two values. Justify its shape by some arguments, if possible.

---

*Data from A. H. S. Ang and W. H. Tang (1975), *Probability Concepts in Engineering Planning and Design*, Vol. 1; *Basic Principles*, Wiley, New York.
†Benjamin and Cornell (1976), p. 20.

Problem 5.9 Do Example 5.1 using the maximum entropy distribution based on moments. Plot the density function.

Problem 5.10 You wish to compare two designs on the basis of reliability. The following lives (given in hours) have been obtained from samples of eight and six from designs A and B, respectively. Both designs are expected to have a Weibull distribution.

Design A	Design B
1300	1841
1859	3307
903	6451
1880	4729
2894	922
2237	2604
2453	
1527	

a. The specified life is 500 hr. Which design would you choose?
b. A customer wants to know the probability that this machine will survive for 100 additional hours if it has survived up to a life of 2000 hr.

Repeat parts (a) and (b) on the assumption that nothing is known of the anticipated probability distributions.

Problem 5.11 A polynomial can be fitted to a density function using the sample moments.* The polynomial is represented by

$$f(x) \underset{\sim}{\sim} \sum_{j=1}^{m} a_m x^j$$

The coefficients are evaluated by minimizing the squared deviation.

$$U = \int_{\ell}^{u} \left[f(x) - \sum_{j=1}^{m} a_j x^j \right]^2 dx = \text{minimum}$$

Setting the derivatives with respect to a_j equal to zero gives

---

*M. Kendall and A. Stuart (1977), *The Advanced Theory of Statistics*, Vol. 1: *Distribution Theory*, 4th ed., Macmillan, New York.

$$\int_{\ell}^{u} \left[f(x) - \sum_{j=1}^{m} a_j x^j \right] x^i \, dx = 0$$

or

$$\int_{\ell}^{u} x^i f(x) \, dx = \sum_{j=1}^{m} a_j \int_{\ell}^{u} x^{i+j} \, dx, \quad i = 1, m$$

The left-hand side represent the ith moment about the origin, which can be estimated from the ith sample moment, m_i. We now have a linear set of simultaneous equations

$$\sum_{j=1}^{m} \frac{1}{i+j+1} (u^{i+j+1} - \ell^{i+j+1}) a_j = m_i, \quad i = 1, m$$

that can be solved for the a_j's. Do the problem of Example 5.3 using this method and compare the plotted density functions.

Problem 5.12 It has been stated in this chapter that the use of the maximum entropy method based on moments has dubious validity for small samples because the higher moments become meaningless. This is not strictly true. What actually happens is that the distribution "overfits" the sample. Find a sample size of 60 or more, and break it up into separate samples of 20. Generate for each new sample a maximum entropy density function based on moments and compare plots. Repeat using the maximum entropy distribution based on ranks. In what sense is the first method overfitting?

Problem 5.13 Sometimes statistical information is available only in the form of frequency histograms, or in equivalent tabular form. Section 4.2 suggested an approximate method of estimating moments from data in this form, and these could be used to fit an analytical distribution using the method of moments, or a maximum entropy distribution could be estimated. An alternative approach would be to adjust the parameters of a theoretical distribution so as to minimize the chi-square goodness-of-fit test value, using a nonlinear programming algorithm, such as those available in the companion volume, *Optimal Engineering Design*. A convenient source of such histogram data is the *Metals Handbook*.* Try the two techniques noted above using a histogram from this, or some other source.

Metals Handbook, American Society for Metals, Vol. 1, Metal Parks, Ohio.

Problem 5.14 Discuss how the method of maximum likelihood could be used to estimate parameters for a beta distribution based on a sample of observed values.

Problem 5.15 Show that the result is a normal distribution with the ends truncated if we assume that the maximum entropy distribution based on moments has parameters defined as follows.

$$\lambda_1 = \frac{\mu}{\sigma^2} \ , \quad \lambda_2 = -\frac{1}{2\sigma^2} \ ,$$

$$\lambda_3 = \lambda_4 = \cdots = \lambda_m = 0$$

where

μ = mean

σ^2 = variance

Problem 5.16 There are many goodness-of-fit tests, and it is essentially arbitrary which one is used. Describe a method based directly on the concept of mean-rank plotting that would be applicable to any distribution.

Problem 5.17 One could argue that the use of mean-rank plotting with variable transformations to get a straight-line fit to a specified theoretical distribution is incorrect because the transformations of the mean rank give incorrect weight to the "errors" in the least-squared approximation. For example, when fitting the Weibull function, the transformation of the mean rank is

$$y_i = \ln \ln \frac{1}{1 - [i/(n+1)]}$$

Suggest an alternative algorithm for obtaining the best-fitting values of the distribution parameters, but still using the mean-rank approximation to sample values of the cumulative distribution function.

Problem 5.18 Suggest an algorithm for deciding if a straight-line mean-rank plot "really is" a straight line. In other words, does the sample really come from a population represented by the proposed theoretical distribution?

Problem 5.19 The data set below is a sample of measured values of resistance of 10-kΩ resistors.

a. Use HIST to generate a histogram and PLOTPL to generate a superimposed density function. Repeat for at least four values of the number of histogram bars. Discuss your results.

b. Use SMOM to obtain the first four moments of the sample, and PLOTPL and MEP1 to generate a density function. Discuss your results.

10.14	9.99	10.17	9.97
10.01	10.02	9.98	10.10
10.00	9.96	9.95	10.03
10.03	10.10	9.97	10.05
10.05	9.96	9.99	10.08
9.98	9.98	9.96	10.00
9.98	9.97	10.03	10.15
10.00	10.02	10.02	10.11
10.06	10.00	10.04	10.08
9.92	9.99	9.95	10.12
10.08	9.96	9.98	10.04
10.08	9.96	9.98	10.04
10.00	10.13	10.05	9.99
9.95	9.92	9.96	9.93
9.97	10.01	10.02	10.00
10.04	9.95	9.94	9.97
9.95	10.16	10.03	10.07
10.05	9.99	9.91	9.99
10.00	10.02	9.99	9.95
9.96	9.97	10.14	10.01
10.24	10.04	10.01	10.01
9.99	10.03	10.03	10.00
10.02	9.91	10.07	9.93
9.94	9.97	10.02	10.00
10.16	10.16	10.08	10.00
10.10	10.00	10.03	9.96
10.00	9.99	10.11	10.09
10.21	9.99	10.18	10.10
10.04	10.01	9.98	9.96
10.02	10.00	10.02	10.18
9.95	10.14	10.02	9.94
10.11	10.01	10.03	10.17
9.96	9.99	10.01	9.95
10.02	10.13	10.01	10.00
9.96	10.03	10.04	9.98
9.99	10.03	9.99	
10.00	9.95	10.00	
9.98	10.09	10.18	
9.99	10.04	10.00	
9.99	10.02	10.02	

Problem 5.20 Use the data set of Problem 5.19 to generate a Weibull distribution mean-rank plot. Use MRWEIB.

a. Obtain a plot using the first 10 values.

b. Obtain a plot using the first 50 values.
c. Obtain a plot using all values.
d. Plot the density function generated in part (c), using FWEIB and PLOTPL.
e. Discuss your results, comparing them with those of Problem 5.19.

Problem 5.21 Use the data set of Problem 5.19 to generate a maximum entropy distribution based on sample moments. Use MEP1. Obtain a plot using HIST, ENTRPF, and PLOTPL so that the density function from MEP 1 is superimposed on the histogram from the data (use the best histogram from Problem 5.19). Compare with the results of Problems 5.19 and 5.20. An interesting extension would be to generate two curves, one using 50 values, the other using all values.

Problem 5.22 Use both goodness-to-fit tests to compare the Weibull and maximum entropy methods for fitting the data of Problem 5.19.

REFERENCES

Abramson, N. (1963), *Information Theory and Coding*, McGraw-Hill, New York.

Benjamin, J. R., and C. A. Cornell (1970), *Probability, Statistics, and Decision for Civil Engineers,* McGraw-Hill, New York.

Dunham, J. W., G. N. Brekke, and G. N. Thompson (1972), "Live Loads on Floors in Buildings," *Building Materials and Structures*, National Bureau of Standards, Washington, D.C., p. 22.

Griffeath, D. S. (1972), Computer Solution of the Discrete Maximum Entropy Problem, *Technometrics*, Vol. 14, No. 4, pp. 891-897.

Guest, P. G. (1961), *Numerical Methods of Curve Fitting*, Cambridge University Press, Cambridge.

Gumbel, E. J. (1958), *Statistics of Extremes*, Columbia University Press, New York.

Guttman, I., S. S. Wilks, and J. S. Hunter (1971), *Introductory Engineering Statistics*, 2nd ed., Wiley, New York.

Hahn, G. J., and S. S. Shapiro (1967), *Statistical Models in Engineering*, Wiley, New York.

Haugen, E. B. (1980), *Probabilistic Mechanical Design,* Wiley-Interscience, New York.

Jaynes, E. T. (1957), Information Theory and Statistical Mechanics, *Phys. Rev.*, Vol. 106, pp. 620-630.

Jones, D. S. (1979), *Elementary Information Theory*, Clarendon Press, Oxford.

Kempthorne, O. (1971), Probability, Statistics and the Knowledge Business, in *Foundations of Statistical Inference*, V. P. Godambe and D. A. Sprott, eds., Holt, Rinehart and Winston of Canada, Toronto.

Kendall, M., and A. Stuart (1979), *The Advanced Theory of Statistics*, Vol. 2: *Inference and Relationship*, 4th ed., Griffin, London.

Kies, J. A., H. L. Smith, H. E. Romine, and M. Bernstein (1965), Fracture Testing of Weldments, *ASTM Spec. Publ. No. 381*, pp. 328-356.

Lipson, C., and N. J. Sheth (1973), *Statistical Design and Analysis of Engineering Experiments*, McGraw-Hill, New York.

Sackman, H. (1975), *Delphi Critique*, Lexington Books, D.C. Heath and Co., Lexington, Mass.

Siddall, J. N. (1982), *Optimal Engineering Design: Principles and Applications*, Marcel Dekker, New York.

Siddall, J. N., and A. Badawy (1979), Use in Probabilistic Design of the Maximum Entropy Distribution Based on Ranked Data, *ASME Trans., J. Mech. Des.*, Vol. 102, No. 3, pp. 460-468.

Siddall, J. N., and Y. Diab (1975), The Use in Probabilistic Design of Probability Curves Generated by Maximizing the Shannon Entropy Function Constrained by Moments, *ASME Trans., J. Eng. Ind.*, Vol. 97, Ser. V, No. 3, pp. 843-852.

Tribus, M. (1969), *Rational Descriptions, Decisions, and Design*, Pergamon, New York.

6

Probabilistic Analysis

If [the experimenter] wanted a green-eyed pig with curly hair and six toes and if this event had a non-zero probability, then the Monte Carlo experimenter, unlike the agriculturist, could immediately produce the animal. [Kahn, 1954]

Why is a probabilistic approach an important component of modern design theory? • *Probability distributions conceived as a codification of the designer's uncertainty of judgment* • *The importance of codification of uncertainty for consistent communication and analytical rigor* • *The basic mathematical problem* • *A study of mathematical techniques and their comparative advantages* • *Practical examples*

6.1 INTRODUCTION

Generally speaking, probabilistic design can be defined as design in which the engineer codifies uncertainty by probability distributions. Traditionally, uncertainty in knowledge about material properties, loadings, flow rates, heat rates, and the like, as well as about the mathematical model itself, has been taken care of by using factors of safety. If a specification requires a certain value for a design characteristic, such as pressure drop in a ventilation system, the designer applies a factor of safety to the predicted value of pressure drop to take care of the inability to make exact predictions. Evaluating this uncertainty is a design decision, not part of the modeling of the device's performance.

Let us consider another example to illustrate the probabilistic approach and demonstrate that it is part of the design decision-making process. As part of some agricultural equipment, an engineer is de-

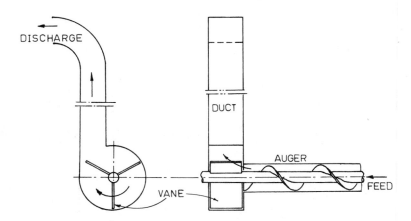

FIG. 6.1 Concept of a grain slinger.

signing a grain slinger, which is a rotary device using impeller blades
to throw grain some vertical distance in a chute. The grain is given
kinetic energy by impact from the impeller blades as it is introduced
axially to the slinger, as illustrated in Fig. 6.1. An important design
characteristic is grain damage, measured in percent of kernels broken
by impact on the blades. An experimental relationship has been obtained giving the percent of grain damage as a function of relative impact velocity and grain moisture content. However, for given values

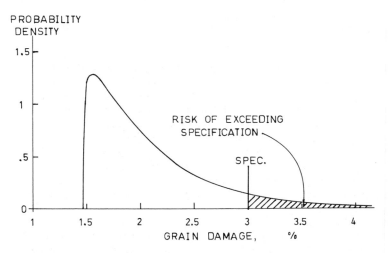

FIG. 6.2 Probability density function for grain damage.

of impact velocity and moisture content, there is a significant spread
in grain damage, represented by the probability density function of
Fig. 6.2. This function is defined such that if we wish to estimate
the probability of the damage being greater than a given amount, say
3%, it is given by the area under the curve past this value. This is
the shaded area in Fig. 6.2. Thus for a given design, the designer
can tell the user the *risk level*, or probability, of exceeding a specified grain damage. In our illustration there is a 7.2% risk of exceeding a 3% damage level. It may help to visualize the concept of risk by
stating that, in 1000 field trials, the damage level of 3% will be exceeded about 72 times. *It is a design decision* as to whether or not this is
an acceptable risk level, and a design procedure in codifying it by use
of probability density functions.

If there is some doubt as to whether the level of uncertainty
associated with grain damage is really typical of most engineering
work, consider the more familiar example of yield stress of steel.
Figure 6.3 shows the probability density of HR 1045 steel (Mischke,
1971), with a range of 40% of the mean. The fatigue strengths of

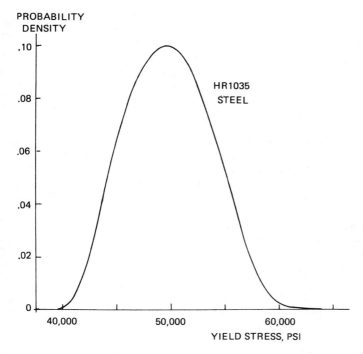

FIG. 6.3 Density function for yield strength of HR 1045 steel.

metals have a much higher spread. Furthermore, we do in fact use relatively large safety factors in design, so the uncertainty must be there somewhere.

The designer in our example also wishes to predict the power required to operate the slinger. He is concerned about the risk of not having sufficient power available. A mathematical model is available which relates power to flow rate, diameter, shaft speed, and the coefficient of friction of grain on steel. The general functional relationship is

$$P = g(Q, D, N, \mu) \tag{6.1.1}$$

The designer realizes that considerable uncertainty is associated with the engine speed, which is poorly controlled, and the coefficient of friction. He codifies this uncertainty in two probability density functions, generated by sampling values of engine speed and coefficient of friction. He is now faced with the problem of determining the probability distribution of power, which is a function of four variables, two of which are random. This is the basic mathematical problem of probabilistic analysis, and techniques are described below for solving it.

The skeptic may now argue that this is all very nice but not realistic about the concept of a factor of safety. It is really a *factor of ignorance*. We use it when we have no sample data from which to generate density functions. We have nothing to go on but crude engineering judgment about the degree of our ignorance—expressed by a factor of safety. This brings us to the concept of subjective probability and subjective probability density functions.

Subjective probability is defined as a measure of one's personal belief that an event will occur, based on experience and judgment. We have seen in Chapter 5 that probabilities associated with random variables can be codified by subjective probability density functions. Lacking any sample data, a designer can still sketch a meaningful subjective density curve. Returning to our example of the grain slinger, we assume now that the designer wishes to estimate the probability that the user will be satisfied. He now recognizes that different farmers have different standards for grain damage. He cannot persuade management to spend money on market research, so he sketches his best estimate of the probability density function for the farmer's tolerable level of damage, based on his experience in meeting farmers, working in the field, and talking to colleagues. A hypothetical result is shown in Fig. 6.4, combined with Fig. 6.2. If we define d and s as the random variables for actual damage level and tolerable damage level, respectively, then risk of an unsatisfied user is expressed by

$$\text{Risk of unsatisfactory design} = P(m < 0) \tag{6.1.2}$$

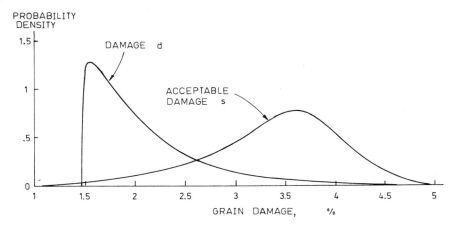

FIG. 6.4 Probability density function for allowable grain damage.

where m = s − d. The notation P(x) represents the probability of event x occurring. The variable m is, of course, a random variable which is a simple function of s and d, and we can evaluate (6.1.2) once the distribution of m is obtained. *The designer's judgment about uncertainty has thus been codified in a way that encourages consistency and rigor and permits communication of it to other people.*

This subjective concept of probability applies to all probability distributions used in engineering, whether or not they are backed up by sample data or relative frequencies. The data are just one kind of information used by the designer in expressing his judgment. The subjective approach also makes the concept of confidence limits meaningless. Even if attempts were made to use them, the level of confidence used is not a meaningful codification of acceptable risk from a designer's point of view.

We have indicated that the basic mathematical problem in probabilistic analysis is the determination of the random nature of a dependent variable, y, which is a known function of a set of random variables, $x_1, x_2, \ldots, x_n$, which have known random natures. The information defining the random variables is not necessarily in the specific form of density functions. This problem is represented graphically in Fig. 6.5. The terms *stochastic simulation, probability modeling,* and *analog prediction under risk* are also used to designate this problem.

An even more general statement of the problem is required when we are concerned with several dependent variables, or design characteristics, y_i.

$$y_i = g_i(x_1, x_2, \ldots, x_n) \tag{6.1.3}$$

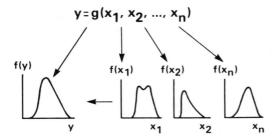

FIG. 6.5 Graphic representation of the problem of probabilistic analysis.

In conjunction with this it may be necessary to determine the probability of combined events associated with the y's. For example, in the design of an evaporator, we may be concerned with the combined probability that the amount of fluid evaporated (y_1) exceeds a specified amount, *and* that the margin of strength of the shell (y_2) exceeds zero.

There are several methods available for this: Monte Carlo simulation, the transformation of variable technique, moment transfers, and the cell technique. All of these are somewhat complex and require use of the computer, and to make their use practicable in design, it is necessary that they be embodied in easy-to-use, standardized, and well-documented subroutines.

6.2 MONTE CARLO SIMULATION

Perhaps the best known and most powerful technique is the Monte Carlo method. The random nature of each independent variable is defined by the cumulative distribution function or the probability density function, most commonly the latter. It may be represented analytically or numerically in any of the ways described in Chapter 5. A member of a theoretical sample is generated from each distribution using one of the techniques described below. Although essentially a simple idea, a theoretical sample is a surprisingly difficult concept to grasp. It is a sample of values of a random variable generated by numerical techniques, which has all of the statistical characteristics of the population represented by the given distribution function. It is in every sense analogous to observing a sample of values of the random variable, and the theoretical sample could be used to recreate approximately the original density function, in just the same fashion and using the same techniques described in Chapter 5 for a real sample.

This set of values from the theoretical samples for the x's is substituted into (6.1.3) to give a member of a sample of values for the y_i. This is literally a theoretical simulation of a field measurement of an actual design, which is a member of a sample of a population of like designs. The procedure is repeated a sufficient number of times to give a meaningful sample of values for y_i. The probabilities of events associated with y_i may be estimated using frequencies in the generated theoretical sample: for example, that the rate of fluid evaporated in an evaporator exceeds a specified amount. The sample may also be used to estimate statistics such as the mean, or to generate a density function. The sample size must be quite large—usually several thousand.

If we are concerned with combined events associated with two or more of the y_i's, and the y_i's are dependent on some common random x's, the events will not be independent. However, we can "observe" the combined event directly for each trial, and thus dependence causes no difficulty.

Returning to our original example of the grain slinger, let us assume that the specifications require that the damage not exceed 2% and the power not exceed 12 hp with a combined probability of 0.90. Thus y_1 corresponds to damage and y_2 to power, and the feasibility statement is

$$P(y_1 < 2 \text{ and } y_2 < 12) \geq 0.90 \qquad (6.2.1)$$

The Monte Carlo simulation pretends that each trial is an actual field measurement of one slinger, and each imaginary slinger varies randomly from all others in accordance with the characteristics of the density functions for power and damage.

Returning now to the problem of generating a theoretical sample, we first need to have some understanding of the concept of a random number generator. A simple device for this purpose would be to mark 1000 chips with numbers 0.000, 0.001, 0.002, ..., 1.000. If these are then well mixed in a container, any chip would be equally likely to be drawn; and a sample drawn with replacement would represent a uniform distribution between 0 and 1. In other words, for a sample of 200, about 20 would fall between 0 and 0.1, and the same for other equal intervals.

Random numbers can be generated by the digital computer, each having a number of digits up to the word size of the computer. They are called pseudo-random numbers because they are theoretically predictable; but a sample has the statistical characteristics of a uniform distribution. Information on random number generation can be found in Tocher (1963) and Forsythe et al. (1977). The algorithms are quite short, and a simple example is

$$x_{i+1} = ax_i + c \text{ (modulo m)} \qquad (6.2.2)$$

where m is equal to 2^t for a machine with t-digit binary integers. The quantities a and c are constants selected to have certain desirable values related to m. The initial value of x is arbitrary and a string of numbers is generated in one program using the recurrence formula above, even if there is more than one call. The number of random numbers that can be generated before the cycle is repeated is finite, and equal to m. Therefore, in any one simulation this number of trials should not be exceeded.

A random number between 0 and 1, called r, can be transformed to a random number x between a specified interval ℓ to u by

$$x = \ell + r(u - \ell) \qquad (6.2.3)$$

A set of values of x would represent a theoretical sample from the uniform distribution

$$f(x) = \frac{1}{u - \ell}, \quad \ell \leqslant x \leqslant u$$

$$= 0, \qquad \text{elsewhere} \qquad (6.2.4)$$

Subroutine FRAND in Appendix A is a random number generator.

We are now ready to approach the problem of generating a theoretical sample from any distribution. We begin by following a method used by Tocher (1963), and imagine that we are generating a sample

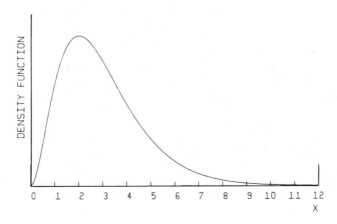

FIG. 6.6. Density function to be sampled.

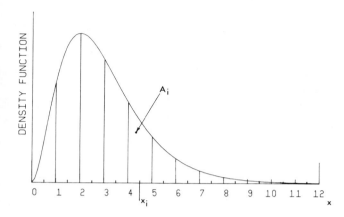

FIG. 6.7 Discretizing of the density function.

by actually using poker-chip-size discs. The algorithm used is as follows. It leads us to the *inverse cumulative distribution method*.

1. A random variable has a known density function, shown in Fig. 6.6.

2. The range of the density function is divided into n equal intervals, as shown in Fig. 6.7.

3. The area of each interval, A_i, is determined. By the definition of a density function, the probability of x occurring in the ith interval is

$$p_i = A_i \qquad\qquad (6.2.5)$$

4. A set of N discs are made, where N is a fairly large number, say 1000. For each interval, $p_i N$ discs are marked with the value of x at the midpoint of the interval, x_i.

   Total number of discs $= p_1 N + p_2 N + \cdots + p_n N$

   $$= N \qquad\qquad (6.2.6)$$

5. We now imagine that we have the density function drawn on a panel set up vertically, and that the x axis and disc diameters are scaled so that the disc diameter just equals the interval width. The discs are now stacked on a ledge along the x axis in front of the curve, in the appropriate interval. This is shown in Fig. 6.8. The discs would represent a histogram corresponding to the density function, except for scale.

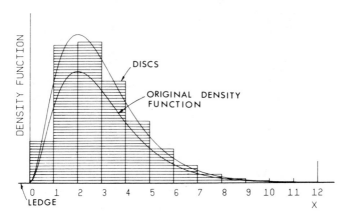

FIG. 6.8 Histogram of discs.

6. If the discs are now mixed, and one drawn at random, the prob-
    ability of drawing one with x_j on it is

$$P(x_j) = \frac{p_j N}{N} = p_j \qquad\qquad (6.2.7)$$

7. If we draw a relatively small sample with replacement, say m, then
    the probability distribution of the sample should approximate the
    distribution of all N. This, in turn, represents the variable x
    discretized. So the number of discs drawn from the ith interval
    will tend to be $p_i m$.
8. If we stack the small sample on the ledge in front of the density
    function curve, it should look much the same as before, but to a
    different scale and somewhat irregular. This is suggested in
    Fig. 6.9.
9. Thus the sample m is *a theoretical sample* of the variable x, simu-
    lating a set of m measurements of x.
10. The procedure can be computerized by using hypothetical discs.
    Serial numbers from 1 to N are assigned to each disc, so that
    numbers 1 to k_1 would apply to the set marked x_1, the numbers
    $k_1 + 1$ to k_2 would apply to the set marked x_2, and so on.
11. One number is called from a source of uniformly distributed ran-
    dom numbers, designated r, which will be between 0 and 1. The
    number rN can be thought of as the serial number of a randomly
    drawn disc. The serial numbers are scanned and the interval
    located in which rN lies. The midpoint value of x for this inter-
    val is the value on the disc, and the sample value of x.

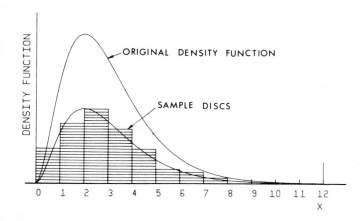

FIG. 6.9 Histogram of sample discs.

12. This can be refined still further if we plot serial numbers versus x, as shown in Fig. 6.10. This is clearly a cumulative frequency polygon for x, and if we divide ordinates by N, it becomes an approximate cumulative distribution function. So now we can forget about discs and use the random number r directly with the cumulative distribution function. Using a table look-up subprogram such as FTABLE in Appendix A, we can match r to an interpolated value of x.

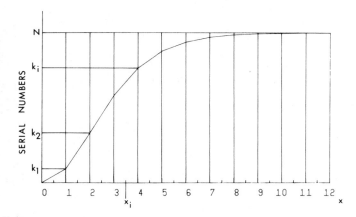

FIG. 6.10 Plot of serial numbers versus x-intervals.

13.   We have in effect

$$F(x) = r \qquad\qquad (6.2.8)$$

and we can now also forget about discretization of x and simply
solve the expression (6.2.8) for x. If $F(x)$ is in analytical form,
it may be possible to invert the function, or we could use a Newton-Raphson technique to solve for x for a given r, or an interpolation technique. Function FINVRT can be used.

We have derived the inverse cumulative distribution method in a
rather long-winded fashion in order to illustrate the concept of a
theoretical sample. It actually derives immediately from the fact that
the cumulative distribution function for an parent population has itself
a uniform distribution, no matter what the parent population distribution is. Figure 6.11 helps to understand the proof of this. We let

$$y = F_x(x)$$

and let x' and y' be any corresponding values of x and y. It follows
that

$$F_y(y') = P[F_x(x) \leqslant y'] = P(x \leqslant x')$$

$$= F_x(x') = y'$$

If $F_y(y) = y$, the distribution of y must be uniform.
     If the density function is available in either analytical or numeri-
cal form, the *rejection method* can be used to obtain a theoretical
sample. It is a faster method than the previous one, but requires
more random numbers. The density function must be bounded; and
if a theoretical distribution is used with one or both tails running to
infinity, it must be arbitrarily bounded so that the tails are negligible. We determine the maximum value of $f(x)$ as a.* The algorithm
is to generate two random numbers between 0 and 1 from a uniform
distribution, designated r_1 and r_2. The choice of a value for x is

$$x = r_1(u - \ell) + \ell \qquad\qquad (6.2.9)$$

where u and ℓ are the bounds. If

$$r_2 \leqslant \frac{f(x)}{a} \qquad\qquad (6.2.10)$$

---

*This can be done using function AMAX.

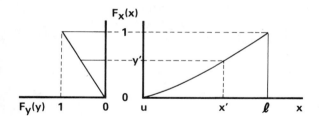

FIG. 6.11 Illustration to prove that the distribution of any cumulative distribution function is uniform.

we accept the trial value of x. Otherwise, we reject it and try again. The basis for the rejection method must be considered intuitively obvious. The number of values of x generated should be proportional to the value of the density function in any given region; so the rejection filters out the uniformly distributed numbers so that this is satisfied.

It is of considerable importance with this method to know the mean number of trials required before a successful one is achieved (i.e., how many rejections can be expected in a simulation). We can write the following probability statement:

P (infinitesimal event of x occurring in dx *and* being accepted)

$$= \frac{dx}{u - \ell} \frac{f(x)}{a}$$

Therefore, the probability of any value tried being accepted is

$$k = \int_{\ell}^{u} \frac{f(x)}{a(u - \ell)} \, dx = \frac{1}{a(u - \ell)}$$

and thus $a(u - \ell)$ trials are required on the average to obtain one accepted value.

EXAMPLE 6.1

It is convenient to illustrate this with the beta distribution having a symmetrical bell shape. If q and r are set equal to 3, (3.4.8) becomes

$$f(x) = \frac{30(x - \ell)^2 (u - x)^2}{(u - \ell)^5}, \quad \ell \leqslant x \leqslant u$$

$$= 0, \qquad \qquad \text{elsewhere} \qquad (6.2.11)$$

We can now try different ranges and determine the number of trials required to obtain 1000 successes, using the following program. It also illustrates the simple coding required for the rejection method.

```
      DIMENSION R(2)
      WRITE(6,4)
      CALL FRANDN(R,1,125)
C     XL=0.0
      DO 6 I=1,3
      NTRIAL=0
      KOUNT=0
      IF(I.EQ.1)XU=30000.
      IF(I.EQ.2)XU=1.
      IF(I.EQ.3)XU=.1
      A=30.*(XU/2.)**4/XU**5
1     CALL FRANDN(R,2,0)
      X=XU*R(1)
      DENSF=30.*X*X*(XU-X)**2/XU**5
      IF(R(2).LE.DENSF/A)GO TO 2
      GO TO 3
2     KOUNT=KOUNT+1
3     NTRIAL=NTRIAL+1
      IF(KOUNT.LT.1000)GO TO 1
      WRITE(6,5)XU,A,NTRIAL
4     FORMAT(1H1,5X,5HRANGE,8X,1HA,7X,6HTRIALS,//)
5     FORMAT(1H ,4X,F7.1,2X,E10.4,2X,I6)
6     CONTINUE
      STOP
      END
```

RANGE	A	TRIALS
30000.0	.6250E-04	1935
1.0	.1875E+01	1840
.1	.1875E+02	1883

The rejection method can also be applied to a discrete random variable. A random value of x is obtained using (6.2.9), and the nearest discrete value is determined, say x_j. Equation (6.2.10) is then applied to this value:

$$r_2 \leqslant \frac{f(x_j)}{a}$$

Having available a means for generating sample values for a random variable in the computer, it is now possible to outline the *simulation algorithm*. We shall assume that the rejection method is used.

1. Define the density function for each functionally independent random variable in the set defined by the relationship

$$y_i = g_i(x_1, x_2, \ldots, x_n) \tag{6.2.12}$$

We shall assume for the moment that the x's are also stochastically independent.

2. Generate *one* sample value for each of the x's in turn by means of the rejection method.
3. Substitute these values into (6.2.12) and calculate the corresponding value of y_i. This is one sample value for y_i.
4. If a probability function is required for y_i, the values are stored in an array. If only frequency counts are required to estimate probabilities of events associated with y_i, only the count increment needs to be saved. Other sample characteristics could also be calculated from the array, such as median, range, and variance.
5. Has the required number of trials been completed?
    Yes: Continue.
    No: Go to 2.
6. Call an appropriate probability density function generating subroutine, such as MEP1, if this is required; *or* estimate required probabilities directly from frequency counts.

This basic algorithm is very easy to program and the engineer is advised to write the software for simple problems, as will be illustrated in the examples in this section. However, a rather elaborate user-oriented package is provided in Appendix A, called subroutine CARLO.

Estimating the required sample size is a key problem in Monte Carlo simulation. We can get some feeling for this if we look at the particular problem of using simulation to estimate the mean of the dependent variable y. We imagine that we have generated n_s random values of y, and from these we calculate the sample mean.

$$\bar{y} = \frac{1}{n_s} \sum_{i=1}^{n_s} y_i = \sum_{i=1}^{n_s} \frac{y_i}{n_s} \qquad (6.2.13)$$

where the y_i's are the values in the simulated theoretical sample for y. The sample mean can be thought of as a random variable; and we would like its variance to be as small as possible, in order to have a good estimate of μ_y the population mean for y. We shall see in Sec. 6.4 [equations (6.4.12) and (6.4.14)] that the mean of $\bar{y}$ is given by

$$E(\bar{y}) = \mu_y \qquad (6.2.14)$$

and the variance by

$$E[(\overline{y} - \mu_y)^2] = \frac{\sigma_y^2}{n_s} \qquad (6.2.15)$$

where σ_y^2 is the population variance.

Thus we see that the error in the estimation of the mean is inversely proportional to $\sqrt{n_s}$, and to improve the estimate by 2 we must increase the sample size by 4. Or if the sample size is 10,000, the standard deviation of the mean is 1/100 of the standard deviation of the population. This is the first clue that the sample size must be large.

We can further say that $\overline{y}$ will strongly tend to be normally distributed because of the central limit theorem (*but this is not true for* y).

EXAMPLE 6.2

We can get a better feeling for sample size by looking at the density function for the sample mean of a simple example. Our engineering model is a piece of 0.100-in.-thick sheet metal, 5 in. wide; and subject to a nominal bending moment of 30 in. lb. The stress is given by

$$S = \frac{6M}{bt^3} \qquad (6.2.16)$$

where the independent random variables are M the bending moment and t the thickness, and b is the width. The density functions are defined numerically as follows, and plotted in Fig. 6.12.

$$f(t) = [0, 1.25, 2.24, 3.01, 3.53, 3.94, 4.28, 4.55, 4.78,$$
$$4.90, 5.00, 4.92, 4.67, 4.30, 3.83, 3.32, 2.70, 2.12,$$
$$1.63, 1.19, 0.82, 0.51, 0.27, 0.09, 0]$$

$$f(M) = [0, 0.11, 0.36, 0.95, 1.92, 3.05, 4.06, 4.77, 5.0,$$
$$4.83, 4.44, 3.89, 3.23, 2.64, 2.14, 1.74, 1.40, 1.09,$$
$$0.82, 0.69, 0.40, 0.26, 0.13, 0.04, 0.0]$$

These values are unnormalized and will be corrected by subroutine FNORM, although this is not really necessary for the rejection method.

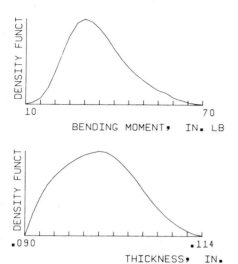

FIG. 6.12 Density functions for thickness and bending moment.

The following program does the simulation to get the sample for S, calculates the mean and standard deviation for S, and determines a normalized frequency histogram for S, all for a sample size of 10,000.

```
      DIMENSION R(2), THICKD(25), FTHICK(25), BMOMD(25), FMOM(25), SI(25
     1), FREQ(24)
      DIMENSION STRESS(10000),XDENS(101),FDENS(101),W1(24),W2(24)
      DATA FTHICK/0.,1.25,2.24,3.01,3.53,3.94,4.28,4.55,4.78,4.90,5.00,4
     1.92,4.67,4.30,3.83,3.32,2.70,2.12,1.63,1.19,.82,.51,.27,.09,0./
      DATA FMOM/0.,.11,.36,.95,1.92,3.05,4.06,4.77,5.00,4.83,4.44,3.89,3
     1.23,2.64,2.14,1.74,1.40,1.09,.82,.69,.40,.26,.13,.04,.0/
      NSAMP=10000
C.... GENERATE DISCRETE VALUES OF VARIABLES
      DO 1 I=1,25
      THICKD(I)=.090+.001*FLOAT(I-1)
      BMOMD(I)=10.+2.5*FLOAT(I-1)
1     CONTINUE
C.... INITIALIZE
      KOUNT1=KOUNT2=0
      SUM=SUMSQ=0.0
      DO 2 I=1,24
      FREQ(I)=0.0
2     CONTINUE
C.... NORMALIZE DENSITY FUNCTIONS
      CALL FNORM (FTHICK,.024,25)
      CALL FNORM(FMOM,60.,25)
C.... INITIALIZE RANDOM NUMBER STRING
      CALL FRANDN(R,1,250)
```

```
C....  SET BOUNDS FOR STRESS - SEE HIST
       SMAX=50000.
       SMIN=50000.
C....  GENERATE THEORETICAL SAMPLE
       AT=FTABLE(THICKD,FTHICK,.100,25)
       AM=FTABLE(BMOMD,FMOM,30.,25)
       DO 7 I=1,NSAMP
4      CALL FRANDN (R,2,0)
       KOUNT1=KOUNT1+1
       IF (KOUNT1.GT.3*NSAMP) STOP
       THICK=R(1)*.024+.090
       DENS=FTABLE(THICKD,FTHICK,THICK,25)
       IF(R(2).GT.DENS/AT)GO TO 4
5      CALL FRANDN (R,2,0)
       IF (KOUNT2.GT.3*NSAMP) STOP
       KOUNT2=KOUNT2+1
       BMOM=R(1)*60.+10.
       DENS=FTABLE(BMOMD,FMOM,BMOM,25)
       IF(R(2).GT.DENS/AM)GO TO 5
C....  CALCULATE SAMPLE VALUE OF STRESS
       STRESS(I)=6.*BMOM/(5.*THICK**3)
C....  CALCULATE SUM AND SUM OF SQUARES
       SUM=SUM+STRESS(I)
       SUMSQ=SUMSQ+STRESS(I)*STRESS(I)
7      CONTINUE
C      CALCULATE SAMPLE MEAN AND VARIANCE
       SMEAN=SUM/FLOAT(NSAMP)
       SVAR=SUMSQ/FLOAT(NSAMP)-SMEAN*SMEAN
       SDEV=SQRT(SVAR)
       WRITE (6,9) SMEAN
       WRITE (6,10) SDEV
       NBARS=24
       LABEL=6HSTRESS
       CALL HIST(STRESS,NSAMP,NBARS,SMAX,SMIN,1,0,6,LABEL,101,S1,
      1FREQ,XDENS,FDENS,W1,W2)
       CALL PLOTPL(1,XDENS,FDENS,101,LABEL,6,SMAX,SMIN)
       XAXIS=0.0
       CALL HTPL(S1,FREQ,0.,24)
9      FORMAT (1H1,6H MEAN=,F7.0)
10     FORMAT (1H0,20H STANDARD DEVIATION=,F7.1,/)
       STOP
       END
```

MEAN= 42113.

STANDARD DEVIATION=13930.5

In Fig. 6.13 the histogram is superimposed on a plot of the density function, derived by smoothing the histogram midpoints. Assuming a normal distribution for the sample mean, and using the sample mean and standard deviation with (6.2.14) and (6.2.15), we can plot the density function for the sample mean for different sample sizes as shown in Fig. 6.14.

It is also useful to show the frequency histogram for different sample sizes; see Fig. 6.15. The conclusion must be that Monte Carlo simulation requires quite large sample sizes, in the range 5000 to 20,000 points, and computer costs are correspondingly high for the method.

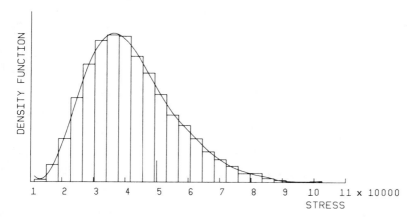

FIG. 6.13 Stress density function from a Monte Carlo simulation.

The problem of requiring very large samples is a serious one with the Monte Carlo method, and a number of techniques are available for improving its effectiveness. We have seen in Example 6.2 that a frequency histogram could be smoothed to connect it to a density function for the dependent variable y. Integrating this for estimates of probabilities of events associated with y could be expected to be better than just counting frequencies of the sample. Perhaps an even better method of generating the density function for y would be to use the maximum entropy method of Sec. 5.5, using sample moments.

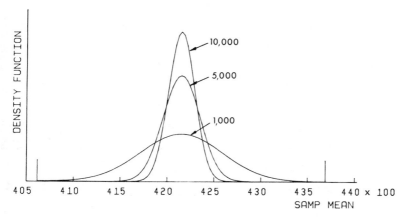

FIG. 6.14 Density function for the mean value of the sample of stress values.

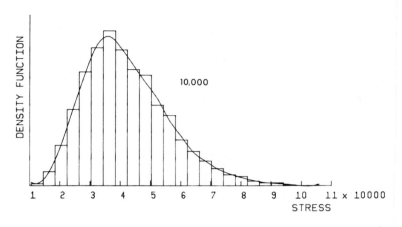

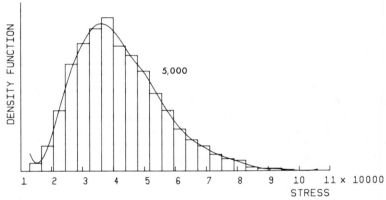

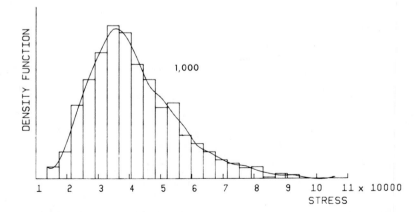

FIG. 6.15 Histograms for different samples sizes.

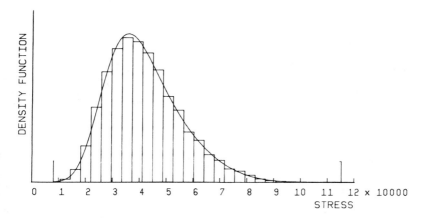

FIG. 6.16 Density function for stress using the maximum entropy
distribution with Monte Carlo.

However, these procedures cannot easily be used for combined events
related to more than one dependent variable, and frequency counts
must be resorted to.

The combination of Monte Carlo with the maximum entropy distribution is shown in the following extension of Example 6.2.

EXAMPLE 6.3

The coding of Example 6.2 is modified so that the first five central moments of the sample of stress are calculated. MEP1 is then
called, and it is shown plotted in Fig. 6.16 together with the
normalized frequency histogram. The bounds were obtained
from Example 6.2.

```
      DIMENSION R(2), THICKD(25), FTHICK(25), BMOMD(25), FMOM(25), S1(25
     1),FREQ(24),SM(5),AL(7),XSPEC(1),PSPEC(1)
      DIMENSION STRESS(10000),XDENS(101),FDENS(101),W1(24),W2(24)
      COMMON/MEP1/KPRINT,TOL,MAXFN
      DATA FTHICK/0.,1.25,2.24,3.01,3.53,3.94,4.28,4.55,4.78,4.90,5.00,4
     1.92,4.67,4.30,3.83,3.32,2.70,2.12,1.63,1.19,.82,.51,.27,.09,0./
      DATA FMOM/0.,.11,.36,.95,1.92,3.05,4.06,4.77,5.00,4.83,4.44,3.89,3
     1.23,2.64,2.14,1.74,1.40,1.09,.82,.69,.40,.26,.13,.04,.0/
      KPRINT=0
      KDATA=0
      MAXFN=300
      NSAMP=10000
C.... GENERATE DISCRETE VALUES OF VARIABLES
      DO 1 I=1,25
      THICKD(I)=.090+.001*FLOAT(I-1)
      BMOMD(I)=10.+2.5*FLOAT(I-1)
1     CONTINUE
C.... INITIALIZE
      KOUNT1=KOUNT2=0
```

```
          SUM=SUMSQ=0.0
          DO 2 I=1,24
          FREQ(I)=0.0
2         CONTINUE
C....  NORMALIZE DENSITY FUNCTIONS
          CALL FNORM (FTHICK,.024,25)
          CALL FNORM(FMOM,60.,25)
C....  INITIALIZE RANDOM NUMBER STRING
          CALL FRANDN(R,1,100)
C....  FIND TRUE BOUNDS OF STRESS
          SMIN=6.*10./(5.*.114**3)
          SMAX=6.*70./(5.*.090**3)
C....  GENERATE THEORETICAL SAMPLE
          AT=FTABLE(THICKD,FTHICK,.100,25)
          AM=FTABLE(BMOMD,FMOM,30.,25)
          DO 7 I=1,NSAMP
4         CALL FRANDN (R,2,0)
          KOUNT1=KOUNT1+1
          IF (KOUNT1.GT.3*NSAMP) STOP
          THICK=R(1)*.024+.090
          DENS=FTABLE(THICKD,FTHICK,THICK,25)
          IF(R(2).GT.DENS/AT)GO TO 4
5         CALL FRANDN (R,2,0)
          IF (KOUNT2.GT.3*NSAMP) STOP
          KOUNT2=KOUNT2+1
          BMOM=R(1)*60.+10.
          DENS=FTABLE(BMOMD,FMOM,BMOM,25)
          IF(R(2).GT.DENS/AM)GO TO 5
C....  CALCULATE SAMPLE VALUE OF STRESS
          STRESS(I)=6.*BMOM/(5.*THICK**3)
7         CONTINUE
          NBARS=24
          LABEL=6HSTRESS
          CALL HIST(STRESS,NSAMP,NBARS,SMAX,SMIN,1,0,6,LABEL,101,S1,
     1FREQ,XDENS,FDENS,W1,W2)
C....  CALCULATE MOMENTS
          CALL SMOM(STRESS,5,NSAMP,SM)
C....  OBTAIN MAX ENTROPY DISTRIBUTION
          CALL MEP1(5,SM,SMIN,SMAX,0,XSPEC,4,0,AL,PSPEC)
          DO 8 I=1,101
8         FDENS(I)=ENTRPF(AL,6,XDENS(I))
          CALL PLOTPL(1,XDENS,FDENS,101,LABEL,6,SMAX,SMIN)
          XAXIS=0.0
          CALL HTPL(S1,FREQ,0.,24)
          STOP
          END
```

We have assumed up to this point that the x's are stochastically independent. If this is not the case, it may be difficult to define the joint density function or functions, but the theoretical sampling by the rejection method is quite straightforward. To illustrate with two stochastically dependent variables, the joint distributions could be defined in two possible ways, $f(x_1, x_2)$ or $f(x_1)$ and $f(x_2|x_1)$. In the first case two random numbers would be used to select a trial value in x_1, x_2, and a third random number would be used for the rejection criterion. In the second case a sample value of x_1 would first be generated from $f(x_1)$, and then given that value in $f(x_2|x_1)$, a sample value of x_2 would be generated.

Another extension of the Monte Carlo method is to use *variance reducing techniques*.* The name is based on the idea that they reduce the variance of characteristic parameters associated with the distribution of the dependent variable. We have seen an illustration of the variance of the mean in Example 6.1. These methods would appear to have rather limited application when used in probabilistic analysis when the engineering modeling is done by functional expressions—the type that we are concerned with here. However, they do have application to complex systems of events, such as PERT networks and estimating the reliability of complex combinations of components and failure modes (Kahn, 1954; Mazumbar, 1975). The latter problem is an important one in engineering design and is discussed in Chapter 9.

The concept of variance reduction is of considerable theoretical interest, and it is possible to show that, by means of some mathematical sleight of hand, it is theoretically possible to have a Monte Carlo simulation with zero variance (i.e., no error in the mean except for numerical round-off error).

It is more general, and indeed more convenient at this time, to assume that all of the x's are stochastically dependent with joint density function $f(x_1, x_2, \ldots, x_n)$. If, in fact, they are not, we can always use some variant of the expression

$$f(x_1, x_2, \ldots, x_n) = f(x_1)f(x_2) \cdots f(x_n) \qquad (6.2.17)$$

Using the general functional relationship

$$y = g(x_1, x_2, \ldots, x_n) \qquad (6.2.18)$$

or, more concisely,

$$y = g(\overline{x}) \qquad (6.2.19)$$

the definition of expected value tells us that

$$\mu_y = E[g(\overline{x})] = \int_R yf(y) \, dy \qquad (6.2.20)$$

We shall see in Sec. 6.3 that the expected value of any function can be be written in the form

$$E[g(\overline{x})] = \int_R \int \cdots \int g(\overline{x})f(\overline{x}) \, \overline{dx} \qquad (6.2.21)$$

---

*The reader is advised to skip this portion on first reading.

and therefore

$$\mu_y = \int_R \int \cdots \int g(\overline{x}) f(\overline{x}) \; \overline{dx} \tag{6.2.22}$$

If we sample from $f(\overline{x})$, the mean of y is estimated by

$$y_m = \frac{1}{n_s} \sum_{i=1}^{n_s} g(\overline{x}_i) \tag{6.2.23}$$

where $\overline{x}_i$ is the ith sample vector. The variance of y_m is, from (6.2.15),

$$V = \frac{1}{n_s} \sigma_y^{\;2} \tag{6.2.24}$$

and since

$$\sigma_y^{\;2} = E[(y - \mu_y)^2] \tag{6.2.25}$$

by analogy to (6.2.21), the variance of y_m becomes

$$V = \frac{1}{n_s} \int_R \int \cdots \int (y - \mu_y)^2 f(\overline{x}) \; \overline{dx} \tag{6.2.26}$$

The basic technique of variance reduction is *importance sampling*, in which an arbitrary "false" density function is substituted for $f(\overline{x})$, designated $f^+(\overline{x})$. Returning to (6.2.22), we multiply top and bottom by $f^+(\overline{x})$.

$$\mu_y = \int_R \int \cdots \int \frac{g(\overline{x}) f(\overline{x})}{f^+(\overline{x})} f^+(\overline{x}) \; \overline{dx} \tag{6.2.27}$$

We let

$$g^+(\overline{x}) = \frac{g(\overline{x}) f(\overline{x})}{f^+(\overline{x})} \tag{6.2.28}$$

and (6.2.27) becomes

$$\mu_y = \int_R \int \cdots \int g^+(\overline{x}) f^+(\overline{x}) \ \overline{dx} \tag{6.2.29}$$

If we now sample for $f^+(\overline{x})$ instead of from $f(\overline{x})$, and calculate $g^+(\overline{x})$ instead of $g(\overline{x})$, then the sample mean is

$$y_m^+ = \frac{1}{n_s} \sum_{i=1}^{n_s} g^+(\overline{x}_i) \tag{6.2.30}$$

and this is a valid estimate of μ_y. The concept of the method is that we hope that y_m^+ will have smaller variance than y_m, by an appropriate choice of $f^+(\overline{x})$. It is important to note that it can only be zero when $f(x)$ or $g(x)$ is zero. The variance of y_m^+ is, by analogy to (6.2.26),

$$v^+ = \frac{1}{n_s} \int_R \int \cdots \int [g^+(\overline{x}) - \mu_y]^2 f^+(\overline{x}) \ \overline{dx} \tag{6.2.31}$$

$$= \frac{1}{n_s} \left\{ \int_R \int \cdots \int \frac{y^2 [f(\overline{x})]^2}{f^+(\overline{x})} \ \overline{dx} - \mu_y^2 \right\} \tag{6.2.32}$$

The ideal choice of $f^+(\overline{x})$ is obtained using calculus.

$$\frac{\partial}{\partial [f^+(\overline{x})]} \left\{ v^+ + \lambda \left[\int_R \int \cdots \int f^+(\overline{x}) \ \overline{dx} - 1 \right] \right\} = 0 \tag{6.2.33}$$

where λ is a Lagrangian multiplier. This can be solved by using the calculus of variations. The solution is omitted here.

$$f^+(\overline{x}) = \frac{|g(\overline{x})| f(\overline{x})}{\int_R \int \cdots \int |g(\overline{x})| f(\overline{x}) \ \overline{dx}} \tag{6.2.34}$$

If $g(\overline{x})$ is positive everywhere, as it commonly is, the denominator becomes μ_y, and (6.2.34) has the form

$$f^+(\overline{x}) = \frac{g(\overline{x}) f(\overline{x})}{\mu_y} \tag{6.2.35}$$

If this is substituted into (6.2.32), the result is that V^+ becomes zero.

This tells that the ideal choice of the false density function is one in which the original density function $f(\bar{x})$ is weighted by $g(\bar{x})$. This would be simple enough if we did not have to normalize it. We thus cannot use a rough estimate of μ_y in (6.2.35). An optimization technique may be possible if $f(\bar{x})$ consists of theoretical functions with parameters, represented as follows:

$$f(\bar{x}) = f_1(x_1; \bar{\alpha}_1) f_2(x_2; \bar{\alpha}_2) \cdots f_n(x_n; \bar{\alpha}_n) \qquad (6.2.36)$$

In this expression, the parameters would be known constants. The false density function, $f^+(\bar{x})$, is given the same form as above, but the parameters are now treated as variables. We want to minimize the variance given by (6.2.32), and it will now have the form

$$V^+ = \frac{1}{n_s} \left\{ \int_R \int \cdots \int \frac{y^2 [f(\bar{x})]^2}{f_1(x_1; \bar{\alpha}_1) f_2(x_2; \bar{\alpha}_2) \cdots f_n(x_n; \bar{\alpha}_n)} \, d\bar{x} - \mu_y^2 \right\}$$

$$(6.2.37)$$

A nonlinear programming technique [see Siddall (1982)] might be used to minimize this expression by adjusting the α's. The quantity μ_y^2 can be dropped from (6.2.37) since it would not affect the minimization. Mazumbar (1975) used this approach when $f(\bar{x})$ was a single-variable exponential function, in which an analytical solution is possible. There is at present no available evidence that this technique is generally successful, and there does not seem to be any general way of selecting $f^+(\bar{x})$ that will ensure variance reduction.

Another technique of variance reduction is called *systematic sampling*. To use it, we require that

$$f(\bar{x}) = f(x_1) f(x_2) \cdots f(x_n) \qquad (6.2.38)$$

or, if there are only two variables, we may have

$$f(\bar{x}) = f(x_1) f(x_2 | x_1) \qquad (6.2.39)$$

For illustration, we shall assume that we have two stochastically independent variables. Instead of generating theoretical sample values for x, based on uniformly distributed random numbers, we calculate a set of numbers, R_{1i}, based on

$$R_{1i} = \frac{i - 1/2}{n_s}, \quad i = 1, 2, \ldots, n_s \qquad (6.2.40)$$

For each R_{1i}, we get the corresponding value of x by solving

$$\int_{\ell_i}^{x_{1i}} f(x_1) \, dx_1 = R_{1i}, \quad i = 1, 2, \ldots, n_s \tag{6.2.41}$$

The usual random theoretical sample is calculated for x_2, and then the systematic samples values for x_1 and the random values for x_2 are combined to give corresponding sample values for $g(x_1, x_2)$. These are used to estimate μ_y, given by

$$y_m = \frac{1}{n_s} \sum_{i=1}^{n_s} g_i(x_1, x_2) \tag{6.2.42}$$

To prove the validity of the method, we must first demonstrate that this is a valid estimate of μ_y, and then show that it is a better estimate. In order to do this, it will be convenient to define the following notation and relationships.

$$E_x(\cdot) = \text{expected value of the argument with respect to } f(x)$$

$$\mu_{y|x_1} = E_{x_2|x_1}(y)$$

$$= \int_{R_2} y f(x_2|x_1) \, dx_2, \quad \text{a function of } x_1 \tag{6.2.43}$$

$$E_{x_1}(\mu_{y|x_1}) = \int_{R_2} \mu_{y|x_1} f(x_1) \, dx_1$$

$$= \int_R \int y f(x_2|x_1) f(x_1) \, dx_1 \, dx_2$$

$$= \int_R \int y f(x_1, x_2) \, dx_1 \, dx_2$$

$$= E_y(y) = \mu_y \tag{6.2.44}$$

$$E_{x_2 | x_1}(y^2) = \int_R y^2 f(x_2 | x_1) \, dx_2 , \text{ a function of } x_1 \qquad (6.2.45)$$

$$E_{x_1}[E_{x_2 | x_1}(y^2)] = \int_R \int y^2 f(x_2 | x_1) f(x_1) \, dx_1 \, dx_2$$

$$= \int_R \int y^2 f(x_1, x_2) \, dx_1 \, dx_2$$

$$= E_y(y^2) \qquad (6.2.46)$$

Equation (6.4.8) is the basis for (6.2.44) and (6.2.46).

We begin our demonstration of the validity of the method by noting that, since

$$\mu_y = E_{x_1}(\mu_{y | x_1})$$

we could estimate μ_y by using samples values of $\mu_{y | x_1}$, designated as $\mu_{y | x_{1i}}$.

$$\bar{y}_m = \frac{1}{n_s} \sum_{i=1}^{n_s} \mu_{y | x_{1i}} \qquad (6.2.47)$$

If we now examine Fig. 6.17, we see that

$$\frac{1}{n_s} = F(x_{1,i+1}) - F(x_{1,i}) \stackrel{\sim}{\approx} f(x_{1i}) \, \Delta x_{1i} \qquad (6.2.48)$$

Substituting this into (6.2.47) gives

$$\bar{y}_m = \sum_{i=1}^{n_s} \mu_{y | x_{1i}} f(x_{1i}) \, \Delta x_{1i} \stackrel{\sim}{\approx} \int_{R_1} \mu_{y | x_{1i}} f(x_1) \, dx_1 \qquad (6.2.49)$$

$$\stackrel{\sim}{\approx} \int_{R_1} \mu_{y | x_1} f(x_1) \, dx_1 \qquad (6.2.50)$$

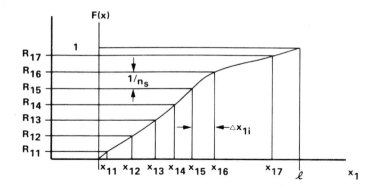

FIG. 6.17 Illustration of systematic sampling for a sample size of 7 based on the cumulative distribution function.

By by (6.2.44), this right-hand side is μ_y, thus confirming the estimate.

We would intuitively expect systematic sampling to provide a valid theoretical sample of any random variable, because the set of numbers R_{1i} would have all of the characteristics of a uniform distribution and (6.2.41) thus corresponds to the inverse distribution method of generating a theoretical sample.

It must still be demonstrated that the variance is reduced by this technique; and even if it is, the cost per trial may be significantly higher unless (6.2.41) is easily solved. The proof of variance reduction is as follows.

The variance of y_m, the estimate of μ_y in this method, is

$$V_s = E_y[(y_m - \bar{y}_m)^2] \tag{6.2.51}$$

Using (6.2.42) and (6.2.47) gives

$$V_s = E_y\left\{\left[\frac{1}{n_s}\sum_{i=1}^{n_s} g_i(x_1, x_2) - \frac{1}{n_s}\sum_{i=1}^{n_s} \mu_y|x_{1i}\right]^2\right\}$$

$$= \frac{1}{n_s^2}\sum_{i=1}^{n_s} E_y\{[g_i(x_1, x_2) - \mu_y|x_{1i}]^2\} \tag{6.2.52}$$

We next define $\sigma_y|x_{1i}^2$ as

$$\sigma_y|_{x_{1i}}^2 = E_y\{[g_i(x_1, x_2) - \mu_y|_{x_{1i}}]^2\}$$

$$= \int_R \int [g_i(x_1, x_2) - \mu_y|_{x_{1i}}]^2 f(x_1, x_2) \, dx_1 \, dx_2$$

$$(6.2.53)$$

Thus (6.2.52) becomes

$$V_s = \frac{1}{n_s} \sum_{i=1}^{n_s} \sigma_y|_{x_{1i}}^2$$

$$(6.2.54)$$

Using (6.2.48), this can be approximated by

$$V_s \approx \frac{1}{n_s} \sum_{i=1}^{n_s} \sigma_y|_{x_{1i}}^2 f(x_{1i}) \, \Delta x_{1i}$$

$$\approx \frac{1}{n_s} \int_{R_1} \sigma_y|_{x_1}^2 f(x_1) \, dx_1$$

$$= \frac{1}{n_s} E_{x_1}(\sigma_y|_{x_1}^2)$$

$$(6.2.55)$$

We need to compare this with the expression for V when the regular Monte Carlo method is used. It was

$$V = \frac{1}{n_s} E_y[(y - \mu_y)^2]$$

$$(6.2.56)$$

And this can easily be reduced to the form

$$V = \frac{1}{n_s} [E_y(y^2) - \mu_y^2]$$

$$(6.2.57)$$

The first term is modified using (6.2.46).

$$V = \frac{1}{n_s} E_{x_1} [E_{x_2|x_1}(y^2) - \mu_y^{\;2}]$$ (6.2.58)

It is useful to operate on $E_{x_2|x_1}(y^2)$.

$$E_{x_2|x_1}(y^2) = E_{x_2|x_1}(y^2 - \mu_{y|x_1}^{\;2}) + \mu_{y|x_1}^{\;2}$$ (6.2.59)

This expression is valid, since

$$E_{x_2|x_1}(\mu_{y|x_1}^{\;2}) = \mu_{y|x_1}^{\;2}$$ (6.2.60)

Continuing to operate on (6.2.59) gives

$$E_{x_2|x_1}(y^2) = E_{x_2|x_1}(y^2 - 2y\mu_{y|x_1} + \mu_{y|x_1}^{\;2}) + \mu_{y|x_1}^{\;2}$$

$$= E_{x_2|x_1}[(y - \mu_{y|x_1})^2] + \mu_{y|x_1}^{\;2}$$

$$= \sigma_{y|x_1}^{\;2} + \mu_{y|x_1}^{\;2}$$ (6.2.61)

Returning to (6.2.58), we have

$$V = \frac{1}{n_s} E_{x_1}(\sigma_{y|x_1}^{\;2} + \mu_{y|x_1}^{\;2} - \mu_y^{\;2})$$ (6.2.62)

We can now compare V and V_s, and we see that the variance reduction, when systematic sampling is used, is

$$V - V_s = \frac{1}{n_s} E_{x_1}(\mu_{y|x_1}^{\;2} - \mu_y^{\;2})$$ (6.2.63)

One more operation is needed to prove that this is a positive quantity. Since

$$E_{x_1}(\mu_{y|x_1}) = \mu_y$$

then

$$V - V_s = \frac{1}{n_s} E_{x_1} (\mu_{y|x_1}^2 - 2\mu_{y|x_1}\mu_y + \mu_y^2)$$

$$= \frac{1}{n_s} E_{x_1} [(\mu_{y|x_1} - \mu_y)^2] \qquad (6.2.64)$$

We have thus demonstrated that the systematic sampling method does indeed reduce the variance.

Variance-reducing techniques are not general-purpose methods. Except in special circumstances, one is never sure in importance sampling whether or not variance reduction has been achieved. Systematic sampling does always reduce variance, but the effect is relatively small and may not be worth the effort. There are many other techniques described in the literature (Hammersley and Handscomb, 1967; Kahn, 1954; Rubinstein, 1981) which may be applicable in special situations. If we want to generate the complete density function for y, these methods should be used with caution. There are a considerable number of other variance-reducing techniques described in the literature cited above. They are based on *sampling theory* from statistical theory; see, for example, Raj (1968) or Kendall and Stuart (1976). Variance-reducing techniques are more commonly applied to systems of discrete events, and the discussion of them is resumed in Chapter 9. Another extension of the Monte Carlo method is the *regionalization technique*, proposed by Scott and Walker (1976).

6.3 TRANSFORMATION OF VARIABLES

The transformation of variable technique is theoretically exact in that it gives explicit analytical solutions to the general problem of probabilistic analysis. However, these still must be evaluated numerically.

We shall begin with only one independent variable, so that the problem reduces to

$$y = g(x) \qquad (6.3.1)$$

We can use the infinitesimal event to relate the density function of y to that of x, as illustrated in Fig. 6.18. Then events of x lying in dx and y in dy must occur together and have the same probability. Therefore,

$$f(y) \, dy = f(x) \, dx \qquad (6.3.2)$$

or

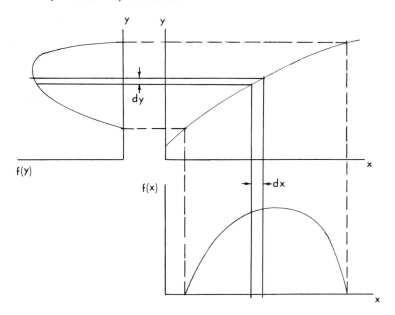

FIG. 6.18 Relating the infinitesimal event in x to that in y.

$$f(y) = \frac{f(x)}{\left| dy/dx \right|} \qquad\qquad (6.3.3)$$

The absolute value sign is added to ensure that $f(y)$ is always positive.

Similar arguments can be used to relate cumulative distribution functions. The argument based on Fig. 6.18 assumes that $g(x)$ is monotonic. If there is a multiple-valued relationship between x and y, the same logic can be applied to expand (6.3.3).

EXAMPLE 6.4

One pound mass of air is compressed adiabatically from 15 psia and 80°F to 75 psia. We wish to predict the new temperature of the air. However, the new pressure of 75 psia is only a nominal figure, due to control variability and instrument error. The probability density function for this pressure is defined numerically as follows:

$$p_2 = [70.0,\ 70.5,\ 71.0,\ 71.5,\ 72.0,\ 72.5,\ 73.0,\ 73.5,\ 74.0,\ 74.5,$$
$$75.0,\ 75.5,\ 76.0,\ 76.5,\ 77.0,\ 77.5,\ 78.0,\ 78.5,\ 79.0,\ 79.5,$$
$$80.0,\ 80.5,\ 81.0,\ 81.5,\ 82.0]$$

$$f(p_2) = [0.0, \ 0.82, \ 1.41, \ 1.87, \ 2.20, \ 2.46, \ 2.66, \ 2.81, \ 2.92, \ 2.98,$$
$$3.00 \ \ 2.90, \ 2.70, \ 2.45, \ 2.15, \ 1.78, \ 1.42, \ 1.11, \ 0.83, \ 0.60,$$
$$0.42, \ 0.26, \ 0.15, \ 0.07, \ 0.0]$$

The relationship for an ideal gas is

$$T_2 = T_1 \left(\frac{p_2}{p_1} \right)^{(k-1)/k} \tag{6.3.4}$$

where k is 1.4 and the temperatures are in °R. The required derivative is

$$\frac{dT_2}{dp_2} = \frac{0.2857T_1}{p_1^{0.2857}} p_2^{-0.7143} \tag{6.3.5}$$

and the density function for T_2 is given by (6.3.3).

$$f(T_2) = \frac{f(p_2)}{dT_2/dp_2} \tag{6.3.6}$$

Substituting for T_1 and p_1, we get

$$f(T_2) = 0.014051p_2^{0.7143}f(p_2) \tag{6.3.7}$$

This illustrates one difficulty with the method—the density function for T_2 is provided as a function of p_2 rather than of T_2. To define $f(T_2)$ adequately, we need a set of equally spaced values of T_2 and the corresponding values of $f(T_2)$. In this example, this is achieved most easily by inverting g(x), or using (6.3.4).

$$p_2 = p_1 \left(\frac{T_2}{T_1} \right)^{1/0.2857} \tag{6.3.8}$$

The algorithm is as follows:

1. Determine the bounds of T_2, using the bounds of p_2 and (6.3.4).
2. Determine a set of equally spaced discrete values of T_2, designated T_{2i}.

3. Determine the corresponding set of values of p_2, using (6.3.8)—designated p_{2i}.
4. Calculate the values of $f(p_2)$ at station i by applying interpolation to the input data, using FTABLE—designated $f(p_{2i})$.
5. Calculate the ith values of $f(T_2)$, using (6.3.7)—designated $f(T_{2i})$.
6. Repeat for all values of i.

It will be useful to extend the program and ask for the probability that the temperature will be less than 855°R, which is the nominal value of T_2.

The algorithm is illustrated in the following computer program.

```
      DIMENSION P2D(25),FP2D(25),T2I(101),P2I(101),FP2I(101),
     1FT2I(101),W(31)
      DATA P2D/70.0,70.5,71.0,71.5,72.0,72.5,73.0,73.5,74.0,74.5,75.0,
     175.5,76.0,76.5,77.0,77.5,78.0,78.5,79.0,79.5,80.0,80.5,81.0,81.5,
     282.0/
      DATA FP2D/0.0,.82,1.41,1.87,2.20,2.46,2.66,2.81,2.92,2.98,3.00,
     12.90,2.70,2.45,2.15,1.78,1.42,1.11,.83,.60,.42,.26,.15,.07,0.0/
      T1=540.
      P1=15.
C.... NORMALIZE DENSITY FUNCTION FOR P2
      RANGE=P2D(25)-P2D(1)
      CALL FNORM(FP2D,RANGE,25)
C.... DETERMINE BOUNDS OF T2
      T2MAX=T1*(P2D(25)/P1)**.2857
      T2MIN=T1*(P2D(1)/P1)**.2857
C.... DETERMINE DISCRETE VALUES OF T2 AND FT2
      RANGE=T2MAX-T2MIN
      DEL=RANGE/100.
      DO 1 I=1,101
      T2I(I)=T2MIN+DEL*FLOAT(I-1)
      P2I(I)=P1*(T2I(I)/T1)**(1./.2857)
      TEMP=P2I(I)
      FP2I(I)=FTABLE(P2D,FP2D,TEMP,25)
1     FT2I(I)=15.**.2857/(.2857*540.)*P2I(I)**.7143*FP2I(I)
C.... OBTAIN PROBABILITY OF EVENT THAT T2 IS LESS THAN 855 DEG R
      FT2I(1)=FT2I(101)=0.0
      PROB=CUM1(T2I,FT2I,101,855.,31,W)
C.... OUTPUT
      WRITE(6,2)PROB
2     FORMAT(//,47H PROBABILITY THAT T2 IS LESS THAN 855 DEG R IS ,
     1F8.5)
C.... PLOT DENSITY FUNCTION FOR P2
      P2MAX=82.
      P2MIN=70.
      CALL PLOTPL(1,P2D,FP2D,25,5HPRESS,5,P2MAX,P2MIN)
      CALL PLOTPL(0,P2D,FP2D,25,5HPRESS,5,P2MAX,P2MIN)
      STOP
      END
```

The input and output density functions are plotted in Fig. 6.19.

It was fortuitous in this example that it was possible to invert $g(x)$. If this is not possible, two alternative techniques could be

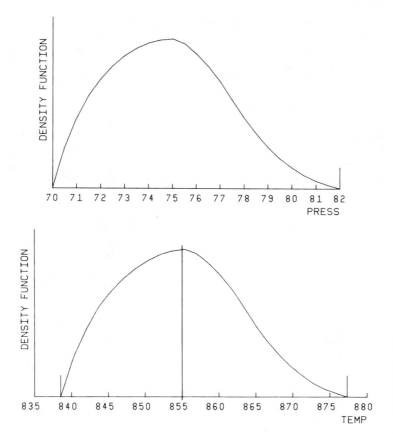

FIG. 6.19 Density functions for Example 6.4.

used. The inversion could be done numerically at each station by a
root-finding technique, or the following algorithm could be used.

1. Determine the set of discrete values of $f(y)$, corresponding to
    the given values for x and $f(x)$, using (6.3.3). Designate these
    $f_i(y)$.
2. Determine the corresponding set of values of y, using (6.3.1).
    Designate these y_i.
3. Determine the bounds of y from the set y_i.
4. Determine a set of equally spaced values of y, using the bounds,
    and designated y_i'. The number of stations need not be the
    same as before.

5. Use interpolation (FTABLE) to determine the set of $f(y)$ corresponding to y_i', designated $f_i'(y)$.
6. Repeat for all values of i.

The next stage in developing the transformation of variable technique is to extend it to two variables. We also require two *dependent* variables; and the second one can be arbitrary. For this reason, the method is sometimes called the *auxiliary variable technique*. We let y_1 be the original dependent variable, and y_2 the auxiliary variable. So we now have

$$y_1 = g_1(x_1, x_2) \qquad (6.3.9)$$

$$y_2 = g_2(x_1, x_2) \qquad (6.3.10)$$

The volume of a region dR of the joint density function $f(y_1, y_2)$ must be the same as the corresponding region of the function $f(x_1, x_2)$, since both represent the probability of the same event. Thus

$$f(x_1, x_2) \, dx_1 \, dx_2 = f(y_1, y_2) \, dy_1 \, dy_2 \qquad (6.3.11)$$

the theory of calculus tells us that

$$dx_1 \, dx_2 = \frac{dy_1 \, dy_2}{J} \qquad (6.3.12)$$

where J is the Jacobian of the transformation and is given by the determinant

$$J = \begin{vmatrix} \dfrac{\partial y_1}{\partial x_1} & \dfrac{\partial y_1}{\partial x_2} \\[2ex] \dfrac{\partial y_2}{\partial x_1} & \dfrac{\partial y_2}{\partial x_2} \end{vmatrix} \qquad (6.3.13)$$

Thus we know that

$$f(y_1, y_2) = \frac{f(x_1, x_2)}{|J|} \qquad (6.3.14)$$

As before, the absolute value of the Jacobian must be used to ensure that $f(y_1, y_2)$ is positive everywhere. If x_1 and x_2 are stochastically independent, this becomes

$$f(y_1, y_2) = \frac{f(x_1)f(x_2)}{|J|} \qquad (6.3.15)$$

We can now obtain the required $f(y_1)$ by using $(3.5.13)$. We shall also begin to use a more precise notation for density functions to avoid ambiguity.

$$f_{y_1}(y_1) = \int_R f_{y_1, y_2}(y_1, y_2) \, dy_2 \qquad (6.3.16)$$

Using $(6.3.15)$ with $(6.3.16)$ gives

$$f_{y_1}(y_1) = \frac{\int_R f_{x_1, x_2}(x_1, x_2)}{|J|} \, dy_2 \qquad (6.3.17)$$

If the x's are independent, this becomes

$$f_{y_1}(y_1) = \int_R \frac{f_{x_1}(x_1) f_{x_2}(x_2)}{|J|} \, dy_2 \qquad (6.3.18)$$

Clearly, the integration cannot be performed unless the functions $f_{x_1}(x_1)$ and $f_{x_2}(x_2)$ are set up as function of y_2. The usual practice is to let y_2 equal one of the independent variables, say x_1, thus making trivial the inversion of one of the functions. We shall represent the inversion of the second one by

$$x_2 = g_1^I(y_1, y_2) \qquad (6.3.19)$$

The variable x_1 does not appear in the argument of the inverted function because we have made it equal y_2. Now $(6.3.18)$ has the form

$$f_{y_1}(y_1) = \int_R \frac{f_{x_1}(y_2) f_{x_2}[g_1^I(y_1, y_2)]}{|J|} \, dy_2 \qquad (6.3.20)$$

Recall that y_2 is actually x_1, and $|J|$ is a function of x_1 and x_2. The Jacobian can be evaluated for given values of y_1 and y_2, using the same inversion values of x_1 and x_2.

For a few simple cases, the inversion is directly analytical, as shown in Table 6.1.

TABLE 6.1 Inversion of a Few Simple Cases

g_1	g_1^{I}	J	
$a_1 x_1 + a_2 x_2$	$\dfrac{y_1 - a_1 y_2}{a_2}$	$- a_2$	(6.3.21)
$\dfrac{x_1}{x_2}$	$\dfrac{y_2}{y_1}$	$\dfrac{x_1}{x_2^{2}}$	(6.3.22)
$x_1 x_2$	$\dfrac{y_1}{y_2}$	$- x_1$	(6.3.23)

In the general case, the inversion might have to be made by an appeal to a numerical technique such as the Newton-Raphson method, applied to the equation

$$y_1 = g_1(x_1, x_2) = g_1(y_2, x_2) \tag{6.3.24}$$

Values of x_2 would be determined corresponding to input values of y_2. (actually x_1) and y_1.

Having achieved an inversion, the algorithm implied by (6.3.20) is quite straightforward.

1. Define a sequence of n station values of y_1, sufficient to define it numerically over its range, designated by y_{1i}.
2. Define a sequence of m values of x_1 over its range, designated x_{ij}.
3. Set i = 1; loop on i to item 11.
4. Set j - 1; loop on j to item 9.
5. Solve $y_{1i} = g_1(x_{ij}, x_2)$ for x_2 [equation (6.3.24)], giving x_{2ij}.
6. Calculate $f_{x_1}(x_{ij})$ and $f_{x_2}(x_{2ij})$.
7. Calculate the value of the Jacobian corresponding to x_{ij} and x_{2ij}.
8. Calculate the argument of the integral in (6.3.20).
9. If j < m, then set j = j + 1 and go to 5;
   else, continue.
10. Determine $f_{y_1}(y_{1i})$, using (6.3.20).
11. If i < n, then set i = i + 1 and go to 4;
    else, continue.
12. All values of $f_{y_1}(y_1)$ completed—output.

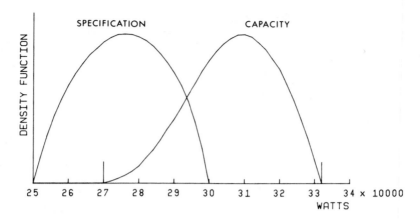

FIG. 6.20 Density functions for Example 6.5.

EXAMPLE 6.5

A heat exchanger is being designed and the probability density
function for the capacity Q in watts has been estimated as shown
in Fig. 6.20. The heat exchanger is being designed for a market
with a variable specification, which has been subjectively estimated to have a density function as shown in Fig. 6.20. We wish
to determine the probability that the heat exchanger will be acceptable to any given user.

The first step in the solution is to define a dependent variable, m (corresponding to y_1), as the *specification margin*, given
by

$$m = C - R \qquad\qquad\qquad (6.3.25)$$

where

C = capacity, W, corresponding to x_1

R = customer requirement, W, corresponding to x_2

Using (6.3.21), the inverse function is (we shall not bother with
y_2, assuming that it equals x_1)

$$R = \frac{m - C}{(-1)}$$

$$= C - m \qquad\qquad\qquad (6.3.26)$$

The partial derivatives are obtained from (6.3.25) and C = C.

$$\frac{\partial m}{\partial C} = 1 \ , \quad \frac{\partial m}{\partial R} = -1$$

$$\frac{\partial C}{\partial C} = 1 \ , \quad \frac{\partial C}{\partial R} = 0$$

and the Jacobian has the value

$$|\mathbf{J}| = \begin{vmatrix} 1 & -1 \\ 1 & 0 \end{vmatrix} = 1$$

The required probability is given by P(m > 0).

The algorithm is embodied in the following computer program, with the results shown.

```
      DIMENSION M(50),C(32),RDISC(26),FM(50),FC(32),FRDISC(26),ARG(32)
      DIMENSION W(41)
      REAL M,MMAX,MMIN
      DATA FC/0.00,0.04,0.09,0.15,0.23,0.34,0.50,0.66,0.86,1.09,1.29,
     1 1.53,1.79,2.02,2.27,2.48,2.66,2.81,2.92,2.99,3.00,2.95,2.83,2.70,
     2 2.50,2.30,2.01,1.69,1.31,0.91,0.46,0.00/
      DATA FRDISC/0.0,0.51,0.97,1.36,1.69,1.98,2.21,2.41,2.59,2.71,2.84,
     1 2.93 ,2.99,3.00,2.99,2.94,2.87,2.76,2.61,2.44,2.23,1.98,1.65,1.24
     2 ,0.70,0.00/
C.....THESE DO LOOPS GENERATE DATA FOR C AND RDISC
      DO 7 I=1,26
7     RDISC(I)=250000.+2000.*FLOAT(I-1)
      DO 8 I=1,32
8     C(I)=270000.+2000.*FLOAT(I-1)
      RANGEC=C(32)-C(1)
      AREA=FSIMP(FC,RANGEC,32)
      DO 10 I=1,32
10    FC(I)=FC(I)/AREA
      RANGER=RDISC(26)-RDISC(1)
      AREA=FSIMP(FRDISC,RANGER,26)
      DO 11 I=1,26
11    FRDISC(I)=FRDISC(I)/AREA
C.....DEFINE VALUES FOR M
      NUMBM=51
      MMAX=332000.-250000.
      MMIN=270000.-300000.
      RANGEM=MMAX-MMIN
      DELM=RANGEM/49.
      M(1)=MMIN
      DO 1 I=2,50
1     M(I)=M(I-1)+DELM
C.....OBTAIN VALUES FOR DENSITY FUNCTION OF M, FM
      DO 3 I=1,50
      DO 2 J=1,32
C.....CALCULATE LOCAL VALUES OF R FROM INVERSION EQUATION
      RIJ=C(J)-M(I)
C.....CALCULATE LOCAL VALUES OF DENSITY FUNCTIONS FC AND FR
C.....FC IS DIRECT FROM INPUT DATA
      FR=0.0
      IF(RIJ.GE.RDISC(1).AND.RIJ.LE.RDISC(26))FR=FTABLE(RDISC,FRDISC,
     1RIJ,26)
C.....CALCULATE ARGUMENT OF INTEGRAL USING (6.3.17)
```

```
2        ARG(J)=FC(J)*FR
C.....OBTAIN LOCAL VALUE OF DENSITY FUNCTION FOR M, FM
3        FM(I)=FSIMP(ARG,RANGEC,32)
C.....CHECK AREA OF FM
         AREA=FSIMP(FM,RANGEM,50)
         WRITE(6,9)AREA
9        FORMAT(//,12H AREA OF FM=,E12.5)
C
C.....OBTAIN REQUIRED PROBABILITY
         PROB=CUM1(M,FM,50,0.0,41,W)
         PROB=1.0-PROB
C.....OUTPUT
         WRITE(6,6)PROB
6        FORMAT(//,35H PROBABILITY OF A SATISFIED USER IS,E12.4)
         CALL PLOTPL(1,M,FM,50,6HMARGIN,6,MMAX,MMIN)
         CALL PLOTPL(0,M,FM,50,6HMARGIN,6,MMAX,MMIN)
         STOP
         END
```

AREA OF FM=   .10095E+01

PROBABILITY OF A SATISFIED USER IS   .9642E+00

The density function for m is shown in Fig. 6.21.

An interesting extension of the problem is to imagine that the designer decides to publicize for the final product a specification of 290,000 W. She estimates that this will "frighten off" those in the potential market above this specification and all below 260,000 W, giving a new and truncated density function for R. The probability of now having a satisfied customer can be calculated by essentially the same computer program, except that $f(R)$ will be be defined by the following array, which must be renormalized.

R = [260, 262, 264, 266, 268, 270, 272, 274, 276, 278, 280, 282,

   284, 286, 288, 290] × 1000

$f(R)$ = [1.98, 2.21, 2.41, 2.59, 2.71, 2.84, 2.93, 2.99, 3.00,

   2.99, 2.94, 2.87, 2.76, 2.61, 2.44, 2.33]

The solution is shown in Fig. 6.22, and gives a probability of 0.9816.

Returning to the general formulation,

$$y_1 = g_1(x_1, x_2)$$

$$y_2 = g_2(x_1, x_2)$$

(6.3.27)

it is possible to have a situation where y_2 actually exists, and we may be concerned with the probability of a combined event associated with

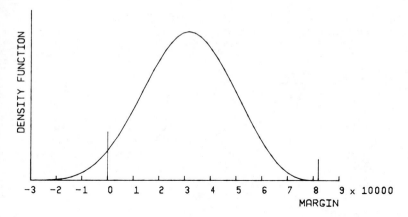

FIG. 6.21 Density function for the specification margin.

both y_1 and y_2. We can get a combined or joint density function from (6.3.14) or (6.3.15). Using the latter, we have

$$f(y_1, y_2) = \frac{f(x_1)f(x_2)}{|J|} \qquad (6.3.28)$$

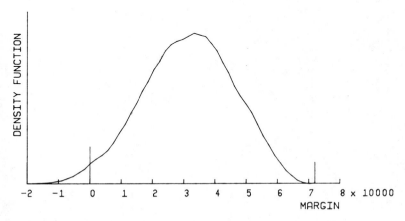

FIG. 6.22 Probability of a satisfied user with a published specification of 260,000W.

We must be able to invert both $g_1(x_1, x_2)$ and $g_2(x_1, x_2)$ to get $f(y_1, y_2)$ in the following usable form:

$$f(y_1, y_2) = \frac{f_{x_1}(y_1, y_2) f_{x_2}(y_1, y_2)}{|J|} \qquad (6.3.29)$$

The required inversion means that, in general, we must solve (6.3.27) simultaneously to obtain x_1 and x_2, corresponding to preestablished discrete values of y_1 and y_2. The initial point could best be solved using the least-squared-residual-error approach with one of the optimization methods given in Siddall (1982), similar to the solution technique of (5.5.44). Solution of adjacent points, for which the current point would be a good start, would probably be done quickly by Newton's method (Forsythe et al., 1977; Kelly, 1967).

Returning to (6.3.29), its solution has the numerical result of defining a grid of values of y_1 and y_2, with a joint density function value attached to each, as suggested in Fig. 6.23. The discrete values of the joint density function could be evaluated at the intersection points on the grid, and numerical integration used to determine the probability of y_1 and y_2 being less, say, than given specified values y_{1s} and y_{2s}.

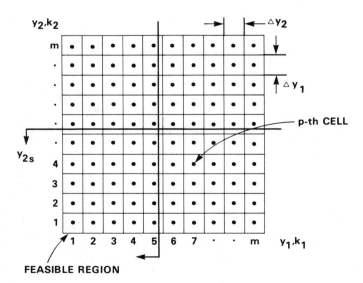

FIG. 6.23 Discretization for y_1 and y_2. The integers along axes correspond to k_1.

$$P(\text{feasibility}) = \int_{\ell_1}^{y_{1s}} \int_{\ell_2}^{y_{2s}} f(y_1, y_2) \, dy_1 \, dy_2 \qquad (6.3.30)$$

However, multiple numerical integration is difficult, and a more practical approach is to define the density functions at the midpoints of the cells in Fig. 6.23. The cell becomes a sort of normalized histogram bar, and the joint density function is treated as constant over the region of the cell. The calculation now is

$$P(\text{feasibility}) = \sum_p f_p \, \Delta y_1 \, \Delta y_2 \qquad (6.3.31)$$

where the summation is for all cells with midpoints in the feasible region and f_p is the value of the $f(y_1, y_2)$ at the midpoint of the pth cell.

It is also possible to obtain marginal density functions for y_1 or y_2. Essentially the same algorithm is used as in the case of an auxiliary variable, but now both functions must be inverted, so that the basic expression for the algorithm becomes

$$f_{y_1}(y_1) = \int_R \frac{f_{x_1}(y_1, y_2) f_{x_2}(y_1, y_2)}{|J|} \, dy_2 \qquad (6.3.32)$$

The inversion process would be the same as for the joint density function, described above.

Before generalizing further with the transformation method, we should identify the basic mathematical requirements that must be satisfied before the method can be used.

1. There must be a single-valued relations between $\bar{y}$ and $\bar{x}$ within the bounds of all the density functions.
2. All of the functions

$$y_i = g_i(\bar{x})$$

must be continuous and have continuous first derivatives.

Everything that has been said about two variables can be extended to n variables.

$$y_i = g_i(x_1, x_2, \ldots, x_n), \quad i = i, m \qquad (6.3.33)$$

where k of the y_i's are real and the remainder are auxiliary variables. Now k simultaneous expressions must be inverted at each discrete value. In principle, this extension causes no difficulty, but the data processing load can become rather heavy, particularly for k greater than 2 or 3.

The *generalized transformation of variables method* is described in the algorith that follows. It explicitly depends only on (6.3.33) and the general transformation equation

$$f(\bar{y}) = \frac{f_x(\bar{y})}{|J|} \qquad (6.3.34)$$

All integrations become summations, and no attempt is made to use numerical quadrature. The following quantities are defined:

n = number of independent variables

m = number of intervals for each variable

p = address of a cell, defined below

k_i = interval index of a cell corresponding to y_i, $1 \leqslant k_i \leqslant m$

Δy_i = discretization interval

The first requirement is to arrange a convenient way of coding the address p of each cell for computer storage, with a correspinding decoding algorithm, which identifies the k_i's associated with any p. This coding of cells was suggested by Scott and Walker (1976). The address p is simply coded as a base 10 number ranging from 1 to n_c, where n_c is the total number of cells.

$$n_c = m^n \qquad (6.3.35)$$

We then conceive of the address as being coded in an m-base number, C_p, which has the form

$$C_p = q_n q_{n-1} \cdots q_1 \qquad (6.3.36)$$

where the digit q_i represents the index value of the discretized y_i. The conversion

$$q_i = k_i - 1 \qquad (6.3.37)$$

is necessary since the digits in C_p must range from 0 to $m - 1$. The m-based number C_p is never actually recorded. The algorithm for decoding p in order to get the k_i's is as follows:

1. i = 1, sum = 0.
2. Do until i = n.

3. II = n − i + 1.
4. If II = n, go to 6,
    Else continue.
5. sum = sum + $q_{II+1}m^{II}$.
6. $q_{II} = (p - sum - 1)/m^{II-1}$.
7. $k_{II} = q_{II} + 1$.
8. i = i + 1.
9. End of do loop.

The number of cells is limited by the integer word size of the computer used. This varies from 23,767 for an 8-bit computer to 2^{59} for a 64-bit computer. However, it is always possible to use two or more words to represent p. The codification listed above is most easily visualized for m equal to 10, but is valid for any m. Its basis is essentially intuitive.

The vector of the coordinates of the midpoint of each cell is used to represent any point occurring in a cell, and the density function at this point is assumed constant over the cell. Thus the probability of any point falling in a given cell is estimated by

$$P_p = P(\overline{y} \text{ in pth cell})$$

$$= f(\overline{y}_p) \, \Delta y_1 \, \Delta y_2 \cdots \Delta y_n \tag{6.3.38}$$

where $f(y_p)$ is the value of the joint density function at the midpoint of cell p.

The algorithm is designed to provide the probability of all y's satisfying upper and/or lower specifications, and the marginal density functions for all the real y's. Recall that, in general, we have k real y's, and the remaining n − k auxiliary variables can conveniently be defined as

$$y_i = x_j, \text{ for } i = k+1 \text{ to } n; j - 1 \text{ to } n - k \tag{6.3.39}$$

The algorithm is as follows:

User Input

1. n = number of variables
2. m = grid size
3. y_{iu} = upper specifications or bound for y_i, i = 1, n
4. $y_{i\ell}$ = lower specification or bound for y_i, i = 1, n
5. x_{iu} = upper bound for x_i, i = 1, n
6. $x_{i\ell}$ = lower bound for x_i, i = 1, n
7. Subprogram defining all functions $g_i(\overline{x})$, including auxiliary functions
8. Subprogram defining $f(\overline{x})$

Algorithm

1. Set up grid and calculate midpoint values y_{ik}, using input item 7.
2. Determine Δy_i's.
3. Calculate number of cells n_c, using (6.3.35)
4. Convert specifications y_{iu} and $y_{i\ell}$ to integer values of k corresponding to y_{ik} nearest above $y_{i\ell}$ and nearest below y_{iu}.
5. Scan each cell.
    Do loop until $p = n_c$.
6.   Decode p to obtain k_i, i = 1, n, for cell p.
7.   Check if cell contains an event inside specifications by comparing k's in item 4 to k_i in item 6.
            Yes: Go to 14.
            No: Go to 8.
8.   Invert all $g_i(\bar{x})$ to get values of all x_i's for cell p.
9.   Determine absolute value of Jacobian. The matrix of partial derivatives is calculated by numerical approximations using forward differences.
10.   Determine midpoint value of $f(\bar{y})$ for cell p using (6.3.34) and input item 8.
11.   Calculate probability associated with cell p using (6.3.38), p_p.
12.   Add p_p to sum of previous values to give total p_{spec}.
13. $p = p + 1$.
14. End of do loop.
15. $p_{spec} = 1 - p_{spec}$.

Output

1. p_{spec} = probability that all specifications are satisfied.

It should be noted that the algorithm actually calculates the probability that all specifications are *not* satisfied, since this usually involves fewer cells.

The algorithm above does not show logic for estimating the marginal density functions for the y's. If this is required, all cells must be evaluated and a frequency histogram is effectively obtained for each y_i that can then be used to develop a density function.

$$f_{ik_i} = \text{sum of all } p_p\text{'s for cells corresponding to } k_i$$

where f_{ik_i} is the histogram frequency for y_i at station k_i. This can be understood more easily by reference to Fig. 6.23, where f_{17} would be evaluated by adding the p_p's in all cells in the column above it. And f_{23} would be evaluated by adding the p_p's in all cells in its row.

The general algorithm for n functionally independent variables and m functionally dependent variables, and any value for k, and for both joint and marginal density functions, is embodied in subroutine TRANSF. An example of its applications follows.

EXAMPLE 6.6

A hollow round shaft is being designed for which the assumed modes of failure are yielding in shear and buckling. The design torque is 800,000 in. lb. The critical stress for buckling is modeled by

$$\tau_{cr} = 0.7E \left(\frac{d_o - d_i}{2d_o} \right)^{3/2} \text{ psi}$$ (6.3.40)

where

E = modulus of elasticity, psi

d_0 = outside diameter, in.

d_i = inside diameter, in.

The predicted shear stress is modeled by

$$\tau = \frac{16Td_o}{\pi(d_o^4 - d_i^4)}$$ (6.3.41)

where T is the torque. The designer uses 23,000 psi as the critical stress for yielding, designated S.

On this basis, the feasibility expressions for the design are

$$\phi_1 = S - \frac{16Td_o}{\pi(d_o^4 - d_i^4)} \geqslant 0$$ (6.3.42)

$$\phi_2 = 0.70E \left(\frac{d_o - d_i}{2d_o} \right)^{3/2} - \frac{16Td_o}{\pi(d_o^4 - d_i^4)} \geqslant 0$$ (6.3.43)

The designer finds an optimum solution for minimum material as follows:*

$d_o = 12.908$ in., $d_i = 12.634$ in.

---

*See Siddall (1982) for optimization procedures.

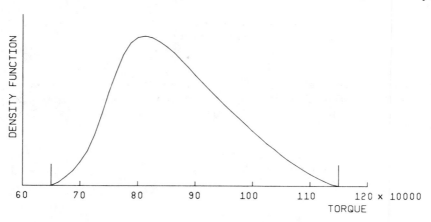

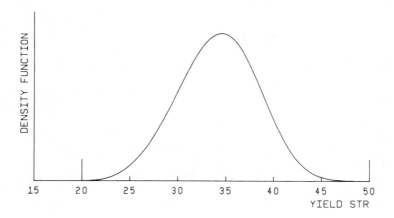

FIG. 6.24 Density functions for yield stress and torque.

He then begins to worry about whether or not he has adequately
taken care of uncertainty. He has used a guaranteed minimum
yield stress for the steel; but has not taken into consideration
the effects of corrosion or the welding process. He has used a
conservative estimate of the torque, but he begins to realize that
over long service, there is a possibility of it going higher. So
he estimates probability density functions for the critical stress
in yielding and the torque, shown in Fig. 6.24.

He also notes that the manufacturing process is to roll sheet
steel into the required circle and seam-weld it; and realizes that
this procedure implies that a better choice of a functionally inde-

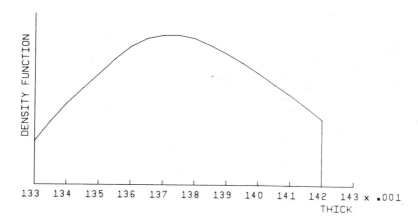

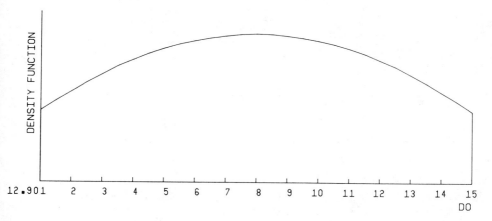

FIG. 6.25 Density functions for outside diameter and thickness.

pendent random variable would be the thickness t, rather than
the inside diameter. He determines tolerances on the sheet metal
from product data, and tolerances on the outside diameter from
the production department. Using this information, he estimates
that these two variables will have density functions as shown in
Fig. 6.25. They are truncated because the design tolerances are
tighter than the natural bounds arising from the manufacturing
processes.

It is of considerable interest to see a plot of the feasible
region in the coordinates of d_0 and t. This is the shaded area
of Fig. 6.26. The design point is at a "corner," so that both
failure modes are "active," or designed to occur simultaneously.

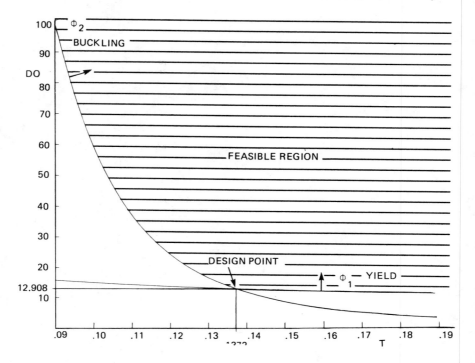

FIG. 6.26 Feasible region for design of a tube in torsion.

This is characteristic of this type of optimization problem. However, it does mean that both modes are at risk of occurring.

The problem has now become one of determining the joint random nature of ϕ_1 and ϕ_2, which are functions of the stochastically independent random variables S, T, d_0, and t. We shall assume that E is determinate with a value of 30×10^6 psi. Finally, having the density function, we can estimate $P(\phi_1 \leq 0$ and $\phi_2 \leq 0)$ to get the probability of no failure of the tube.

Before calling on TRANSF, we must have expressions for the partial derivatives $\partial \phi_1 / \partial S$, $\partial \phi_1 / \partial T$, and so on. Or, failing this, we must use finite-difference estimates of the partial derivatives. Both are available in TRANSF. The latter approach is usually more attractive to engineers, both from the point of view of engineering time and risk of engineering error. The algebra of obtaining analytical partial derivatives can be very messy, and often impossible. The calling program for TRANSF and the results are as follows:

```
C       X(1) = OUTSIDE DIAMETER
C       X(2) = THICKNESS
C       X(3) = YIELD STRESS*1.E3
C       X(4) = TORQUE
C       Y(1) = PHI(1)
C       Y(2) = PHI(2)
C       Y(3) = X(1)
C       Y(4) = X(2)
        DIMENSION SLOW(4),SHIGH(4),XLOW(4),XUPP(4),YLOW(4),YUPP(4),
       1YDENS(2,100),FY(2,100)
        DIMENSION DIFF(4),Y(4,20),ISHIGH(4),ISLOW(4),KVALUE(4),X(4),
       1W1(20),W2(20),W3(4,4),DELY(4),FREQ(2,20),YI(21),FREQI(20),
       2DENS(100),FDENS(100)
        DIMENSION ARG(51),FUNC(51)
        DIMENSION RMAX(4),RMIN(4),XSTRT(4),PHI(8),PSI(1),W(56)
        COMMON/OPT/NU,XL(4),XU(4)
        COMMON/TCHECK/TLIMIT
        COMMON/YSCALE/ISCALE
        ISCALE=1
        DATA XLOW/12.901,.133,20.0,650000./
        DATA XUPP/12.915,.142,50.0,1150000./
        CALL SECOND(T0)
        TLIMIT=500.
        N=4
        M=15
        MINT=100
        ISPEC=1
        NDENS=NDIM=2
        DELX=1.E-3
        SLOW(1)=0.0
        SLOW(2)=0.0
        YUPP(3)=12.915
        YUPP(4)=.142
        YLOW(3)=12.901
        YLOW(4)=.133
        X(3)=XUPP(3)
        X(4)=XLOW(4)
        X(1)=XUPP(1)
        X(2)=XUPP(2)
        YUPP(1)=FUN(X,1)
        X(3)=XLOW(3)
        X(4)=XUPP(4)
        X(1)=XLOW(1)
        X(2)=XLOW(2)
        YLOW(1)=FUN(X,1)
C.... FIND BOUNDS ON Y(2) BY OPTIMIZATION
C       SEE COMPANION VOLUME - *OPTIMAL ENGINEERING DESIGN*
        DO 9 I=1,4
        RMAX(I)=XUPP(I)
        RMIN(I)=XLOW(I)
9       XSTRT(I)=(RMAX(I)+RMIN(I))/2.
        DO 11 I=1,4
        XL(I)=XLOW(I)
11      XU(I)=XUPP(I)
        NU=1
        CALL SEEK(4,8,0,1,RMAX,RMIN,XSTRT,X,U,PHI,PSI,NVIOL,W)
        YLOW(2)=U
        NU=2
        DO 10 I=1,4
10      XSTRT(I)=(RMAX(I)+RMIN(I))/2.
        CALL SEEK(4,8,0,1,RMAX,RMIN,XSTRT,X,U,PHI,PSI,NVIOL,W)
        YUPP(2)=-U
        WRITE(6,100)YLOW(2),YUPP(2)
100     FORMAT(* YLOW(2)= *,E12.5,*  YUPP(2)= *,E12.5)
        SLOW(3)=XLOW(1)
```

```
        SHIGH(3)=XUPP(1)
        SLOW(4)=XLOW(2)
        SHIGH(4)=XUPP(2)
        SHIGH(1)=YUPP(1)
        SHIGH(2)=YUPP(2)
        CALL TRANSF(N,M,ISPEC,SLOW,SHIGH,NDENS,NDIM,MINT,DELX,XLOW,
       1XUPP,YLOW,YUPP,PSPEC,YDENS,FY,DIFF,Y,ISHIGH,ISLOW,KVALUE,
       2X,W1,W2,W3,DELY,FREQ,YI,FREQI,DENS,FDENS)
        WRITE(6,1)PSPEC
1       FORMAT (///,* PROBABILITY OF NO YIELDING AND NO BUCKLING IS*,F8.5)
        CALL SECOND(T1)
        TIME=T1-T0
        WRITE(6,1000)TIME
1000    FORMAT(///,* TIME FOR TRANSFORMATION OF VAR METHOD=*,F8.2)
        DO 2 I=1,100
        DO 2 I=1,100
        DENS(I)=YDENS(1,I)
2       FDENS(I)=FY(1,I)
C.... CALCULATE PROBABILITY OF NO YIELDING PHI(1))
        DO 4 I=1,51
        ARG(I)=YUPP(1)*FLOAT(I-1)/50.
4       FUNC(I)=FTABLE(DENS,FDENS,ARG(I),100)
        PNOY=FSIMP(FUNC,YUPP(1),51)
        WRITE(6,5)PNOY
5       FORMAT(//,* PROBABILITY OF NO YIELDING IS*,F8.5)
        CALL PLOTPL(1,DENS,FDENS,100,3HPHI,3,YUPP(1),YLOW(1))
        DO 3 I=1,100
        DENS(I)=YDENS(2,I)
3       FDENS(I)=FY(2,I)
C.... CALCULATE PROBABILITY OF NO BUCKLING
        DO 6 I=1,51
        ARG(I)=YUPP(2)*FLOAT(I-1)/50.
6       FUNC(I)=FTABLE(DENS,FDENS,ARG(I),100)
        PNOB=FSIMP(FUNC,YUPP(2),51)
        WRITE(6,7)PNOB
7       FORMAT(//,* PROBABILITY OF NO BUCKLING IS*,F8.5)
        DO 8 I=1,100
8       FDENS(I)=FDENS(I)/2.
        CALL PLOTPL(1,DENS,FDENS,100,3HPHI,3,YUPP(2),YLOW(2))
        CALL PLOTPL(0,DENS,FDENS,100,3HPHI,3,YUPP(2),YLOW(2))
        STOP
        END


        FUNCTION DENSIT(N,X)
        DIMENSION X(1),DOUT(29),FDO(29),THICK(37),FTHICK(37),
       1TORQ(41),FTORQ(41)
        DATA ISTART/0/
        DATA FDO/2.38,2.72,3.03,3.34,3.62,3.89,4.11,4.32,4.49,4.64,4.76,
       14.85,4.92,4.98,5.00,4.98,4.92,4.85,4.76,4.64,4.49,4.32,4.11,3.98,
       23.62,3.34,3.03,2.72,2.38/
        DATA FTHICK/2.05,2.47,2.95,3.33,3.76,4.11,4.48,4.81,5.16,5.51,
       15.86,6.14,6.46,6.63,6.84,6.91,7.00,7.00,6.99,6.91,6.86,6.70,6.51,
       26.34,6.15,5.95,5.72,5.50,5.26,4.99,4.71,4.49,4.22,3.95,3.69,3.40,
       33.10/
        DATA FTORQ/0.0,.14,.39,.70,1.13,1.58,2.29,3.30,4.41,5.21,6.10,
       16.65,6.92,7.00,6.92,6.74,6.51,6.22,5.90,5.56,5.20,4.85,4.51,4.17,
       23.81,3.50,3.21,2.90,2.57,2.29,1.99,1.73,1.46,1.19,1.00,.76,.56,
       3.39,.22,.09,0.0/
        IF (ISTART.EQ.1)GO TO 2
C.... GENERATE DISCRETE VALUES
        DO 1 I=1,29
1       DOUT(I)=12.901+FLOAT(I-1)*.0005
        DO 4 I=1,41
4       TORQ(I)=650000.+FLOAT(I-1)*12500.
```

```
      DO 5 I=1,37
5     THICK(I)=.133+FLOAT(I-1)*.00025
C.... NORMALIZE DENSITY FUNCTIONS
      CALL FNORM(FDO,.014,29)
      CALL FNORM(FTHICK,.009,37)
      CALL FNORM(FTORQ,500000.,41)
C
      ISTART=1
2     IF(X(1).LT.12.901.OR.X(1).GT.12.915.OR.X(2).LT..133.OR.X(2).
     1GT..142.OR.X(3).LT.20..OR.X(3).GT.50..OR.X(4).LT.650000..OR.
     2X(4).GT.1150000.)GO TO 3
      TEMP=(X(3)-20.)/16.
      FYIELD=3.93/16.*TEMP**2.93*EXP(-TEMP**3.93)
      DENSIT=FYIELD*FTABLE(DOUT,FDO,X(1),29)*FTABLE(THICK,FTHICK,X(2),
     137)*FTABLE(TORQ,FTORQ,X(4),41)
      RETURN
3     DENSIT=0.0
      RETURN
      END


      FUNCTION FUN(X,I,N)
      DIMENSION X(1)
      IF(X(2).LT.0.0)X(2)=.001
      GO TO (1,2,3,4),I
1     YSTR=X(3)*1.E3
      FUN=YSTR-16.*X(4)*X(1)/(3.14159*(X(1)**4-(X(1)-2.*X(2))**4))
      RETURN
2     FUN=0.70*30.E6*(X(2)/X(1))**1.5-16.*X(4)*X(1)/(3.14159*(X(1)**4-(X
     1(1)-2.*X(2))**4))
      RETURN
3     FUN=X(1)
      RETURN
4     FUN=X(2)
      RETURN
      END


      SUBROUTINE INVRT(L,N,X)
      DIMENSION X(1)
      COMMON/IN/NVAR,YCELL(20)
      X(1)=YCELL(3)
      X(2)=YCELL(4)
      TEMP=3.14159*(X(1)**4-(X(1)-2.*X(2))**4)/(16.*X(1))
      X(4)=(0.7*30.E6*(X(2)/X(1))**1.5-YCELL(2))*TEMP
      YSTR=YCELL(1)+X(4)/TEMP
      X(3)=YSTR*1.E-3
      RETURN
      END


      SUBROUTINE UREAL(X,U)
      DIMENSION X(1)
      COMMON/OPT/NU,XL(4),XU(4)
      GO TO (1,2),NU
1     U=FUN(X,2)
      RETURN
2     U=-FUN(X,2)
      RETURN
      END


      SUBROUTINE CONST(X,NCONS,PHI)
      DIMENSION X(1),PHI(1)
      COMMON/OPT/NU,XLOW(4),XUPP(4)
      DO 1 I=1,4
1     PHI(I)=XUPP(I)-X(I)
```

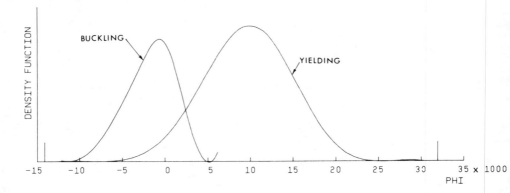

FIG. 6.27 Marginal density functions for Example 6.6.

```
          DO 2 I=1,4
     2    PHI(I+4)=X(I)-XLOW(I)
          RETURN
          END
```

```
PROBABILITY OF NO YIELDING AND NO BUCKLING IS   .29260
PROBABILITY OF NO YIELDING IS   .97059
PROBABILITY OF NO BUCKLING IS   .30181
```

Figure 6.27 shows the marginal density functions for ϕ_1 and ϕ_2.

6.4 MOMENT RELATIONSHIPS

6.4.1 Introduction

It is possible to obtain moment transfer, rather than direct density function transfers, between the functionally independent and dependent variables. These have great utility in probabilistic analysis.

Sometimes we are interested only in the mean value of a design characteristic; generating the complete density function would be a waste of effort. If we do want the density function, the moment relationships make it possible to use the maximum entropy method to obtain it. Both exact analytical relationships and approximate numerical expressions are available to make the moment transfers.

6.4.2 Analytical Relationships

The moments of the functionally dependent variable can be derived from the density functions of the independent variables by direct integration. We shall begin with one variable. We know

$$y = g(x) \tag{6.4.1}$$

and $f(x)$. By definition

$$\mu_y = \int_R yf(y) \, dy \tag{6.4.2}$$

and recalling (6.3.2), we have

$$f(y) \, dy = f(x) \, dx \tag{6.4.3}$$

We multiply both sides by y and integrate.

$$\int_R yf(y) \, dy = \int_R yf(x) \, dx \tag{6.4.4}$$

Equation (6.4.2) has appeared on the left side, giving

$$\mu_y = \int_R g(x)f(x) \, dx \tag{6.4.5}$$

We can use a similar approach to get higher central moments.

$$c_y^{(j)} = \int_R [g(x) - \mu_y]^j f(x) \, dx \tag{6.4.6}$$

where $c_y^{(j)}$ is the j central moment of y. These integral forms are easily evaluated numerically.

These expressions can be generalized in order to work with the multivariable function.

$$y = g(x_1, x_2, \ldots, x_n) \tag{6.4.7}$$

The approach is again the same. We shall revert to a more concise notation.

$$\mu_y = \int_R \int \cdots \int g(\overline{x})f(\overline{x}) \, \overline{dx} \tag{6.4.8}$$

$$c_y^{(j)} = \int_R \int \cdots \int [g(\overline{x}) - \mu_y]^j f(\overline{x}) \, \overline{dx} \tag{6.4.9}$$

where $f(\overline{x})$ is the joint density function of $x_1, x_2, \ldots, x_n$.

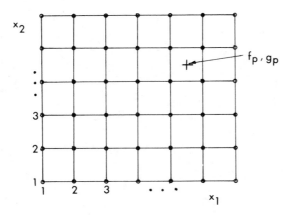

FIG. 6.28 Discretization for approximate moment transfer based on integral expressions.

Numerical integration of these expressions would be difficult, except perhaps for two variables; however, adequate accuracy is possible by using a cell or discretization technique. Each independent variable x_i is discretized into m cells, as indicated in Fig. 6.28. The number of cells, n_c, equals m^n. The algorithm uses a coded word, varying from 1 to n_c, to identify each cell, which is decoded into k, ℓ, ..., the discretization indices are the variables x_1, x_2, ..., x_n. This coded word is convenient to store in the computer, and makes possible the use of a large number of cells. More details are given in Sec. 6.6. For each cell the function values f_p and g_p are calculated, where the subscript p represents the code word or cell identifier. The algorithm to evaluate (6.4.8) and (6.4.9) is as follows.

1. Calculate midpoint values of x's for each cell.
2. p = 1.
3. Do until $p = n_c$.
4. Decode p to give cell discretization index values of x_1, x_2, ..., x_n.
5. Calculate values at midpoints for $g(\bar{x})$ and $f(\bar{f})$, designated g_p and f_p.
6. Calculate A_p where

$$A_p = f_p \, \Delta x_1 \, \Delta x_2 \cdots \Delta x_n$$

7. Obtain μ_y by successive summation.

$$\mu_y = \sum_p g_p A_p$$

8. $p = p + 1$
9. End of do loop.
10. $p = 1$.
11. Do until $p = n_c$.
12.   Obtain all $c_y^{(j)}$ by successive summation.

$$c_y^{(j)} = \sum_p (g_p - \mu_y)^j A_p$$

13. $p = p + 1$.
14. End of do loop.

Subroutine TRANSM is available for calculating moments of the dependent variable from the independent variable density functions, using the techniques described above.

It is useful at this point to examine a special case with quite simple results. We let the function relating y to $x_1, x_2, \ldots, x_n$ be

$$y = a_1 x_1 + a_2 x_2 + \cdots + a_n x_n \tag{6.4.10}$$

and we require that the x's be *stochastically independent*. Applying (6.4.8) to this gives

$$\mu_y = a_1 \int_R \int \cdots \int x_1 f(\bar{x}) \, \overline{dx}$$

$$+ a_2 \int_R \int \cdots \int x_2 f(\bar{x}) \, \overline{dx}$$

$$+ \cdots + a_n \int \int \cdots \int x_n f(\bar{x}) \, \overline{dx} \tag{6.4.11}$$

Referring back to (3.5.15), we see that the ith term is μ_i, the mean of x_i.

$$\mu_y = a_1 \mu_1 + a_2 \mu_2 + \cdots + a_n \mu_n \tag{6.4.12}$$

This is actually valid even with stochastic dependence, but the higher moment relationships are not. These are developed using (6.4.9).

$$c_y^{(j)} = \int_R \int \cdots \int (a_1 x_1 + a_2 x_2 + \cdots + a_n x_n - \mu_y)^j f(\bar{x}) \, \overline{dx}$$

$$= \int_R \int \cdots \int (a_1 x_1 + \cdots + a_n x_n - a_1 \mu_1 - a_2 \mu_2$$

$$- \cdots - a_n \mu_n)^j f(\bar{x}) \qquad\qquad (6.4.14)$$

$$= \int_R \int \cdots \int [a_1^j (x_1 - \mu_1)^j + a_2^j (x_2 - \mu_2)^j$$

$$+ \cdots + a_n^j (x_n - \mu_n)^j + \text{terms containing products of}$$

$$\text{the x's]} f(\bar{x}) \, \overline{dx} \qquad (6.4.15)$$

The terms containing products of the x's disappear because of the stochastic independence (see Sec. 3.5), and the expression becomes

$$c_y^{(j)} = a_1^j c_1^{(j)} + a_2^j c_2^{(j)} + \cdots + a_n^j c_n^{(j)} \qquad\qquad (6.4.16)$$

where $c_i^{(j)}$ is the jth central moment of x_i. If stochastic dependence exists, the cross moments could be included in (6.4.16). It can be shown that if the x's have a normal distribution, then y will also.

6.4.3 Taylor's Series Approximations

In the previous analytical approach, moments of the dependent variable y were obtained as a function of, or by processing, *density functions* of the functionally independent variables x_1, x_2, ..., x_n. We shall now examine an approximate method by which the moments of y are estimated directly from *the moments* of the x's (Hahn and Shapiro, 1967). These could be sample moments or, alternatively, moments calculated from density functions.

We shall begin with the single-variable case

$$y = g(x) \qquad\qquad (6.4.17)$$

The function g(x) is expanded in a Taylor's series about μ_x.

$$g(x) = g(\mu_x) + \frac{dg(\mu_x)}{dx}(x - \mu_x) + \frac{d^2 g(\mu_x)}{dx^2} \frac{(x - \mu_x)^2}{2} + \cdots$$

$$(6.4.18)$$

If the expected value of both sides is taken, we obtain an expression for μ_y.

$$\mu_y = E[g(x)] = g(\mu_x) + \frac{d^2 g(\mu_x)}{dx^2} \frac{c_2}{2} + \frac{d^3 g(\mu_x)}{dx^3} \frac{c_3}{3} + \cdots$$

(6.4.19)

A similar expression can be obtained for σ_y^2 and higher central moments

$$\sigma_y^2 = E\{[g(x) - \mu_y]^2\}$$

$$= E[g(x)^2 - 2g(x)\mu_y + \mu_y^2]$$

$$= E[g(x)^2] - E[2g(x)\mu_y] + E(\mu_y^2) \qquad (6.4.20)$$

$$= E[g(x)^2] - 2\mu_y^2 + \mu_y^2 \qquad (6.4.21)$$

$$= E[g(x)^2] - \mu_y^2 \qquad (6.4.22)$$

The first term can be expanded by using an expansion similar to (6.4.19).

$$E[g(x)^2] = g(\mu_x)^2 + \frac{d^2[g(\mu_x)^2]}{dx^2} \frac{c_2}{2} + \frac{d^3[g(\mu_x)^2]}{dx^3} \frac{c_3}{3} + \cdots$$

(6.4.23)

We thus can obtain μ_y and σ_y^2 to any desired degree of accuracy by using sufficient terms of the series. In many applications only two terms of (6.4.23) are needed, giving the more convenient expression

$$\sigma_y^2 = g(\mu_x)^2 + \left\{ \left[\frac{dg(\mu_x)}{dx} \right]^2 + g(\mu_x) \frac{d^2 g(\mu_x)}{dx^2} \right\} c_2 - \mu_y^2$$

(6.4.24)

Higher moments could be similarly approximated, but it is more useful to proceed to Taylor's series approximation for multivariable relationships. Higher moments for a single-variable relationship then become a special case.

In the general case we have

$$y = g(\bar{x}) \tag{6.4.25}$$

The Taylor's series expansion of this function is

$$y = g(\bar{\mu}) + \sum_{i=1}^{n} \frac{\partial g(\bar{\mu})}{\partial x_i} (x_i - \mu_i) + \frac{1}{2} \left[\sum_{i=1}^{n} \frac{\partial^2 g(\bar{\mu})}{\partial x_i^2} (x_i - \mu_i)^2 \right.$$

$$\left. + 2 \sum_{i=1}^{n} \sum_{j=1}^{i-1} \frac{\partial^2 g(\bar{\mu})}{\partial x_i \partial x_j} (x_i - \mu_i)(x_j - \mu_j) \right] + \cdots \tag{6.4.26}$$

The mean can be estimated by taking the expected value of both sides.

$$\mu_y = g(\mu) + \frac{1}{2} \sum_{i=1}^{n} \frac{\partial^2 g(\bar{\mu})}{\partial x_i^2} c_{2i}$$

$$+ 2 \sum_{i=1}^{n} \sum_{j=1}^{i-1} \frac{\partial^2 g(\bar{\mu})}{\partial x_i \partial x_j} E[(x_i - \mu_i)(x_j - \mu_j)] \tag{6.4.27}$$

where higher terms are now dropped; and we let c_{2i} represent the second central moment of x_i.

The second central moment of y is defined by

$$c_{2y} = E[(y - \mu_y)^2] \tag{6.4.28}$$

Expanding the right sides gives

$$c_{2y} = E(y^2 - 2y\mu_y + \mu_y^2)$$

$$= E(y^2) - \mu_y^2 \tag{6.4.29}$$

The expressions for y and μ_y in (6.4.26) and (6.4.27) can be substituted into (6.4.29). The result, after dropping fourth-degree terms, is

$$c_{2y} = \sum_{i=1}^{n} \left[\frac{\partial g(\bar{\mu})}{\partial x_i} \right]^2 c_{2i}$$

$$+ 2 \sum_{i=1}^{n} \sum_{j=1}^{i-1} \frac{\partial g(\bar{\mu})}{\partial x_i} \frac{\partial g(\bar{\mu})}{\partial x_j} E[(x_i - \mu_i)(x_j - \mu_j)]$$

$$+ \sum_{i=1}^{n} \frac{\partial g(\bar{\mu})}{\partial x_i} \frac{\partial^2 g(\bar{\mu})}{\partial x_i^2} c_{3i}$$

$$+ \sum_{i=1}^{n} \sum_{j=1}^{n} \left\{ \frac{\partial g(\bar{\mu})}{\partial x_i} \frac{\partial^2 g(\bar{\mu})}{\partial x_j^2} E[(x_i - \mu_i)(x_j - \mu_j)^2] \right\}$$
$$i \neq j$$

$$+ 2 \sum_{i=1}^{n} \sum_{j=1}^{n} \left\{ \frac{\partial g(\bar{\mu})}{\partial x_i} \frac{\partial^2 g(\mu)}{\partial x_i \partial x_j} E[(x_i - \mu_i)^2(x_j - \mu_j)] \right\}$$
$$i \neq j$$

$$+ 2 \sum_{i=1}^{n} \sum_{j=1}^{n} \sum_{k=1}^{n} \left\{ \frac{\partial g(\bar{\mu})}{\partial x_i} \frac{\partial^2 g(\bar{\mu})}{\partial x_j \partial x_k} E[(x_i - \mu_i)(x_j - \mu_j) \right.$$
$$i \neq j \quad j \neq k$$

$$\left. \times (x_k - \mu_k)] \right\} \tag{6.4.30}$$

The third and fourth moment approximations follow from a similar algebraic treatment.

$$c_{3y} = \sum_{i=1}^{n} \left[\frac{\partial g(\bar{\mu})}{\partial x_i} \right]^3 c_{3i}$$

$$+ 3 \sum_{i=1}^{n} \sum_{\substack{j=1 \\ i \neq j}}^{n} \left\{ \left[\frac{\partial g(\bar{\mu})}{\partial x_i} \right]^2 \frac{\partial g(\bar{\mu})}{\partial x_j} E[(x_i - \mu_i)^2 (x_j - \mu_j)] \right\}$$

$$+ 6 \sum_{i=1}^{n} \sum_{j=1}^{i-1} \sum_{k=1}^{j-1} \left\{ \frac{\partial g(\bar{\mu})}{\partial x_i} \frac{\partial g(\bar{\mu})}{\partial x_j} \frac{\partial g(\bar{\mu})}{\partial x_k} E[(x_i - \mu_i) \right.$$

$$\left. (x_j - \mu_j)(x_k - \mu_k)] \right\} \tag{6.4.31}$$

$$c_{4y} = \sum_{i=1}^{n} \left[\frac{\partial g(\bar{\mu})}{\partial x_i} \right]^4 c_{4i}$$

$$+ 4 \sum_{i=1}^{n} \sum_{\substack{j=1 \\ i \neq j}}^{n} \left\{ \left[\frac{\partial g(\bar{\mu})}{\partial x_i} \right]^3 \frac{\partial g(\bar{\mu})}{\partial x_j} E[(x_i - \mu_i)^3 (x_j - \mu_j)] \right\}$$

$$+ 6 \sum_{i=1}^{n} \sum_{j=1}^{i-1} \left\{ \left[\frac{\partial g(\bar{\mu})}{\partial x_j} \right]^2 \left[\frac{\partial g(\bar{\mu})}{\partial x_j} \right]^2 E[(x_i - \mu_i)^2 (x_j - \mu_j)^2] \right\}$$

$$+ 12 \sum_{i=1}^{n} \sum_{j=1}^{n} \sum_{\substack{k=1 \\ i \neq j \neq k}}^{n} \left\{ \left[\frac{\partial g(\bar{\mu})}{\partial x_i} \right]^2 \frac{\partial g(\bar{\mu})}{\partial x_j} \frac{\partial g(\bar{\mu})}{\partial x_k} \right.$$

$$\times E[(x_i - \mu_i)^2 (x_j - \mu_j)(x_k - \mu_k)]$$

$$+ 24 \sum_{i=1}^{n} \sum_{j=1}^{i-1} \sum_{k=1}^{j-1} \sum_{\ell=1}^{k-1} \left\{ \frac{\partial g(\bar{\mu})}{\partial x_i} \frac{\partial g(\bar{\mu})}{\partial x_j} \frac{\partial g(\bar{\mu})}{\partial x_k} \frac{\partial g(\bar{\mu})}{\partial x_\ell} \right.$$

$$\left. \times E[(x_i - \mu_i)(x_j - \mu_j)(x_k - \mu_k)(x_\ell - \mu_\ell)] \right\} \tag{6.4.32}$$

In these expressions the expected values that are functions of more than one variable will be simplified if the variables are independent. Consider the following two cases.

$$E[(x_i - \mu_i)(x_j - \mu_j)]$$

$$= \int_R \int (x_i - \mu_i)(x_j - \mu_j)f(x_i)f(x_j)\, dx_i\, dx_j$$

$$= \int_{R_i} (x_i - \mu_i)f(x_i)\, dx_i \int_{R_i} (x_j - \mu_j)f(x_j)\, dx_j$$

$$= 0 \tag{6.4.33}$$

and

$$E[(x_i - \mu_i)^2(x_j - \mu_j)^2]$$

$$= \int_R \int (x_i - \mu_i)^2(x_j - \mu_j)^2 f(x_i)f(x_j)\, dx_i\, dx_j$$

$$= \int_{R_i} (x_i - \mu_i)^2 f(x_i)\, dx_i \int_{R_j} (x_j - \mu_j)^2 f(x_j)\, dx_j$$

$$= c_{2i} c_{2j} \tag{6.4.34}$$

It is apparent that any expected value containing an argument with a factor of the type $(x_i - \mu_i)$ must be zero. All such factors must be of higher degree than 1.

Although we could obtain approximations of higher moments, the algebra becomes rather complex. Moment transfers using the expressions of this section are coded in subroutine MOMENT.

6.4.4 Probabilistic Analysis Using Moment Transfers

Having established techniques for transferring moments from functionally independent to dependent variables, we are now ready to apply these to the central problem of this chapter—the determination of the density function for y, given the general relationship

$$y = g(x_1, x_2, \ldots, x_n)$$

In some cases the moments of the x's can be estimated directly from the data using the expressions in Sec. 4.2. Alternatively, the inde-

pendent variable may have a density function established by one of
the methods of Chapter 5, and in this event the moments can be determined using the integral expressions in Sec. 4.1.

Before using the maximum entropy distribution technique of Sec.
5.5. to generate a density function for y, we must have some estimate
of the bounds of the density function for y. In some kinds of engineering modeling it may be quite practicable to obtain the bounds of
y directly from the bounds of the x's, using the functional relationship $g(x_1, x_2, \ldots, x_n)$. This requires that we have an estimate
of the bounds of those x's that had moments estimated directly from
sample data.

If bounds cannot be estimated directly, an approximate choice
can be made based on the moments. A normal distribution has an area
of 0.999534 enclosed within bounds of $\mu - 3.5\sigma \leqslant \mu + 3.5\sigma$. For many
distributions that are not too far off symmetry, this would provide a
reasonable first guess at the bounds. The third moment can be used
to determine if the distribution is significantly skewed to the right or
left, and the upper or lower bound can be corresponding increased.*
This logic is incorporated in subroutines MANAL and MOMINT, the
software embodiment of the moment transfer methods. MANAL uses
moment transfers based on Taylor's series approximations, while
MOMINT uses moment transfers based on analytical relationships, and
has the advantage of not requiring partial derivatives.

EXAMPLE 6.7

Example 6.2 was solved by the Monte Carlo method. The solution
is repeated here using moment transfers with the maximum entropy distribution. The engineering modeling relationship is

$$S = \frac{6M}{bt^3}$$

and the density function for M and t are shown in Fig. 6.12.
We require the following derivatives for the use of MANAL:

$$\frac{\partial S}{\partial M} = \frac{6}{bt^3}, \quad \frac{\partial^2 S}{\partial M^2} = 0$$

$$\frac{\partial S}{\partial t} = -\frac{18M}{bt^4}, \quad \frac{\partial^2 S}{\partial t^2} = \frac{72M}{bt^5}$$

---

*This idea originated with T. G. Pal.

The bounds of S were obtained from Example 6.2. The calling programs for both MANAL and MOMINT are given below.

```
C.... EXAMPLE 6.7A
      DIMENSION YP(1),CUM(1),AL(8)
      DIMENSION XUPP(2),XLOW(2),CC(6),DELX(2),XGRID(2,40),KVALUE(2),X(2)
      COMMON/MOMINT/IPRINT,IDATA,NBND
      COMMON/MEP1/KPRINT,TOL,MAXFN
C.... X(1) = MOMENT
C     X(2) = THICKNESS
      IDATA=1
      MAXFN=300
      KSTART=4
      YMAX=6.*70./(5.*.090**3)
      YMIN=6.*10./(5.*.114**3)
      N=2
      M=25
      NMOM=4
      XUPP(1)=70.
      XLOW(1)=10.
      XUPP(2)=.114
      XLOW(2)=.090
      NL=NMOM+1
      NYP=0
      CALL MOMINT(N,M,NMOM,XUPP,XLOW,YMIN,YMAX,NYP,YP,KSTART,AL,CUM,
     1DELX,XGRID,KVALUE,X)
      CALL ENTPL(YMIN,YMAX,NL,AL,6,6HSTRESS)
      CALL MEPOUT(AL,NL,NYP,YP,CUM)
      STOP
      END




      FUNCTION FUN(X,N)
      DIMENSION X(1)
      FUN=6.*X(1)/(5.*X(2)**3)
      RETURN
      END




      FUNCTION DENSIT(N,X)
      DIMENSION X(1),THICK(25),FTHICK(25),BMOM(25),FMOM(25)
      DATA FTHICK/0.0,1.25,2.24,3.01,3.53,3.94,4.28,4.55,4.78,4.90,5.00,
     14.92,4.67,4.30,3.83,3.32,2.70,2.12,1.63,1.19,.82,.51,.27,.09,0.0/
      DATA FMOM/0.0,.11,.36,.95,1.92,3.05,4.06,4.77,5.00,4.83,4.44,3.89,
     13.23,2.64,2.14,1.74,1.40,1.09,.82,.69,.40,.26,.13,.04,0.0/
      DATA NCALL/0/
      IF(NCALL.EQ.1)GO TO 2
      DO 1 I=1,25
      THICK(I)=.090+.001*FLOAT(I-1)
      BMOM(I)=10.+2.5*FLOAT(I-1)
1     CONTINUE
C.... NORMALIZE DENSITY FUNCTION
      CALL FNORM(FTHICK,.024,25)
      CALL FNORM(FMOM,60.,25)
      NCALL=1
C.... CALCULATE JOINT DENSITY FUNCTION AT X
2     DENSIT=FTABLE(THICK,FTHICK,X(2),25)*FTABLE(BMOM,FMOM,X(1),25)
      RETURN
      END
```

```
C....  EXAMPLE 6.7B
       DIMENSION YP(1),CUM(1),AL(6),CM(2,4)
       DIMENSION XUPP(2),XLOW(2),CC(6)
       DIMENSION ARG(101),DENS(101),W(101)
       COMMON/MANAL/IPRINT,IDATA,NBND
       COMMON/MEP1/KPRINT,TOL,MAXFN
C....  X(1) = MOMENT
C      X(2) = THICKNESS
       IDATA=1
       MAXFN=300
       KSTART=4
       YMAX=6.*70./(5.*.090**3)
       YMIN=6.*10./(5.*.114**3)
       N=2
       XUPP(1)=70.
       XLOW(1)=10.
       XUPP(2)=.114
       XLOW(2)=.090
       NL=5
       NYP=0
       DO 1 I=1,101
       ARG(I)=10.+60.*FLOAT(I-1)/100.
1      DENS(I)=DENSF(1,ARG(I))
       CALL CMOM(ARG,DENS,101,4,CC,W)
       DO 2 I=1,4
2      CM(1,I)=CC(I)
       DO 3 I=1,101
       ARG(I)=.090+.024*FLOAT(I-1)/100.
3      DENS(I)=DENSF(2,ARG(I))
       CALL CMOM(ARG,DENS,101,4,CC,W)
       DO 4 I=1,4
4      CM(2,I)=CC(I)
       CALL MANAL(N,CM,YMIN,YMAX,NYP,YP,KSTART,AL,CUM)
       CALL ENTPL(YMIN,YMAX,NL,AL,6,6HSTRESS)
       CALL MEPOUT(AL,NL,NYP,YP,CUM)
       STOP
       END



       FUNCTION DENSF(J,X)
       DIMENSION THICK(25),FTHICK(25),BMOM(25),FMOM(25)
       DATA FTHICK/0.0,1.25,2.24,3.01,3.53,3.94,4.28,4.55,4.78,4.90,5.00,
      14.92,4.67,4.30,3.83,3.32,2.70,2.12,1.63,1.19,.82,.51,.27,.09,0.0/
       DATA FMOM/0.0,.11,.36,.95,1.92,3.05,4.06,4.77,5.00,4.83,4.44,3.89,
      13.23,2.64,2.14,1.74,1.40,1.09,.82,.69,.40,.26,.13,.04,0.0/
       DATA NCALL/0/
       IF(NCALL.EQ.1)GO TO 2
       DO 1 I=1,25
       THICK(I)=.090+.001*FLOAT(I-1)
       BMOM(I)=10.+2.5*FLOAT(I-1)
1      CONTINUE
C....  NORMALIZE DENSITY FUNCTION
       CALL FNORM(FTHICK,.024,25)
       CALL FNORM(FMOM,60.,25)
       NCALL=1
C....  CALCULATE JOINT DENSITY FUNCTION AT X
2      CONTINUE
       GO TO (3,4),J
3      DENSF=FTABLE(BMOM,FMOM,X,25)
       RETURN
4      DENSF=FTABLE(THICK,FTHICK,X,25)
       RETURN
       END
```

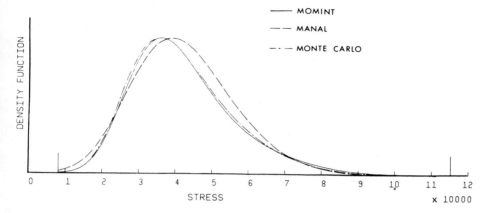

FIG. 6.29 Results of Example 6.7, comparing the moment transfer method with the Monte Carlo method.

```
SUBROUTINE DERV(FUN,DE1,DE2,N,X)
DIMENSION DE1(1),DE2(1),X(1)
FUN=6.*X(1)/(5.*X(2)**3)
DE1(1)=6./(5.*X(2)**3)
DE1(2)=-18.*X(1)/(5.*X(2)**4)
DE2(1)=0.0
DE2(2)=72.*X(1)/(5.*X(2)**5)
RETURN
END
```

The results of these two moment transfer methods are plotted on Fig. 6.29, together with the results from Example 6.3.

6.5 OPERATIONAL TECHNIQUES

Fourier and Mellin transforms can be used for probabilistic analysis when the functional relationship has a general polynomial form (i.e., combinations of sums and products of the variables). This approach has considerable theoretical interest, but in the author's opinion, it is not as practical a technique as others given in this chapter.

In probability theory, the fourier transform of a density function is commonly called the *characteristic function*. It is defined by

$$\phi(u) = \int_R e^{iux} f(x) \ dx \qquad (6.5.1)$$

where u is an arbitrary transform variable and i is $\sqrt{-1}$.

The characteristic function has the interesting property that moments about the origin can be derived directly from it. If we take the first derivative of $\phi(u)$, we get

$$\frac{d\phi}{du} = i \int_R xe^{iux} f(x) \, dx \tag{6.5.2}$$

If this is evaluated at $u = 0$, the results is

$$\frac{d\phi(0)}{du} = i \int_R xf(x) \, dx \tag{6.5.3}$$

It follows that

$$\mu = \frac{1}{i} \frac{d\phi(0)}{du} \tag{6.5.4}$$

Similar arguments lead to

$$m_j = \frac{1}{i^j} \frac{d^j \phi(0)}{du^j} \tag{6.5.5}$$

where m_j is the j moment about the origin.

We now wish to see how Fourier transforms can be used to obtain the probability distribution of the sum of a set of stochastically independent random variables. We begin with the known function

$$y = a_1 x_1 + a_2 x_2 + \cdots + a_n x_n \tag{6.5.6}$$

and known density function for the x's. We also assume that we can determine the Fourier transforms of the x's, designated $\phi_i(u)$ for x_i. The Fourier transform for y is

$$\phi_y(u) = \int_R e^{iuy} f(y) \, dy \tag{6.5.7}$$

$$= \int_R e^{iu(a_1 x_1 + a_2 x_2 + \cdots + a_n x_n)} f(y) \, dy$$

$$= \int_R e^{iua_1 x_1} e^{iua_2 x_2} \cdots e^{iua_n x_n} f(y) \, dy \tag{6.5.8}$$

By the use of infinitesimal events, we have

$$f(y) \, dy = f(x_1) \, dx_1 \, f(x_2) \, dx_2 \cdots f(x_n) \, dx_n \qquad (6.5.9)$$

Subsituting this into (6.5.8) gives

$$\phi_y(u) = \int_R e^{iua_1x_1} f(x_1) \, dx_1 \int_R e^{iua_2x_2} f(x_2) \, dx_2 \cdots$$

$$\int_R e^{iua_nx_n} f(x_n) \, dx_n \qquad (6.5.10)$$

We also know by infinitesimal events that

$$f(x_i) \, dx_i = f(a_ix_i) \, d(a_ix_i) \qquad (6.5.11)$$

so (6.5.10) reduces to the simple product rule

$$\phi_y(u) = \phi_1(a_1x_1)\phi_2(a_2x_2) \cdots \phi_n(a_nx_n) \qquad (6.5.12)$$

We require the inverse Fourier transform to obtain $f(y)$, which formally is

$$f(y) = \frac{1}{2\pi} \int_{-\infty}^{\infty} e^{-iuy} \phi_y(u) \, du \qquad (6.5.13)$$

In a few special cases, both direct and inverse Fourier transforms can be obtained analytically. However, in general, numerical techniques are required.

The Mellin transform can be used in a similar way for products of random variables. It is defined as

$$\mu(v) = \int_0^{\infty} x^{v-1} f(x) \, dx \qquad (6.5.14)$$

We thus require that the variable be nonnegative. However, this requirement can be circumvented. Our known function now is

$$y = x_1x_2 \cdots x_n \qquad (6.5.15)$$

The Mellin transform for y is

$$M_y(v) = \int_0^\infty y^{v-1} f(y) \, dy$$

$$= \int_0^\infty x_1^{v-1} x_2^{v-1} \cdots x_n^{v-1} f(y) \, dy \qquad (6.5.16)$$

If we again use (6.5.9) to replace $f(y)$, we obtain

$$M_y(v) = \int_0^\infty x_1^{v-1} f(x_1) \, dx_1 \int_0^\infty x_2^{v-1} f(x_2) \, dx_2 \cdots$$

$$\int_0^\infty x_n^{v-1} f(x_n) \, dx_n \qquad (6.5.17)$$

Using the definition of Mellin transforms, this becomes

$$M_y(v) = M_{x_1}(v) M_{x_2}(v) \cdots M_{x_n}(v) \qquad (6.5.18)$$

The procedure can be extended to quotients and powers of random numbers. The density function $f(y)$ is given by the inverse transform of $M_y(v)$, which is

$$f(y) = \frac{1}{2\pi i} \int_{C-i\infty}^{C+i\infty} x^{-v} M_y(v) \, dv$$

where C is an arbitrary constant.

General numerical techniques for inverting Fourier and Mellin transforms are apparently not yet available (Springer, 1979). For a more complete treatment of this approach, the reader is referred to Springer (1979).

6.6 THE INDEPENDENT VARIABLE CELL TECHNIQUE

This cell technique is a discretization method somewhat similar to the general version of the transformation of variable method described in

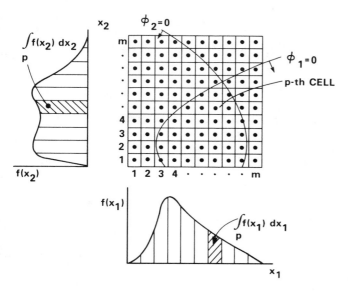

FIG. 6.30 Illustration of discretization.

Sec. 6.3. However, we now work directly with a discretization of the functionally independent variables, rather than the dependent variables. This is illustrated for two variables in Fig. 6.30. The following quantities are defined.

n = number of functionally independent variables

m = number of intervals for each variable

p = address of a cell, defined below

k_i = interval index of a cell corresponding to $x_i - 1 \leqslant k_i \leqslant m$

Δx_1 = discretization interval

The first requirement is to arrange a convenient way of coding the address, p, of each cell for computer storage, with corresponding decoding algorithm, which identifies the k_i's associated with any p. The procedure was described in Sec. 6.3, but is repeated here for convenience.

The address p is simply coded as a base 10 number ranging from 1 to n_c, where n_c is the total number of cells.

$$n_c = m^n \tag{6.6.1}$$

We then conceive of the address as being coded in an m-base number, C_p, which has the form

$$C_p = q_n q_{n-1} \cdots q_1 \tag{6.6.2}$$

where the digit q_i represents the index value of the discretized x_i. The conversion

$$q_i = k_i - 1 \tag{6.6.3}$$

is necessary since the digits in C_p must range from 0 to $m - 1$. The m-based number C_p is never actually recorded. The algorithm for decoding p in order to get the k_i's is as follows.

1. $i = 1$, sum $= 0$.
2. Do until $i = n$.
3. $II = n - i + 1$.
4.   If $II = n$, go to 6.
5.   sum $=$ sum $+ q_{II+1} m^{II}$.
6. $q_{II} = (p - $ sum $- 1)/m^{II-1}$.
7. $k_{II} = q_{II} + 1$.
8. $i = i + 1$.
9. End of do loop.

The number of cells is limited by the integer word size of the computer used. However, it is always possible to use two or more words to represent p. The codification listed above is most easily visualized for m equal to 10, but is valid for any m.

The vector of the coordinates of the midpoint of each cell is used to represent any point occurring in a cell. Thus the probability of any point falling in a given cell is

$$P_p = P(\bar{x} \text{ in pth cell}) = \int_p f(\bar{x}) \, \overline{dx} \tag{6.6.4}$$

in which the joint density function $f(\bar{x})$ is integrated over the region of the pth cell. In the illustration, where x_1 and x_2 are assumed stochastically independent, we have

$$P_p = \int_p f(x_1) \, dx_1 \int_p f(x_2) \, dx_2 \tag{6.6.5}$$

Let us now assume that we have two design characteristics,

$$y_1 = g_1(x_1, x_2)$$

$$y_2 = g_2(x_1, x_2)$$

(6.6.6)

Our general concern in this chapter is to obtain the density functions for y_1 and y_2, but this is difficult to do in this method. However, we are usually only interested in events associated with values of the y's, rather than explicitly knowing the marginal density functions. This is quite easily done, even for joint events. If the requirements are that

$$y_1 \leqslant y_{1S} \quad \text{and} \quad y_2 \leqslant y_{2S}$$

(6.6.7)

then it is convenient to define two new functions,

$$\phi_1 = y_{1S} - y_1 \geqslant 0$$

$$\phi_2 = y_{2S} - y_2 \geqslant 0$$

(6.6.8)

If ϕ_1 and ϕ_2 were set equal to zero, they would plot as lines in Fig. 6.30 and the would enclose a feasible region in which both specifications are satisfied. We are thus concerned with evaluating

$$P(\text{feasibility}) = P(\phi_1 \geqslant 1 \text{ and } \phi_2 \geqslant 0)$$

(6.6.9)

The algorithm for this is quite straightforward:

1. Scan through all cells from 1 to n_c.
2. Decode the cell number p to get k_i, i = 1,n.
3. Obtain corresponding values of x_i, i = 1,n.
4. Evaluate ϕ_1 and ϕ_2 for the cell and determine if the cell is feasible.
5. For feasible cells calculate p_p.
6. P(feasibility) = Σ p_p for all feasible cells.

Rather than integrating to calculate p_p as in (6.6.5), its value could be estimated by using the midpoint values of $f(x_1)$ and $f(x_2)$.

$$p_p = f(x_{1p})f(x_{2p}) \, \Delta x_1 \, \Delta x_2$$

(6.6.10)

It commonly would be more efficient to calculate probabilities at infeasible cells, since the probability of feasibility usually exceeds 0.5.

Extension of the method to more than two variables, and more than two feasibility functions is quite straightforward. The use of random specifications y_{1S}, y_{2S}, and so on, would also cause no real difficulty.

The method is more directly comparable to the generalized transformation of variable technique. The latter requires an inversion of the function at every feasible (or infeasible) cell, whereas this method requires evaluation of the functions at all cells. In complicated sets of functions, an inversion could be considerably more time consuming than an evaluation. On the other hand, there are many cases when inversion is simple, since the auxiliary variables represent very simple functions. This method has the advantage of accepting multivalued functions, but it does not easily yield explicit density functions for the y's. The method is embodied in subroutine CELLV.

6.7 COMPARISON OF METHODS

Choice of the best method for a given analysis is rather difficult. It depends on the nature and complexity of the problem and on what is required from the analysis.

The following tables summarize, in some respects rather crudely, the relative characteristics and advantages of the four methods.

Requirements of the Problem

Method	Probability of combined events	Marginal density functions	Density function of y in analytical form
Monte Carlo	Yes	Yes	No
Transformation of variables	Yes	Yes	No
Moment transfers	No	No	Yes
Cell method	Yes	No	No

Complexity of the Problem

Method	Multi-valued functions	Very complex functions	Discontinuous functions	Discrete events	Analytical derivation required	Stochastic dependence of x's
Monte Carlo	Yes	High	Yes	Yes	No	Yes, but somewhat more difficult
Transformation of variables	No	Low	No	No	Yes/no	Yes
Moment transfers	Yes	Medium	No	No	Yes/no	Possible, but difficult
Cell method	Yes	High	Yes	No	No	Yes

Accuracy, Complexity, and Computer Time

Method	Uses sample moments of x's	Algorithm complexity	Accuracy	Computer time
Monte Carlo	No	Low	Good	High
Transformation of variables	No	High	Fair	High
Moment transfer	Yes for MANAL	High	Good	Low
Cell method	No	Low to medium	Good	Low

One particularly interesting point is the fact that the moment transfer method, which estimates y moments by a Taylor's series approximation, avoids the inevitable bias resulting from representing the distributions of the x's by theoretical distributions such as the normal or Weibull. The moments of the x's are estimated directly from the data, thus both avoiding bias and simplifying the solution.

6.8 ANALYSIS UNCERTAINTY

We have so far ignored a rather important source of uncertainty, and one rather difficult to cope with—analysis error. If y represents a

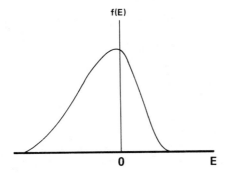

FIG. 6.31 Density function for analysis error.

dependent variable or design characteristic as calculated from the mathematical model, then we define a true value for y as

$$y^t = y - E \tag{6.8.1}$$

where E is a random variable representing the analysis error for y. A subjective density function can be sketched for E which might have a form like that indicated in Fig. 6.31. It will tend to lie mainly in the negative region, since analyses are commonly set up to be conservative. The distribution of y^t can be obtained from (6.8.1), using one of the methods described above.

PROBLEMS

Problem 6.1 This problem is intended to assist in understanding the concept of a theoretical distribution. Generate a theoretical sample with 100 elements from the following distribution for yield stress of a steel in kips/in.² (1 kip = 1000 lb):

$$f(S) = \frac{3.52}{7.68} \left(\frac{S - 37.3}{7.68} \right)^{2.52} \exp\left[-\left(\frac{S - 37.3}{7.68} \right)^{3.52} \right]$$

(What type of analytical distribution is this?) Use the sample to regenerate the probability density function, using the program developed for Problem 5.1. Plot the result on top of a plot for the given analytical function and compare. Do the same for sample sizes of 1000, 2000, 5000, and 10,000.

Problem 6.2 Give an algorithm for applying the inverse cumulative distribution method of generating a theoretical sample to a discrete random variable.

Problem 6.3 Repeat Example 6.1, using the inverse cumulative distribution method, and compare the results with Example 6.1.

Problem 6.4 Generate a theoretical sample for

$$f(x) = 10^{-3} \exp(-10^{-3}x)$$

using the inverse cumulative distribution method. Compare the results with the population density function for sample sizes of 1000, 3000, 5000, 7000, and 9000.

Problem 6.5 We might be inclined to apply the transformation of variable method to a problem such as the following.

$$y = x_1 x_2 + \frac{x_1}{x_2}$$

We could let $w_1 = x_1 x_2$, and $w_2 = x_1/x_2$. The method could be applied quite easily to obtain $f(w_1)$ and $f(w_2)$. What would be wrong with again applying it to

$$y = w_1 + w_2$$

to obtain $f(y)$? Give the correct algorithm.

Problem 6.6 A high-strength steel has a yield stress, S, with density function defined by

$$f(S) = \frac{1.41}{6.48} \left(\frac{S - 92.2}{6.48} \right)^{0.41} \exp\left[-\left(\frac{S - 92.2}{6.48} \right)^{1.41} \right]$$

where S is in kips/in.^{2}. It is used in a wide-flange steel beam having a section modulus (I/y) of 70.7 in.^{3}. The beam is subjected to a bending moment of 6.71×10^6 in. lb. If only the yield stress is considered a random variable, what is the probability of yield failure in simple bending?

Problem 6.7 All textbook examples are, through necessity of space, oversimplified. Discuss some of the additional realities and uncertainties that should be considered in Example 6.6.

Problem 6.8 Develop an algorithm to give $f(y)$, similar to that used in Example 6.5, for the case when y and x have a multivalued relationship, such as that shown in Fig. 6.32. Assume that $f(x)$ is defined numerically and $g(x)$ is defined analytically.

Problem 6.9 Discuss the problem of estimating $f(y)$ when $y = x_1/x_2$, $f(x_1)$ and $f(x_2)$ are given, and x_2 has bounds 0 and 1. What methods might have difficulties? Which is likely to be the best method?

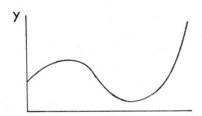

y

x   FIG. 6.32 Multivalued function.

Problem 6.10 Use function FINVRT to compare a systematic sample (see Sec. 6.3) with a random theoretical sample of the same size. Use any density function and compare the first four moments calculated from the two samples.

Problem 6.11 A pump is connected to a straight-line section of pipe. We wish to determine the probability density function for the flow rate. We assume that the following are known:

1. The function relationship for the pump between head H (ft) flow rate Q (ft^3/sec). Let this be

   $$Q = g_1(H)$$

2. The pressure drop in the pipe is given by

   $$H = \frac{2fV^2L}{gD} \text{ ft}$$

   where

   > f = friction factor
   >
   > V = velocity, ft/sec
   >
   > D = internal pipe diameter, ft
   >
   > g = acceleration due to gravity
   >
   > L = pipelength, ft

3. The friction factor is given by

   $$\frac{1}{\sqrt{f}} = a \log_{10} (Re \sqrt{f}) + b$$

   where a and b are empirical parameters and Re is the Reynolds number.

   $$Re = \frac{DV\rho}{\mu}$$

   where

   > ρ = water density, lb/ft^2
   >
   > μ = water viscosity, lb/sec ft

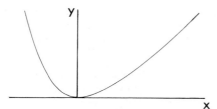

X   FIG. 6.33 Problem 6.12.

4. The following quantities are known and determinate: L, g. The
    following quantities are random but completely defined: a,b,
    normal with known parameters; D, Weibull with known parameters.
5. ρ and μ are known functions of the temperature t (°F):

$$\rho = g_2(t), \quad \mu = g_3(t)$$

The density function of t is known from frequency data, but cannot be represented satisfactorily by a theoretical distribution.

Explain in logical, step-by-step fashion how you would solve this
problem.

Problem 6.12 If the function y = g(x) has the form shown in Fig.
6.33, determine the relationship between f(y) and f(x), and make a
sketch similar to Fig. 6.18.

Problem 6.13 Extend Example 6.4 by obtaining the mean and mode of
the dependent and independent variables. Compare with the nominal
values.

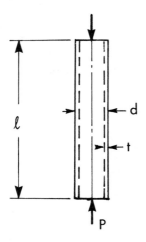

FIG. 6.34 Euler's column.

Problem 6.14 You are designing a Euler-type column, as shown in Fig. 6.34. The critical stress in a pin-ended Euler column with a tubular cross section is

$$S_{cr} = \frac{\pi^2 E}{(\ell/\rho)^2}$$

where

ℓ = column length = 120 ± 0.375, in.

E = Young's modulus = $31 \times 10^6 \pm 10^4$, psi

ρ = radius of gyration = $(I_z/A)^{1/2}$

I_z = moment of inertia $\simeq \pi r^3 t$

d = outer diameter = 2.572 ± 0.015 in.

t = wall thickness = 0.051 ± 0.006 in.

A = cross-sectional area = $2\pi r t$, in^2

The applied load is a normal random variable, P $\sim$ (6070, 200) lb. The column length ℓ, outer diameter d, and thickness t are all normal random variables as well as the Young's modulus of elasticity E. The tolerances are given $\pm 3\sigma$.

a. Use the Monte Carlo technique and maximum entropy principle to calculate the probability of failure of this design, neglecting the effect of eccentricity. Compare the results of both methods.
b. Show analytically how the effect of eccentricity of the outer and inner diameters may be included in the analysis. (Outline briefly without repeating the solution on the computer.)

Problem 6.15 A slider-crank mechanism has the nominal configuration shown in Fig. 6.35. It is to be used as a function generator, and the configuration was optimized to minimize the error. The mechanism geometry is defined by the expression

$$S^2 + (2\rho_1 \cos \alpha)S - 2\rho_1\rho_3 \sin \alpha + (\rho_1^2 - \rho_2^2 + \rho_3^2) = 0$$

where

S = slider position

α = crank angle

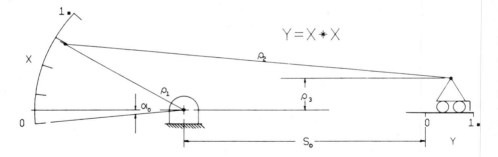

FIG. 6.35 Slider-crank function generator.

S_0 = initial position of slider = 5.2649

α_0 = initial crank angle = $-7.5401°$

α_{max} = final crank angle = $42.4598°$

ρ_1 = 2.8269

ρ_2 = 8.1367

ρ_3 = 0.6892

The error in the value of y is defined as

$$e = y_a - y_t$$

where

y_a = actual position = $S - S_0$

y_t = desired position = $x^2 = [(\alpha - \alpha_0)/50]^2$

The values given for ρ_1, ρ_2, and ρ_3 are nominal, and due to manufacturing errors, they are actually random variables having a normal distribution with standard deviations given by

$$\sigma_i = 0.001 + 0.0003\rho_i$$

Plot a curve showing the error as a function of crank angle; also plot 0.10 and 0.90 fractile lines, as defined in Fig. 6.36.

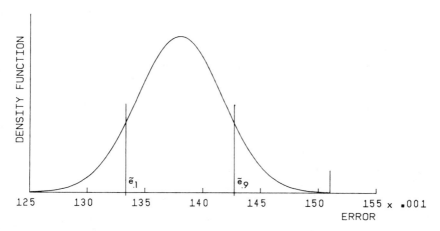

FIG. 6.36 Error density function at some given value of α.

Could f(e) be adequately approximated by a normal distribution for all values of α? If it could, how could we estimate the mean and standard deviation of e?

(This problem was adapted from a paper by Schade.*

Problem 6.16 A bread toaster for use in the home is being designed; and after doing some testing of a prototype, the density function is estimated to determine the percent of toast burned. The designer then consults with the marketing department and makes a subjective estimate of the density function for percent of toast burned that would be acceptable to consumers. Hypothetical curves are defined by the following data arrays.

% burned by toaster = B - [0.0 to 7.0 in increments of 0.2]

f(B) = [0.0, 0.03, 0.12, 0.29, 0.59, 1.18, 1.89, 2.40, 2.71, 2.91,
        2.99, 2.95, 2.73, 2.43, 2.08, 1.73, 1.45, 1.21, 1.00, 0.83,
        0.69, 0.58, 0.48, 0.40, 0.34, 0.30, 0.23, 0.20, 0.17, 0.11,
        0.10, 0.09, 0.05, 0.02, 0.01, 0.0]

---

*R. Schade (1980), Probabilistic Models in Computer Automated Slider-Crank Function Generator Design, *ASME Paper 80-DET-46.*

Acceptable % burned = A = [0.0 to 8.0 in increments of 0.2]

f(A) = [0.0, 0.0, 0.01, 0.05, 0.09, 0.12, 0.20, 0.28, 0.37, 0.50,
        0.63, 0.78, 0.93, 1.11, 1.30, 1.50, 1.71, 1.95, 2.19, 2.40,
        2.56, 2.70, 2.81, 2.90, 2.97, 3.00, 2.99, 2.92, 2.85, 2.74,
        2.61, 2.50, 2.33, 2.16, 1.93, 1.70, 1.41, 1.10, 0.79, 0.40,
        0.0]

Do the following:

a. Plot the given curves.
b. Determine and plot the density function for the specification margin.
c. Determine the probability of a satisfied customer.

Problem 6.17 In the design of bolted connections, it is theoretically possible to relate bolt preload to wrench torque. It is desirable to give bolts as high a preload as possible so as to minimize fatigue and risk of loosening.* Standard practice is to torque the bolts so that the preload is 90% of the nominal yield stress of the bolt material. The wrenching torque is calculated from the expression

$$T = \frac{Fd_m}{2} \frac{\ell + \pi\mu d_m \sec \alpha}{\pi d - \mu\ell \sec a} + 0.625\mu_c dF$$

where

F = preload, lb

d_m = mean bolt diameter, in.

ℓ = thread pitch, in.

μ = coefficient of friction between the threads

μ_c = coefficient of friction between the washer and nut

α = half the thread angle, 30°

d = bold outside diameter, in.

The torque is measured while tightening by means of a torque wrench. Experience has shown that there is a large amount of variation in the

*J. E. Shigley (1963), *Mechanical Engineering Design*, McGraw-Hill, New York.

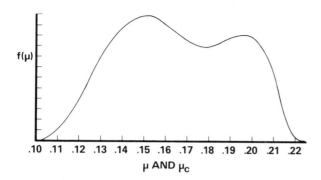

FIG. 6.37 Density function for the coefficient of friction.

actual preload force in the bolt, if nominal values are used in the expression above to predict the necessary torque.

We shall use the specific example of A325 high-strength bolts, 3/4-UNC10, having a yield stress S defined by the following probability density function:

$$f(S) = \frac{1.41}{6.48}\left(\frac{S - 92.2}{6.48}\right)^{0.41} \exp\left[-\left(\frac{S - 92.2}{6.48}\right)^{1.41}\right]$$

where S is in kips/in^2, with a mean value of 98.1 kips/in^2. Other quantities having significant variability are μ, μ_c, and T. Hypothetical but fairly reasonable probability density functions are shown in Fig. 6.37 for the coefficients of friction. It is assumed that these are derived from smoothed frequency data. We shall also assume that it has been found that the actual torque exerted by a worker using a torque wrench is normally distributed with a standard deviation of 5% of the nominal or specified value. The major and minor diameters of the specified bolt are 0.750 and 0.62 in. Determine the probability of the yield strength being exceeded when a bolt is torqued to the nominal value using a torque wrench. Solve the problem using all four of the methods given in this chapter.

Problem 6.18 Two standard I-sections support a load P, as shown in Fig. 6.38. The beams are assumed to be simply supported with length ℓ, and the load concentrated at midspan. The beams have a camber when they come from the rolling mill, as indicated in Fig. 6.39, measured by the quantity ε. The measure of camber is always considered positive, as shown, and the beams are always installed with positive camber. The cambers can be different, as indicated in the

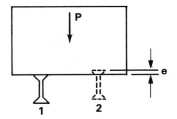

FIG. 6.38 Problem 6.18.

first figure, so that one beam picks up more of the load than the other. Failure is in yielding. You have the following data.

$P = 40,000$ lb

$\ell = 240$ in.

$$f(S_y) = \frac{3.52}{7.68} \left(\frac{S_y - 37.3}{7.68} \right)^{2.52} \exp \left[- \left(\frac{S_y - 37.3}{7.68} \right)^{3.52} \right]$$

where S_y is in kips/in^2 in this expression.

$$f(\varepsilon) = \frac{2.2}{0.295} \left(\frac{\varepsilon}{0.295} \right)^{1.2} e^{-(\varepsilon/0.295)^{2.2}}$$

$E = 30.0 \times 10^6$ psi

Choose a standard steel section that will give an overall probability of failure of 0.05. Use any two methods.

   Note: You should plot the two given random variables to see what they look like. Use their nominal values to get a start on your trials for your selection of the member size.

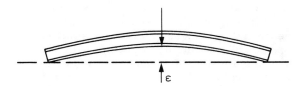

FIG. 6.39 Problem 6.18.

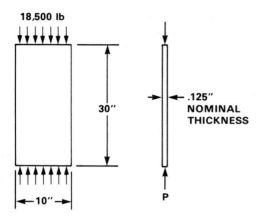

FIG. 6.40 Problem 6.19.

Problem 6.19 A hot-rolled steel sheet having dimensions shown in Fig. 6.40 is subject to a total compression loading of 18,500 lb. The critical buckling stress is given by

$$\sigma_{cr} = 3.62E \left(\frac{t}{b}\right)^2 \text{ psi}$$

where

t = thickness, in.

E = modulus of elasticity = 30.0×10^6 psi

b = width, in.

The thickness has a tolerance from the mill of ±0.010 in. The probability density function for t is a normal distribution with standard deviation equal to one-third the tolerance. What is the probability of failure?

Problem 6.20 The critical stress for an eccentrically loaded column shown in Fig. 6.41 is defined implicitly by the following expression:

$$S_y = \sigma_{cr}\left[(1 + \frac{ec}{\rho^2} \sec \left(\frac{L}{2\rho} \sqrt{\frac{\sigma_{cr}}{E}}\right)\right]$$

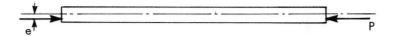

FIG. 6.41   Problem 6.20.

where

S_y = yield stress

e = eccentricity

c = maximum distance from the neutral axis to the extreme point on the section

ρ = radius of gyration

L = equivalent length

E = modulus of elasticity

σ_{cr} = stress level above which buckling will occur

The stress level is calculated by

$$\sigma = \frac{P}{A}$$

where

P = axial load

A = cross-sectional area

There is considerable uncertainty about the value of e for any given column for the particular design in question, and its probability density function f(e) is unknown. All of the other parameters are assumed deterministic, and we want to determine the probability density function for σ_{cr}. Make your own subjective estimate of f(e) and obtain a solution using the transformation of variable method, without using TRANSF. Then obtain a solution with TRANSF.

Problem 6.21 We wish to design a round bar in torsion to resist a torque T having the following probability density function:

$$f(T) = \frac{1}{10\sqrt{2\pi}} \exp\left[-\frac{(T - 100)^2}{2 \times 10^2} \right]$$

where T is in kip in.

The yield stress for the material in kips/in.^{2} has a distribution

$$f(S) = \frac{3.70}{46.46} \left(\frac{S - 157.0}{46.46} \right)^{2.70} \exp\left[-\left(\frac{S - 157.0}{46.46} \right)^{3.70} \right]$$

The bar diameter has a tolerance of ±0.005d, where d is the mean shaft diameter. The diameter distribution is assumed normal with a standard deviation of one-third the tolerance. The probability of no failure is to be 0.999. What is the nominal factor of safety? The stress in a round shaft is predicted by

$$\sigma = \frac{16T}{\pi d^3}$$

Problem 6.22 Suggest an algorithm for determining f(y) when

$$y = \sum_{i=1}^{n} a_i x_i$$

and n is a discrete random variable with known probability mass function f(n), and x_i's are continuous independent random variables with known $f(x_i)$.

Problem 6.23 Write a segment of computer code to embody the *inverse cumulative distribution method* of generating a theoretical sample, calling subprograms from the PROBVAR package if appropriate. Under what circumstances would you use this method rather than the rejection method?

REFERENCES

Forsythe, G. E., M. A. Malcolm, and C. B. Moler (1977), *Computer Methods for Mathematical Computations*, Prentice-Hall, Englewood Cliffs, N.J.

Hahn, G. J., and S. S. Shapiro (1967), *Statistical Models in Engineering*, Wiley, New York.

Hammersley, J. M., and D. C. Handscomb (1967), *Monte Carlo Methods*, Methuen, London.

Kahn, H. (1954), Use of Different Monte Carlo Sampling Techniques, in *Proceedings of Symposium on Monte Carlo Methods*, H. A. Meyer, ed., Wiley, New York.

Kelly, L. G. (1967), *Handbook of Numerical Methods and Applications*, Addison-Wesley, Reading, Mass.

Kendall, M. G., and A. Stuart (1976), *The Advanced Theory of Statistics*, Vol. 3: *Design and Analysis, and Time Series*, 3rd ed., Griffin, London.

Mazumbar, M. (1975), Importance Sampling in Reliability Estimation, in *Reliability and Fault Tree Analysis*, R. E. Barlow, J. B. Fussell, and N. D. Singpurwalla, eds., Society for Industrial and Applied Mathematics, Philadelphia.

Mischke, C. R. (1971), Some Tentative Weibullian Descriptions of the Properties of Steels, Aluminums, and Titaniums, *ASME Publ. No. 71-Vibr-64*.

Raj, D. (1968), *Sampling Theory*, McGraw-Hill, New York.

Rubinstein, R. Y. (1981), *Simulation and the Monte Carlo Method*, Wiley, New York.

Scott, T. R., and T. P. Walker, Jr. (1976), Regionalization: A Method for Generating Joint Density Estimates, *IEEE Trans. Circuits Syst.* Vol. CAS-23, No. 4, pp. 229-234.

Siddall, J. N. (1982), *Optimal Engineering Design: Principles and Applications*, Marcel Dekker, New York.

Springer, M. D. (1979), *The Algebra of Random Variables*, Wiley, New York.

Tocher, K. D. (1963), *The Art of Simulation*, The English Universities Press, London.

7

Sequential Events

Introduction to Markov chains · Monte Carlo simulation of sequential events with an example · The elements of random time function theory

7.1 INTRODUCTION

We are concerned in this chapter with random events that occur sequentially in time. They may be discrete events, or events associated with continuous variables that vary with time in a random fashion. Sets of these events are in general called *random processes* or *stochastic processes*.

The Poisson process described in Sec. 3.3 is one rather simple type of random process when placed in a time context. It consists of a series of independent discrete events, having a constant rate of occurrence. At the other end of the range of complexity of random processes is the *random time function*, described in Sec. 7.4. An intermediate area of complexity is included in the modeling of *Markov processes*, which represent systems of sequential events where the occurrence of any event at a given stage depends, in the probabilistic sense, only on the specific event that has occurred in the immediately previous stage. Markov chains are described briefly in Sec. 7.2.

7.2 MARKOV CHAINS

We are concerned with modeling a systematic sequence of events. Each event can assume one of m different possible states at stage n in the sequence. We require the further limitation that the probabilities associated with each state at stage n depend only on the state that occurred in the previous stage, not on any earlier stages. Finally, we require that the process be *stationary*, so that the probabilities associated with going from one state to another in successive stages do not vary with time. Such a process is also sometimes said to be *homogeneous*.

The general analytical theory for Markov chains is beyond the scope of this book. Information can be found in Kemeny and Snell (1960), Saaty (1961), Parzen (1962), and Rau (1970). In principle, it is always possible to solve problems of this type using Monte Carlo simulation, and an example is given in the following section.

7.3 MONTE CARLO SIMULATION

Sequential event problems can usually be solved by Monte Carlo simulation. Although computer costs tend to be much higher than when Markov chain theory is applied, there is less risk of error in setting up events and misinterpreting probability laws. Monte Carlo simulation may also be applicable in more complex sequential systems that cannot be modeled as Markov chains. More information can be found in Bulgren (1982) and Medhi (1982).

EXAMPLE 7.1 Gravel Truck Queue

One of the commonest types of sequential event problems is the queuing problem. A typical engineering example is a fleet of trucks picking up loads of gravel at a gravel pit and transporting them to a variety of locations. We are concerned with the queue of trucks waiting to be filled at the pit. The *state* is defined as the number of trucks in the queue waiting to be filled; and the *nth stage* is defined as the moment that the nth truck to be serviced on a given day has been filled and leaves. The state is represented by the random variable x, the number of trucks in the queue including the one just about to be filled, and x_n is the value of x at the nth stage. The stage is 1 when the first truck to arrive leaves.

The system behavior depends on two other random variables: the time t_a between successive trucks arriving, and the time t_f to fill a truck. We assume that we know the distributions for these variables. It may be reasonable to model, for example, one variable as the result of a Poisson process, so that the distribution would be exponential, and it would only be necessary to determine the mean time. Or we could generate a more general distribution from observations, using the methods of Chapter 5.

We are interested in determining the probability distribution for queue size at any stage, the probability distribution for wait-

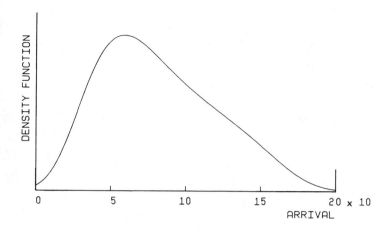

FIG. 7.1 Density function for truck arrival interval. Moments are -0.81772×10^2, 0.15411×10^4, 0.26374×10^5, 0.59137×10^7, and 0.25067×10^9.

ing time for any given truck, and the probability that the maximum queue in any given day will equal a specified value. The last item may be important because of limited queuing space.

There are 47 trucks being serviced at the gravel pit. The interval between arrivals, t_a, has the density function shown in Fig. 7.1; and the filling time, t_f, has the density function shown in Fig. 7.2. These are maximum entropy density functions, and the parameters are given in the computer program that follows the algorithm. The day length is 9 hours.

The algorithm has the following steps:

1. *Input*

t_d = length of working day, sec

n_s = sample size

m = maximum queue size or number of states

$f(t_a)$ = density function for arrival time

$f(t_f)$ = density function for fill time

n_d = number of stages for Monte Carlo trials (it is easier to work with a specified number of trials than a specified time)

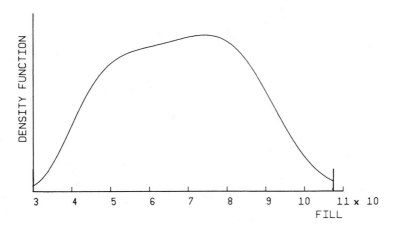

FIG. 7.2 Density function for truck filling time. Moments are 0.78042 × 10^2, 0.27802 × 10^3, 0.89884 × 10^2, 0.16943 × 10^6, 0.10748 × 10^6.

2. *Initialize*
    a. Get maximum values of $f(t_a)$ and $f(t_f)$ for use in the rejection method of sampling.
    b. Define values for x, the queue size or states.
    c. Zero frequency counts.
3. DO WHILE i ≤ n_s to step 22.
4. Begin simulation and initialize.

n_t = number of trucks arriving during a fill = 0

T_a = cumulative arrival times = 0

T_f = cumulative fill times = 0

i_w = flag set at 1 if there is a truck waiting, and at 0 if not = 0

n_f = cumulative number of fills = 0

n_q = current number in queue = 1

tin_1 = arrival time of first truck = 0

i = index for arrival time of first truck = 1

n_{max} = maximum observed queue length in a day = 0

5. IF $n_f = 0$,
    THEN go to 14.
    ELSE continue.
    ENDIF

6. A new stage has occurred; a truck has just left.

    $$n_q = n_q - 1$$

    $$n_f = n_f + 1$$

7. Check for end of day.

    IF $T_f > t_d$,

    THEN exit with error message.
    ELSE continue.
    ENDIF

    Check for completion of number of stages.

    IF $n_f > n_d$,

    THEN go to 22.
    ELSE continue.
    ENDIF

8. Increment frequency count for queue size at current stage.

    $$n_s = \text{current state index} = n_q + 1$$

    $A(n_s, n_f)$ = frequency count for occurrence of n_s state at n_f
    stage; increment by 1

9. Check if current queue size is maximum.

    IF $n_q > n_{max}$,

    THEN $n_{max} = n_q$.

    ELSE continue.
    ENDIF

10. Set flag.

    IF $n_t > 0$ *or* $n_q > 0$,

    THEN $i_w = 1$.

    ELSE $i_w = 0$.

    ENDIF

11. If $n_q = 0$, we must wait until next truck arrives before fill-
    ing begins.
    IF $n_q > 0$,
    THEN go to 14.
    ELSE continue.
    ENDIF.

12. $T_f = T_f + \Delta t$

    $i = n_f + 1$

    $tin_i = T_a$

    $i = i + 1$

13. Record waiting time.

    tw_i = waiting time for ith truck

    $= T_f - tin_i$

    $n_t = 0$

14. Sample fill time and begin a new stage. Generate a trial val-
    ue for fill time, using the rejection method.

    $$T_f = T_f + t_f$$

15. Determine number of arrivals during a fill, but first check
    for *any* arrivals.

16. IF $n_f = 0$,
    THEN go to 18.
    ELSE continue.
    ENDIF

17. IF $T_f > T_a$,
    THEN
    IF $i_w = 1$,
       THEN $n_q = n_q + 1$

              $n_t = n_t + 1$

                 $i = i + 1$
                 $tin_i = T_a$.
    ELSE continue.
    ENDIF
    ELSE $\Delta t = T_a - T_f$
         go to 6.
    ENDIF

18. Generate a trial value for arrival time, using the rejection method.

19. $T_a = T_a + t_a$

20. IF $T_a < T_f$,

    THEN $n_q = n_q + 1$

    $n_t = n_t + 1$

    $i = i + 1$

    $tin_i = T_a$

    go to 18.

    ELSE $\Delta t = T_a - T_f$
    go to 6.

    ENDIF

21. Increment frequency counts for observed value of waiting time and for observed maximum queue size in one day.

22. ENDDO

    A Monte Carlo trial has been completed.

23. Normalize frequencies for queue size and output queue size versus relative frequency (estimated probabilities).

24. Normalize waiting time frequencies and plot waiting time versus relative frequency histogram.

25. Normalize maximum queue size in 1-day frequencies and output as probability estimates.

The foregoing algorithm is coded in the following program, with accompanying output.

```
C.... INPUT
C         TDAY= ESTIMATED TIME IN SECONDS TO STEADY STATE
C         NSAMP= SAMPLE SIZE FOR MONTE CARLO SIMULATION
C         NQUMAX= ESTIMATED MAXIMUM QUE SIZE
C         FUNCA= DENSITY FUNCTION FOR ARRIVAL TIMES - EXTERNAL FUNCTION
C         FUNCF= DENSITY FUNCTION FOR FILL TIMES - EXTERNAL FUNCTION
C.... INTERNAL VARIABLES
C         NSTAGES= EXPECTED NUMBER OF STAGES IN TIME TDAY
C         NTRUCK= NUMBER OF TRUCKS THAT ARRIVE DURING A FILL
C         TA= CUMULATIVE ARRIVAL TIMES
C         TF= CUMULATIVE FILL TIMES
C         NFILLS= NUMBER OF TRUCKS THAT HAVE BEEN FILLED
C         NQUE= CURRENT NUMBER IN QUEUE
C         NSTATE= CURRENT STATE NUMBER
C         IWAIT= 0, NO TRUCKS ARE WAITING WHEN CURRENT TRUCK LEAVES
C              = 1, ONE OR MORE TRUCKS ARE WAITING
C         TFILL= SAMPLED FILL TIME
C         TARRIV= SAMPLE ARRIVAL TIME
C         TIN(I)= ARRIVAL TIME OF I TH TRUCK
C         TWAIT(I)= WAITING TIME OF I TH TRUCK
C         TWMAX= WAITING TIME UPPER BOUND
C         NMAX= MAX OBSERVED QUE LENGTH IN A DAY
```

```
C.... OUTPUT
C        FREQ(I,J)= FREQUENCY COUNT FOR I TH STATE AND J TH STAGE
C        FQDAY(I)= FREQUENCY COUNT FOR MAX OBSERVED QUE IN ONE DAY
         DIMENSION FREQ(10,100), R(2), X(10), TIN(100), TWAIT(100), FREQW(1
        100), WAIT(101), FQDAY(10), DUM(1), XDENS(100), FDENS(100), W1(100)
        2, W2(100)
         COMMON /HIST/ ICALL
         EXTERNAL FUNCA,FUNCF
         ICALL=1
C.... INITIALIZE
C.... CALCULATE DAY LENGTH IN SEC
         TDAY=9.*60.*60.
         NSAMP=5000
         NQUMAX=10
         TWMAX=650.
         NSTAGES=48
         DO 1 I=1,NQUMAX
         FQDAY(I)=0.0
1        CONTINUE
         DO 2 I=1,50
         FREQW(I)=0.0
2        CONTINUE
C.... DEFINE INTERVALS FOR WAITING TIME FREQ COUNT
         DELW=TWMAX/50.
         DO 3 I=1.51
         WAIT(I)=DELW*FLOAT(I-1)
3        CONTINUE
C.... GET MAX VALUES OF GIVEN DENSITY FUNCTIONS
         CALL AMAX (FUNCA,0.0,200.,ARRIVM,X1)
         CALL AMAX (FUNCF,30.,107.5,FILLM,X1)
         DO 4 I=1,NQUMAX
         X(I)=FLOAT(I-1)
         DO 4 J=1,NSTAGES
4        FREQ(I,J)=0.0
C.... BEGIN SIMULATION - A TRIAL IS ONE DAY
         CALL FRAND (R,1,1002246)
         DO 14 ISAMP=1,NSAMP
         NTRUCK=0
         TA=0.0
         TF=0.0
         NQUE=1
         NFILLS=0
         TIN(1)=0.0
         IT=1
         NMAX=0
         IWAIT=0
         IF (NFILLS.EQ.0) GO TO 6
C.... A NEW STAGE HAS OCCURRED, A TRUCK HAS JUST LEFT
5        NQUE=NQUE-1
         NFILLS=NFILLS+1
C.... CHECK FOR END OF DAY
         IF (TF.GT.TDAY) GO TO 30
C.... CHECK FOR COMPLETION OF SPECIFIED NUMBER OF STAGES
         IF (NFILLS.GT.NSTAGES) GO TO 10
C.... INCREMENT FREQUENCY COUNT FOR QUEUE SIZE
         NSTATE=NQUE+1
         FREQ(NSTATE,NFILLS)=FREQ(NSTATE,NFILLS)+1.
C.... CHECK IF CURRENT QUEUE IS MAXIMUM FOR DAY
         IF (NQUE.GT.NMAX) NMAX=NQUE
         IWAIT=0
         IF (NTRUCK.GT.0.OR.NQUE.GT.0) IWAIT=1
C
C.... IF NQUE=0 WE MUST WAIT UNTIL NEXT TRUCK ARRIVES BEFORE
C        FILLING BEGINS
```

```
         IF (NQUE.GT.0) GO TO 6
         TF=TF+DELT
         NQUE=NQUE+1
         IT=IT+1
         TIN(IT)=TA
6        CONTINUE
C....  RECORD WAITING TIME
         TWAIT(NFILLS+1)=TF-TIN(NFILLS+1)
         NTRUCK=0
C....  SAMPLE FILL TIME AND BEGIN A NEW STAGE
C
         CALL FRAND (R,2,0)
         TFILL=R(1)*77.5+30.
         IF (R(2).GT.FUNCF(TFILL)/FILLM) GO TO 6
         TF=TF+TFILL
C....  DETERMINE NUMBER OF ARRIVALS DURING FILL
C....  FIRST CHECK FOR ANY ARRIVALS
         IF (NFILLS.EQ.0) GO TO 8
         IF (TF.GT.TA) GO TO 7
         DELT=TA-TF
         GO TO 5
7        IF (IWAIT.EQ.0) GO TO 8
         NQUE=NQUE+1
         NTRUCK=NTRUCK+1
         IT=IT+1
         TIN(IT)=TA
8        CONTINUE
         CALL FRAND (R,2,0)
         TARRIV=R(1)*200.
         IF (R(2).GT.FUNCA(TARRIV)/ARRIVM) GO TO 8
         TA=TA+TARRIV
         IF (TA.LT.TF) GO TO 9
         DELT=TA-TF
         GO TO 5
9        NQUE=NQUE+1
         NTRUCK=NTRUCK+1
         IT=IT+1
         TIN(IT)=TA
         GO TO 8
10       CONTINUE
C....  OBTAIN WAITING TIME FREQUENCY COUNT
         DO 13 I=1,NSTAGES
         DO 11 J=2,51
         IF (TWAIT(I).LT.WAIT(J)) GO TO 13
11       CONTINUE
         WRITE (6,12)
12       FORMAT (37H UPPER BOUND FOR WAITING TIME TOO LOW)
         STOP
13       FREQW(J-1)=FREQW(J-1)+1.
C....  OBTAIN MAX QUEUE FOR DAY FREQ COUNT
         FQDAY(NMAX+1)=FQDAY(NMAX+1)+1.
C....  OUTPUT TIME USED IN A TRIAL
14       CONTINUE
         WRITE (6,15) TF
15       FORMAT (37H TOTAL TIME USED IN A ONE DAY TRIAL =,E12.5)
C....  NORMALIZE FREQUENCIES FOR QUEUE SIZE
         DO 18 J=1,NSTAGES
         SUM=0.0
         DO 16 I=1,NQUMAX
         SUM=SUM+FREQ(I,J)
16       CONTINUE
         DO 17 I=1,NQUMAX
         FREQ(I,J)=FREQ(I,J)/SUM
```

```
17      CONTINUE
18      CONTINUE
C....   OUTPUT
        WRITE (6,19)
19      FORMAT(//,10X,41HPROBABILITIES FOR QUEUE SIZE AT ANY STAGE,/)
        WRITE (6,20)
20      FORMAT (//,3X,5HSTAGE,25X,10HQUEUE SIZE)
        WRITE (6,21)
21      FORMAT (12X,1H0,6X,1H1,6X,1H2,6X,1H3,6X,1H4,6X,1H5,6X,1H6,6X,1H7,6
       1X,1H8,6X,1H9,/)
        DO 23 J=1,NSTAGES
        WRITE (6,22) J,(FREQ(I,J),I=1,10)
22      FORMAT (2X,I4,3X,10(F6.4,1X))
23      CONTINUE
C....   NORMALIZE MAX QUEUE FOR DAY FREQUENCIES
        SUM=0.0
        DO 24 I=1,NQUMAX
        SUM=SUM+FQDAY(I)
24      CONTINUE
        DO 25 I=1,NQUMAX
        FQDAY(I)=FQDAY(I)/SUM
25      CONTINUE
        WRITE (6,26)
26      FORMAT (//,10X,44HPROBABILITIES FOR MAX QUEUE SIZE FOR ANY DAY,/)
        WRITE (6,27) (FQDAY(I),I=1,NQUMAX)
27      FORMAT (9X,10(F6.4,1X))
C....   OBTAIN WAITING TIME HISTOGRAM
C....   NORMALIZE FREQUENCIES FOR WAITING TIME
        SUM=0.0
        DO 28 I=1,50
        SUM=SUM+FREQW(I)
28      CONTINUE
        DO 29 I=1,50
        FREQW(I)=FREQW(I)/SUM
29      CONTINUE
        CALL HIST (DUM,NSAMP,50,TWMAX,0.0,1,1,9,9HWAIT TIME,100,WAIT,FREQW
       1,XDENS,FDENS,W1,W2)
        STOP
30      WRITE (6,31)
31      FORMAT (59H THE SPECIFIED NUMBER OF STAGES HAS EXCEEDED THE DAY LE
       1NGTH)
        END
        FUNCTION FUNCA (X)
        DIMENSION AL(6)
        DATA AL/-.80546E1,.16154E0,-.26641E-2,.18875E-4,-.57186E-7,.42215E
       1-10/
        FUNCA=ENTRPF(AL,6,X)
        RETURN
        END
        FUNCTION FUNCF (X)
        DIMENSION AL(6)
        DATA AL/-.55593E2,.36043E1,-.99413E-1,.13438E-2,-.88187E-5,.22195E
       1-7/
        FUNCF=ENTRPF(AL,6,X)
        RETURN
        END
```

TOTAL TIME USED IN A ONE DAY TRIAL = .41798E+04

PROBABILITIES FOR QUEUE SIZE AT ANY STAGE

STAGE	QUEUE SIZE									
	0	1	2	3	4	5	6	7	8	9
1	.5958	.3674	.0354	.0014	0.0000	0.0000	0.0000	0.0000	0.0000	0.0000
2	.4868	.4322	.0756	.0050	.0004	0.0000	0.0000	0.0000	0.0000	0.0000
3	.4534	.4432	.0894	.0132	.0008	0.0000	0.0000	0.0000	0.0000	0.0000
4	.4320	.4388	.1104	.0170	.0018	0.0000	0.0000	0.0000	0.0000	0.0000
5	.4260	.4316	.1194	.0200	.0028	.0002	0.0000	0.0000	0.0000	0.0000
6	.4128	.4332	.1252	.0254	.0030	.0004	0.0000	0.0000	0.0000	0.0000
7	.3984	.4296	.1372	.0300	.0044	.0004	0.0000	0.0000	0.0000	0.0000
8	.3936	.4252	.1392	.0358	.0054	.0008	0.0000	0.0000	0.0000	0.0000
9	.3898	.4210	.1494	.0350	.0042	.0004	.0002	0.0000	0.0000	0.0000
10	.3654	.4404	.1484	.0372	.0072	.0010	.0004	0.0000	0.0000	0.0000
11	.3626	.4396	.1488	.0388	.0080	.0016	.0006	0.0000	0.0000	0.0000
12	.3688	.4346	.1458	.0398	.0086	.0022	.0002	0.0000	0.0000	0.0000
13	.3664	.4374	.1448	.0362	.0124	.0022	.0006	0.0000	0.0000	0.0000
14	.3654	.4314	.1488	.0408	.0106	.0028	.0002	0.0000	0.0000	0.0000
15	.3608	.4342	.1490	.0424	.0092	.0038	.0006	0.0000	0.0000	0.0000
16	.3738	.4158	.1550	.0386	.0118	.0044	.0006	0.0000	0.0000	0.0000
17	.3714	.4196	.1486	.0436	.0126	.0032	.0010	0.0000	0.0000	0.0000
18	.3652	.4264	.1454	.0464	.0124	.0032	.0008	.0002	0.0000	0.0000
19	.3546	.4268	.1564	.0452	.0124	.0036	.0010	0.0000	0.0000	0.0000
20	.3652	.4214	.1510	.0458	.0122	.0034	.0010	0.0000	0.0000	0.0000
21	.3590	.4194	.1598	.0434	.0136	.0040	.0006	.0002	0.0000	0.0000
22	.3606	.4212	.1550	.0436	.0142	.0044	.0008	.0002	0.0000	0.0000
23	.3620	.4204	.1526	.0436	.0166	.0038	.0006	.0002	.0002	0.0000
24	.3662	.4138	.1546	.0462	.0132	.0050	.0004	.0006	0.0000	0.0000
25	.3616	.4180	.1554	.0478	.0118	.0036	.0014	.0004	0.0000	0.0000
26	.3646	.4160	.1610	.0408	.0128	.0036	.0006	.0004	.0002	0.0000
27	.3562	.4230	.1586	.0432	.0126	.0052	.0008	.0002	.0002	0.0000
28	.3596	.4230	.1520	.0476	.0122	.0040	.0012	.0002	.0002	0.0000
29	.3612	.4138	.1596	.0478	.0124	.0030	.0018	.0002	.0002	0.0000
30	.3572	.4114	.1646	.0492	.0130	.0030	.0010	.0004	.0002	0.0000
31	.3486	.4256	.1566	.0474	.0158	.0038	.0016	.0004	.0002	0.0000
32	.3412	.4358	.1548	.0468	.0152	.0040	.0014	.0008	0.0000	0.0000
33	.3432	.4338	.1576	.0456	.0136	.0042	.0014	.0006	0.0000	0.0000
34	.3476	.4382	.1458	.0472	.0154	.0034	.0016	.0006	.0002	0.0000
35	.3564	.4206	.1564	.0452	.0146	.0048	.0016	.0004	0.0000	0.0000
36	.3620	.4192	.1480	.0510	.0134	.0052	.0008	.0004	0.0000	0.0000
37	.3568	.4194	.1552	.0484	.0138	.0048	.0014	.0002	0.0000	0.0000
38	.3556	.4228	.1540	.0472	.0136	.0050	.0016	.0002	0.0000	0.0000
39	.3608	.4152	.1506	.0526	.0142	.0046	.0018	.0002	0.0000	0.0000
40	.3616	.4084	.1544	.0528	.0162	.0060	.0004	.0002	0.0000	0.0000
41	.3582	.4198	.1462	.0550	.0154	.0040	.0010	.0004	0.0000	0.0000
42	.3548	.4244	.1496	.0512	.0142	.0044	.0014	0.0000	0.0000	0.0000
43	.3564	.4150	.1566	.0502	.0152	.0062	.0004	0.0000	0.0000	0.0000
44	.3490	.4278	.1528	.0500	.0144	.0052	.0008	0.0000	0.0000	0.0000
45	.3574	.4252	.1478	.0484	.0156	.0054	.0002	0.0000	0.0000	0.0000
46	.3584	.4190	.1500	.0508	.0168	.0042	.0008	0.0000	0.0000	0.0000
47	.3560	.4200	.1510	.0484	.0192	.0038	.0016	0.0000	0.0000	0.0000
48	.3626	.4104	.1524	.0542	.0144	.0046	.0014	0.0000	0.0000	0.0000

PROBABILITIES FOR MAX QUEUE SIZE FOR ANY DAY

0.0000 .0444 .4880 .3218 .1056 .0286 .0100 .0010 .0006 0.0000

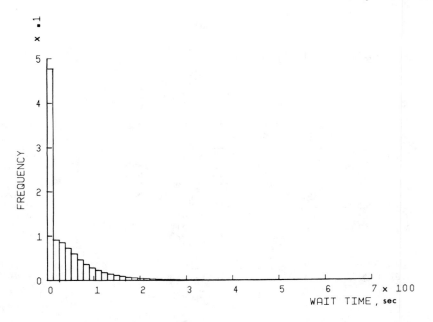

FIG. 7.3 Waiting-time relative frequency histogram for gravel truck queue.

     The theory of Markov chains demonstrates for most common chains that the queue size probability mass function at each stage converges to a constant repetitive set of values. The foregoing results indicate that this has happened at about the tenth stage for four significant figures.

     Because of excessive computer time, the simulation has only been executed for about one-tenth of a full day. For a full day, the steady-state waiting times would swamp out the effect of the initial period, and the relative frequency histogram for waiting time, shown in Fig. 7.3, would be slightly different.

     The probability of a queue size of 6 or larger occurring on any given day is calculated as follows:

$$P(x \geqslant 6) = P(n_{max} = 6 \ or \ n_{max} = 7 \ or \ n_{max} = 8 \ or \ n_{max} = 9)$$

$$= 0.0100 + 0.0010 + 0.0006 + 0.0000$$

$$= 0.0116$$

Thus we would expect this problem to occur roughly once in every 100 days.

7.4 RANDOM TIME FUNCTIONS

A random time function is a random variable in the usual sense that its value cannot be predicted and varies randomly when measured in different devices of a population of similar devices. However, it also varies randomly as a function of time in any given device. This is illustrated in Fig. 7.4. The totality of possible time histories for the whole population is an *ensemble* and they are the result when the same operation is performed by every device for the whole population of devices. An example would be a fleet of the same kind of truck driving over the same stretch of road, where the random variable is the stress at a given point on the frame.

An ordinary random variable can be fully described by its density function or distribution function. However, a random time function may require considerably more information to describe completely; in the most general case, it requires an infinite number of density functions. If we measure the value of x for each device of a sample t_1, as shown in Fig. 7.4, we can construct a density function for x(t). In general, these will not be the same for different values of t_1. Thus we would have in general the density function $f_i[x(t_i)]$ to describe the distribution of x at time t_i. This is called the *first-order probability density function*.

Second-order distributions are defined by the joint probability density function $f_{ij}[x(t_i), x(t_j)]$. Such a density function can be used to determine the probability that, for a particular device, x will

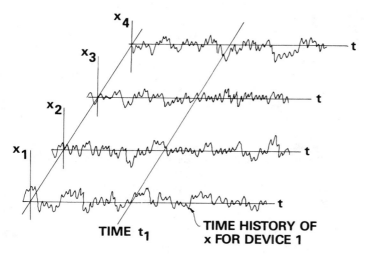

FIG. 7.4 Sample of time histories for variable x in a sample of devices.

have a value between a_1 and b_1 at time t_1, and a value between time a_2 and b_2 at time t_2.

$$P[a_1 \leqslant x(t_1) \leqslant b_1, a_2 \leqslant x(t_2) \leqslant b_2]$$

$$= \int_{a_1}^{b_1} \int_{a_2}^{b_2} f_{1,2}[x(t_1), x(t_2)] \, dx_1 \, dx_2 \qquad (7.4.1)$$

where dx_1 and dx_2 are the infinitesimals associated with $x(t_1)$ and $x(t_2)$, respectively.

A commonly used quantity is the autocorrelation or the first joint moment about the origin.

$$R_{ij} = E[x(t_i)x(t_j)] = \int_{-\infty}^{+\infty} \int_{-\infty}^{+\infty} x(t_i)x(t_j)f[x(t_i), x(t_j)] \, dx_1 \, dx_2$$

$$(7.4.2)$$

A general random process is fully described only by an infinite set of all first-order density functions, all second-order density functions, and all higher-order density functions. Fortunately, many processes may be adequately described by fewer functions if they are *stationary* processes. This occurs if all first-order density functions are the same and all second-order density functions are the same; or, in other words, if the *nature of the randomness does not change with time*. An example would be the fleet of trucks traveling over the same road, and the road is essentially the same in nature over its whole length. If the road changes its nature, say from concrete to old blacktop to gravel, the process would be nonstationary.

In a stationary process, the second-order joint density functions, $f_{ij}[x(t_i), x(t_j)]$, will vary only when $t_j - t_i$ varies. If we call this difference τ, then

$$R_{ij} = R(\tau) = E[x(t)x(t + \tau)] \qquad (7.4.3)$$

The distributions and moments given above are ensemble distributions—for a whole fleet of trucks. Let us look at one truck and define *temporal distributions* for a sample of x for a given time span T. The moments are defined directly from the sample, rather than from the density function. Thus the first temporal moment is

$$\mu^t = \frac{1}{T} \int_{-T/2}^{T/2} x(t) \, dt \qquad (7.4.4)$$

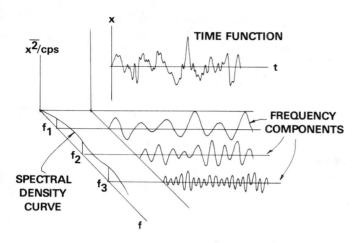

FIG. 7.5 Analysis of a random vibration in terms of spectral density.

and the temporal autocorrelation is

$$R^t(\tau) = \frac{1}{T} \int_{-T/2}^{T/2} x(t)x(t + \tau)\, dt \qquad (7.4.5)$$

Another common way of measuring the random nature of a process is by use of the concept of *spectral density*. It is most easily understood by an explanation of how it is measured. We may imagine that a random time function consists of a superposition of a large number of frequencies, each having randomly varying amplitudes, as shown in Fig. 7.5. The spectral density function or curve gives the amount of each frequency component present, where f is the frequency in cycles per second. The amount of each component is in terms of the mean square value of the component, $\overline{x^2}$. To be meaningful, the spectral density must be given in units of $\overline{x^2}/\text{cps}$, since the frequency varies continuously over the range of components.

Returning to the measurement of the spectral density, a combination of instruments is used as shown schematically in Fig. 7.6. The

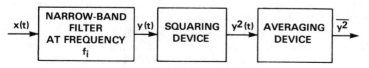

FIG. 7.6 Schematic diagram for spectral density measurement.

time function signal, x(t), is fed into a narrow-band filter which
passes only that component, y(t), of x(t) which has frequencies in a
narrow band about a set frequency, f_i. The signal y(t) is then
squared, and then averaged. If the filter band width is Δf, the
spectral density is given by

$$s(f) = \overline{y^2}/\Delta f \qquad\qquad (7.4.6)$$

It can be shown that the spectral density is related to the autocorrelation function as follows:

$$S(f) = \int_{-\infty}^{\infty} R(\tau)e^{-if\tau}\,dt \qquad\qquad (7.4.7)$$

where $i = \sqrt{-1}$. This is the Fourier transform.

This relationship helps to give some insight into the concept of
the autocorrelation. Idealized "white noise" is shown in the spectral
density curve of Fig. 7.7. All frequency components are present in
equal quantities. The corresponding autocorrelation function is
shown in Fig. 7.8. This means that the value of x at any successive
finite time differences are completely independent. The presence of
so many frequencies has complete "randomized" the value timewise.
If the band width of white noise is restricted, as shown in Fig. 7.9,
the autocorrelation function begins to widen out as indicated in Fig.
7.10. Successive values of x now begin to depend more on previous
values, although the effect dies out as the time difference increases.
If the bandwidth becomes very narrow, we are left with essentially
one frequency, and the autocorrelation function becomes very broad.
There is now great dependence between successive cycles.

Some random stationary processes have the characteristic that
temporal distribution and moments from any sample function are equal
to corresponding distributions and moments for the ensemble. These
are called *ergodic processes* and, of course, are extremely convenient
since we can get all the necessary information to define randomness
from one sample function or time history of sufficient length to adequately sample the variable. To illustrate from our previous example
of a fleet of trucks, the process would become ergodic if only one
truck was used by one driver, or if the variation between drivers and
trucks is negligible compared to the variation in the road.

Stationary and ergodic processes are usually identified by physical arguments, although they can be verified by testing. If a time
sample is available on magnetic tape, it can be checked for stationari-
ness by electrically integrating successive sections to calculate the
mean by (7.4.4). If it is essentially constant, the process may be
stationary. Similarly, the tape can be run past two heads in a re-

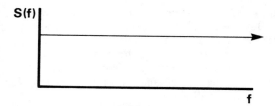

FIG. 7.7 White noise spectral density.

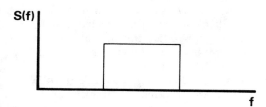

FIG. 7.8 Autocorrelation function for white noise.

FIG. 7.9 Band-limited white noise spectrum.

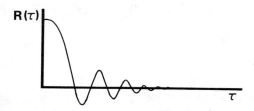

FIG. 7.10 Autocorrelation function of band-limited white noise.

corder, so that the signals for x are registered at time τ apart, where τ will depend on the distance between the heads and the tape speed. Equation (7.4.5) can be electrically evaluated to determine $R(\tau)$. If $R(\tau)$ also remains constant for successive sections, the process is usually assumed to be stationary. A check for ergodicity required comparison of different sample functions, since it assumes that randomness is independent of the sample function considered.

The theory of random processes can be extended to provide a basis for engineering applications such as predicting stresses in members due to imposed random loadings or displacements, design of filters for electrical signals containing undesired frequencies or noise, and design of control systems for chemical or physical processes.

The random time variable can also be discrete, either inherently or because data describing it are obtained by *sampling* at equal time intervals. A similar theory to the one described above is used, and an important application is in control systems using digital computers.

Further consideration of general random processes is beyond the scope of this book, and the reader is referred to texts such as Larson and Shubert (1979), Chatfield (1975), McSchane (1974) and Box and Jenkins (1976).

PROBLEM

Problem 7.1 An operator of a harvester combine for custom harvesting of grain uses the machine for quite prolonged periods in a year. If a machine is not ready to go on any given morning, there is a considerable financial loss. He wishes to predict the probability that a machine will be ready to go on any given day, and the mean number of days in a season of 90 days that the machine will be ready. The life to failure for the machine is a random variable, and its density function has been extimated to be an exponential function.

$$f_L(t_L) = \frac{1}{\mu_L} e^{-t_L/\mu_L} \tag{1}$$

where μ_L is the mean life. The corresponding reliability function is*

$$R_L(t_L) = 1 - F(t_L) = e^{-t_L/\mu_L} \tag{2}$$

---

*The reliability function is defined as the probability that a device will have a given life t, where t is a random variable. For further details, see Chapter 11.

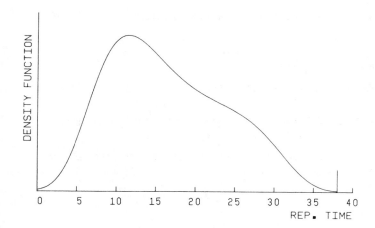

FIG. 7.11 Probability density function for repair time. The moments are 16.489, 55.194, 149.88, and 7148.5.

It is characteristic of exponential reliability functions that life can be assumed to restart at zero time after a repair, or at the beginning of each day, and in fact at any instant when *it is known* that the device is working. The mean life is 45.7 hr and the working day is 10 hr. The probability density function for time to repair, shown in Fig. 7.11, is a maximum entropy function. It may be noted that the upper bound is such that the repair time can never extend beyond 38 hr. Solve the problem using Monte Carlo simulation.

REFERENCES

Box, G. E. P., and G. H. Jenkins (1976), *Time Series Analysis: Forecasting and Control*, rev. ed., Holden-Day, San Francisco.

Bulgren, W. G. (1982), *Discrete System Simulation*, Prentice-Hall, Englewood Cliffs, N.J.

Chatfield, C. (1975), *The Analysis of Time Series: Theory and Practice*, Chapman & Hall, London.

Kemeny, J. G., and J. L. Snell (1960), *Finite Markov Chains*, D. Van Nostrand, Princeton, N.J.

Larson, H. J., and B. O. Shubert (1979), *Probabilistic Models in Engineering Sciences*, Vol. 1: *Random Variables and Stochastic Processes*, Wiley, New York.

McShane, E. J. (1974), *Stochastic Calculus and Stochastic Models*, Academic Press, New York.

Medhi, J. (1982), *Stochastic Processes: Theory and Applications*, Halsted Press, New York.

Parzen, E. (1962), *Stochastic Processes*, Holden-Day, San Francisco.

Rau, J. G. (1970), *Optimization and Probability in System Engineering*, Van Nostrand Reinhold, New York.

Saaty, T. L. (1961), *Elements of Queuing Theory: With Applications*, McGraw-Hill, New York.

8

Order Statistics and Extreme Values

The concept of order statistics and ranks, including their distribu-
tions • Extreme value distributions with applications to design and
to selection of bounds and judgment of outliers

8.1 INTRODUCTION

We have used mean ranks to generate distribution functions in Sec.
5.4, and it may be recalled that the rank is related to the order statistic. The theory of extreme values is also developed from the theory of order statistics, and both topics have important applications in
design.

8.2 ORDER STATISTICS

We saw in Sec. 5.4 an application of order statistics in the plotting of
sample data using mean ranks. In order statistics a sample is ordered
from low to high; and the ith member is treated as a random variable
having different values for every possible sample of the same size
from the same population. The sample must be randomly selected,
and each sample value must be independent of all others. This random variable is called the *ith order statistic*, designated o_i. Each
order statistic has a different distribution with a range effectively
smaller than the parent distribution $f(x)$. This is suggested in Fig.
8.1. The mean of o_i is designated $\bar{o}_i$. It is perhaps not obvious that
o_i is a random variable until we realize that every x_i from different
samples of size n from the same population will have a different value.

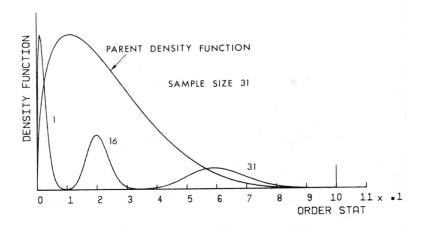

FIG. 8.1 Density functions for order statistics.

The probability density function for o_i can be developed using the infinitesimal event

$$x' \leqslant o_i \leqslant x' + dx \qquad\qquad (8.2.1)$$

illustrated in Fig. 8.2. This is true for any value of x, so the prime can be omitted. It is useful to define an equivalent event related to sample values.

$$P(x \leqslant o_i \leqslant x + dx) = P(i - 1 \text{ occurrences of sample values}$$
$$\text{less than } o_i \text{ \textit{and} one occurrence of a}$$
$$\text{sample value in the interval dx \textit{and}}$$
$$n - i \text{ occurrences of sample values}$$
$$\text{greater than } o_i) \qquad (8.2.2)$$

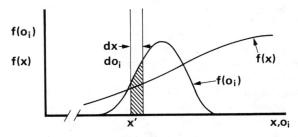

FIG. 8.2 Illustration of infinitesimal event for o_i and x. Vertical scales are different for the two curves.

The probabilities of these three events on the right-hand side are $F_X(o_i)$, $f_X(o_i)$ dx, and $1 - F_X(o_i)$, respectively. The left-hand side can be expressed as $f_{o_i}(o_i)$ do_i, and the multinomial distribution can be used to model the right-hand side. It has the general form (see Sec. 3.3)

$$f(y_1, y_2, \ldots, y_k) = \frac{m!}{y_1! \, y_2! \, \cdots \, y_k!} \, \lambda_1^{y_1} \lambda_2^{y_2} \cdots \lambda_k^{y_k}$$

(8.2.3)

where

$$m = \sum_{i=1}^{k} y_i = \text{number of trials}$$

y_i = random variable representing the number of events occurring of the ith type

λ_i = probability of ith event occurring

In terms of this problem we have three events with probabilities given above; and m equals n. Thus (8.27) becomes

$$f_{o_i}(o_i) \, do_i = \frac{n!}{(i - 1)! \, (n - i)!} [F_X(o_i)]^{i-1} f_X(o_i) \, dx \, [1 - F_X(o_i)]^{n-i}$$

(8.2.4)

We finally get the density function by dividing out dx, noting that do_i and dx are equivalent.

$$f_{o_i}(o_i) = \frac{n!}{(i - 1)! \, (n - i)!} [F_X(o_i)]^{i-1} f_X(o_i) [1 - F_X(o_i)]^{n-i}$$

(8.2.5)

It will be conceptually helpful to illustrate this with a sample of 21 from an exponential distribution.

$$f(o_1) = 21\lambda e^{-21\lambda o_1}$$

$$f(o_{11}) = \frac{21!}{10! \, 10!} \left[1 - e^{-\lambda o_{11}} \right]^{10} \lambda e^{-11\lambda o_{11}}$$

(8.2.6)

$$f(o_{21}) = 21\lambda \left(1 - e^{-\lambda o_{21}} \right)^{20} e^{-\lambda o_{21}}$$

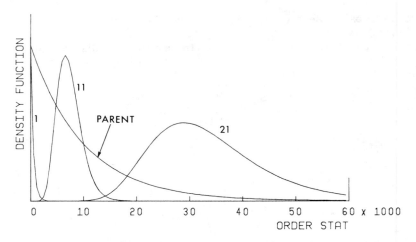

FIG. 8.3 Illustration of density functions of order statistics for a sample size of 21 from an exponential distribution.

These include the upper bound, the lower bound, and the median, and are plotted in Fig. 8.3 for a λ of 10^{-4}.

In Sec. 5.4 we had occasion to use the distribution for the cumulative distribution function of the primary variable x, when x equals an order statistic. It is defined by

$$r_i = F_x(o_i) = \int_{-\infty}^{o_i} f_x(x) \, dx \qquad (8.2.7)$$

This random variable is commonly called the *rank*. Its density function could be derived by arguments similar to those used for o_i. However, we can do it very simply by using (6.3.3).

$$f_y(y) = \frac{f_x(x)}{dy/dx} \qquad (8.2.8)$$

where y and x are random variables related in general by

$$y = g(x) \qquad (8.2.9)$$

In this case r_i corresponds to y, o_i corresponds to x, $F_x(o_i)$ corresponds to $g(x)$, and $dF_x(o_i)/do_i$ corresponds to dy/dx. Applying (8.2.8) gives

$$f_{r_i}(r_i) = \frac{f_{o_i}(o_i)}{dF_x(o_i)/do_i} \qquad (8.2.10)$$

We note that, by definition,

$$\frac{dF_x(o_i)}{do_i} = f_x(o_i) \qquad (8.2.11)$$

and using this with (8.2.5) in (8.2.10) gives

$$f_{f_i}(r_i) = \frac{n!}{(i-1)!\,(n-i)!}[F_x(o_i)]^{i-1}[1 - F_x(o_i)]^{n-i} \qquad (8.2.12)$$

However, by definition,

$$r_i = F_x(o_i)$$

so (8.2.12) becomes

$$f_{r_i}(r_i) = \frac{n!}{(i-1)!\,(n-1)!}\,r_i^{i-1}(1 - r_i)^{n-i} \qquad (8.2.13)$$

with a range of 0 to 1. Thus the rank distribution is independent of the primary distribution. It is illustrated in Fig. 8.4 for various values of i and n. These figures help give some feeling for the randomness associated with mean-rank plotting for different sample sizes. They also show the interesting feature that the central ranks have greater uncertainty by a considerable margin over the upper and lower ranks.

The mean rank, used in Sec. 5.4, is obtained from (8.2.13).

$$\bar{r}_i = \int_0^1 r_i f_{r_i}(r_i)\,dr_i$$

$$= \frac{n!}{(i-1)!\,(n-i)!} \int_0^1 r_i r_i^{i-1}(1 - r_i)^{n-i}\,dr_i$$

$$= \frac{n!}{(i-1)!\,(n-1)!} \int_0^1 r_i^i(1 - r_i)^{n-i}\,dr_i \qquad (8.2.14)$$

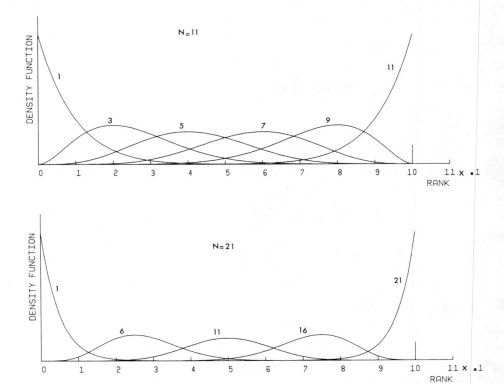

FIG. 8.4 Rank distributions.

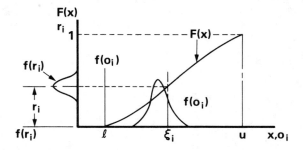

FIG. 8.5 Order statistics and rank density functions.

The standard definition of a beta function reduces this to

$$\bar{r}_i = \frac{n!}{(i-1)!\,(n-1)!}\,B(i+1,\,n-1+1)$$

The standard relationship between the beta function and gamma functions gives

$$\bar{r}_i = \frac{n!}{(i-1)!\,(n-i)!}\,\frac{\Gamma(i+1)\Gamma(n-i+1)}{\Gamma(n+2)}$$

Finally, if k is an integer, $\Gamma(k)$ is $(k-1)!$, and $\bar{r}_i$ becomes

$$\bar{r}_i = \frac{i}{n+1} \tag{8.2.15}$$

The cumulative distribution for the rank can be obtained in analytical form. We let a specific observation of the value of the ith order statistic be ξ_i. In Fig. 8.5, the density functions for the ith order statistic, $f(o_i)$, and the density function for the ith rank, $f(r_i)$, are shown superimposed on the cumulative distribution function for x. The cumulative distribution function of r_i has the value

$$F(r_i) = P[r_i \leqslant r_i(\xi_i)] = P(o_i \leqslant \xi_i) \tag{8.2.16}$$

If we refer to Fig. 8.6, we see that, for i equal to 3, the event of o_3 $\leqslant \xi_3$ occurs if

3 among 9 trials or less than ξ_3

or 4 among 9 trials are less than ξ_3

$$\vdots$$

or 9 among 9 trials are less than ξ_3

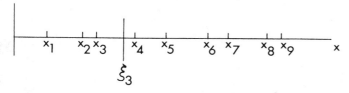

FIG. 8.6 Plot of a sample of nine trials for values of x.

and, in general,

$$P(o_i \leqslant \xi_i) = \sum_{j=1}^{n} P(j \text{ among } n \text{ trials less than } \xi_i) \qquad (8.2.17)$$

However, it is clear that the probability of any single trial being less than ξ_i is r_i. Thus the right-hand side of (8.2.17) is given by the binomial distribution. Combining this with (8.2.16) gives

$$F(r_i) = \sum_{j=i}^{n} \frac{n!}{j!(n-j)!} r_i^{j}(i - r_i)^{n-j} \qquad (8.2.18)$$

This can be conveniently used to obtain the median rank, by letting $F(r_i)$ equal 0.5 and solving for r_i. Some statisticians have argued that the median-rank values should be used in rank plotting, rather than the means, as described in Sec. 5.4. The inversion of (8.2.18) is commonly avoided by generating tabular values of the median, as a function of i and n. Function FINVRT can be used to invert (8.2.18).

Some interesting applications arise in the use of the lowest- and highest-order statistics.

EXAMPLE 8.1 Wooden Beam

A wooden beam is made of six laminated sections, shown in Fig. 8.7. Each section is assumed to have the same stiffness and to equally share the load, which is also assumed determinate. However, the strength of wood is quite variable due to the random

LOAD

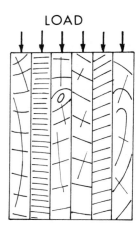

FIG. 8.7 Laminated beam cross section.

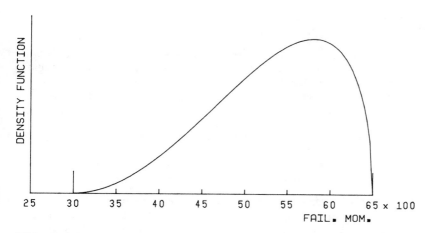

FIG. 8.8 Density function for the bending moment at failure of beam elements.

occurrence of defects, and the beams have a known strength distribution as indicated in Fig. 8.8, designated $f_X(x)$. Each element takes a share, B, of the applied bending moment, and we wish to predict the probability of failure.

A solution can conveniently be obtained by considering that the six elements are an ordered random sample. Failure will be controlled by the first order statistic, or weakest beam element. From (8.2.5) the density function for the strength of the weakest element is then

$$f_W(w) = 6f_X(w)[1 - F_X(w)]^5 \qquad (8.2.19)$$

where w is the failure bending moment of the weakest element, and

$$F_X(w) = \int_\ell^W f_X(x) \, dx \qquad (8.2.20)$$

where ℓ is the lower bound. The probability of failure is

$$P(\text{failure}) = \int_\ell^B f_W(w) \, dw \qquad (8.2.21)$$

EXAMPLE 8.2 Wooden Beam (Cont.)

The beam from Example 8.1 is now assumed to have 25 random loads applied to it, at occasional intervals. The density function $f_y(y)$ for this random variable is also assumed known. We treat the loads as an ordered sample, and our concern is with the largest element. From (8.2.5), the density function for the largest load is

$$f_F(F) = 25[F_y(F)]^{24} f_y(F) \tag{8.2.22}$$

We shall leave the solution of this example to Problem 8.6 and, alternatively, by another method, in Example 9.2.

These examples are closely related to the concept of extreme values; and we shall see in Sec. 8.3 how, with some rather broad assumptions about the primary distribution, it is possible to develop an extreme value distribution independent of the primary distribution.

The density function for the *range* of a random variable can also be derived by similar methods (Guttman et al., 1971). We let the range be

$$W = x_n - x_1 \tag{8.2.23}$$

The corresponding density function is

$$f(W) = n(n-1) \int_R [F_x(x + W) - F_x(x)]^{n-2} f_x(x) f_x(x + w) \, dx$$

$$\tag{8.2.24}$$

This expression can be used to estimate the probability that the range will be less than a specified amount, or greater than a specified amount, or within some interval.

The order statistics are not independent random variables, and any subset r, s, t, u has the following joint density function (David, 1981):

$$f(o_r, o_s, o_t, o_u) = \frac{n!}{(r-1)! \, (s-r-1)! \, (t-s-1)! \, (u-t-1)! \, (n-u)}$$

$$\times [F(o_r)]^{r-1} [F(o_s) - F(o_r)]^{s-r-1} [F(o_t) - F(o_s)]^{t-s-1}$$

$$\times [F(o_u) - F(o_t)]^{u-t-1} [1 - F(o_u)]^{n-u}$$

$$\times f(o_r) f(o_s) f(o_t) f(o_u) \tag{8.2.25}$$

where $r < s < t < u$, but they need not be sequential. The joint density function of all n order statistics reduces to

$$f(o_1, o_2, \ldots, o_n) = n! \; f(o_1) f(o_2) \cdots f(o_n) \qquad (8.2.26)$$

An interesting application of this equation is given in Problem 8.4.

8.3 EXTREME VALUE DISTRIBUTIONS

8.3.1 Introduction

We have seen in Sec. 8.2 that it may be useful to apply the largest and smallest order statistics to problems where we are concerned with extreme values of a small sample. This concept can be extended to situations where we are interested in the extreme value of a large sample.

It is not infrequent in engineering design to be concerned about predicting the maximum value of some quantity. It might be a quantity such as the maximum depth of a pothole in the road, or the maximum weight of boulder dropped into a truck body by a power shovel in a quarry, or the maximum vertical gust velocity encountered by an aircraft, or maximum wind speed acting on a structure over its lifetime. We are, of course, implying random quantities, and can only predict the probability of a selected maximum value not being exceeded.

One approach would be to take a large sample, say 1000 measured values of boulder weights in a quarry, and establish a density function from the data. If we now assume that we wish to estimate the maximum boulder weight to be encountered, with a probability of 0.95 that it will not be exceeded, this can be calculated directly from the density function. The drawback of this method is that the tails of a distribution are usually the most poorly defined part of the curve.

A review of Sec. 8.2 will suggest that order statistics provides a kind of microscopic view of a portion of the region of a random variable, permitting a sort of magnified image of the random nature of a variable in a local region. We shall see how this is done in the following discussion.

Extreme value theory, an extension of order statistics, provides a more accurate estimate of extreme values than does the parent distributions, if the appropriate data can be gathered. We require a set of samples, and each maximum (or minimum) observation in a sample is now taken as a sample of extreme values. It may not be actually necessary to take a complete sample to determine the extreme value in it. In our example, we might keep a record of the largest boulder observed in each day for 1 year, yielding 250 extreme values.

It has been found that, *for some* parent distributions of specific general types, the extreme value distribution can be accurately de-

fined by certain theretical distributins, for large n. It is important
to note that these distributions, commonly called asymptotic distributions, are functions of n; or more specifically, their parameters are
functions of n. In the event that such an extreme value distribution
can be used, we need now know the sample size, or observe all values
of the parent random variable, as long as we know that the sample
size is large. Thus for our potholes, we need not know the number
encountered, or the size of each. We assume that, *for a given interval*, the number of potholes is about constant; and only require some
means of monitoring to obtain the largest in the given interval—always
providing that the pothole population meets the asymptotic requirement. There are three known asymptotic distributions (Gumbel,
1958): types 1, 2, and 3.

8.3.2. Type 1 Asymptotic Distribution

The requirement for the type 1 extreme value distribution is that the
tail of the parent distribution that contains the extreme value be exponential in nature; and this includes the normal, the Weibull, the
exponential, the gamma, and other similarly shaped density functions.
      For *largest values* the parent asymptotic tail must be to the right,
and the distribution has the form

$$f(x_e) = \alpha \exp (-y - e^{-y}) \qquad\qquad (8.3.1)$$

$$F(x_e) = \exp (-e^{-y}) \qquad\qquad (8.3.2)$$

where x_e is the extreme random variable and y is a convenient intermediate parameter, or reduced variate, related to x_e by

$$y = \alpha(x_e - u) \qquad\qquad (8.3.3)$$

and α and u are parameters of the distribution.
      A similar distribution exists for *smallest extreme values*.

$$f(x_e) = \alpha \exp (y - e^{y}) \qquad\qquad (8.3.4)$$

$$F(x_e) = 1 - \exp (-e^{y}) \qquad\qquad (8.3.5)$$

$$y = \alpha(x_e - u) \qquad\qquad (8.3.6)$$

The parent exponential tail must now extend to the left.
      The derivation of these distributions is based on the requirement
that the parent population have a cumulative distribution function with
the upper tail defined by (for an upper extreme)

$$F(x) = 1 - e^{-h(x)} \qquad (8.3.7)$$

where $h(x)$ is a monotonically increasing function of x. For derivation of these extreme value expressions, the reader is referred to Gumbel (1958) or Kendall and Stuart (1977).

The type 1 upper extreme value distribution has a mode, mean, variance, and third central moment as follows:

$$\tilde{x}_e = u$$

$$\mu_e = u + \frac{0.57722}{\alpha} \qquad (8.3.8)$$

$$\sigma_e^2 = \frac{\pi^2}{6\alpha^2}$$

$$c_{e3} = 1.1396\sigma_e^3$$

The parameters u and α can be related to the parent distribution as follows (Gumbel, 1958). For the upper extreme

$$F_x(u) = 1 - \frac{1}{n} \qquad (8.3.9)$$

$$\alpha = nf_x(u) \qquad (8.3.10)$$

and for the lower extreme

$$F_x(u) = \frac{1}{n} \qquad (8.3.11)$$

$$\alpha = nf_x(u) \qquad (8.3.12)$$

The extreme value distribution is used to estimate the probability that a specified value of the random variable will not be exceeded *within a specified number of observations*. This must be specified for extreme values to be meaningful.

The probability that *any* observed value of a random variable x will be less than a specified value x_s is

$$P(x < x_s) = F_x(x_s) \qquad (8.3.13)$$

the probability that *all values in n observations* will be less than x_s is

P(maximum value in n observed values is less than x_s)

$$= P(x < x_s \mid n \text{ trials}) = F_x^n(x_s)$$

$$= F_{o_n}(x_s) \quad \text{for any n} \qquad (8.3.14)$$

$$\underset{\sim}{\approx} F_{x_e}(x_s) \quad \text{for large n}$$

where $F_{x_e}(\cdot)$ is the type 1 extreme value cumulative distribution function.

$F_x(x)$ can be though of as $F_{o_n}(o_n)$ for $i = n = 1$. This is confirmed if we write the expression for the density function for the order statistic (8.2.5) and set $i = n = 1$. It reduces to $f_x(x)$.

In applying extreme value theory it is common practice to use a *specified time interval* rather than a specified number of observations. For example, in aircraft design we may be attempting to predict the probability that a specified gust velocity, x_s, will not be exceeded in 50,000 h. We have recorded maximum values of gust velocity observed each 50 h for 5000 h, in an aircraft in similar service. The number of observations in each 50 h, from which the maximum was extracted, is large and uncounted; and we have no knowledge of the parent distribution except a judgment that the upper tail is exponential. The 100 observations can be used to estimate the parameters of the type 1 distribution (using methods to be discussed below). This gives $f_{x_e}(x_e)$ and $F_{x_e}(x_e)$. The probability that x_s will not be exceeded in 50 h is $F_{x_e}(x_s)$. The probability that x_s will not be exceeded in 50,000 h is, using the product law for combined events,

$$P(x < x_s \mid 50,000 \text{ hr}) = [F_{x_e}(x_s)]^{1000} \qquad (8.3.15)$$

An additional requirement for validity of the asymptotic distribution is that the number of observations in each 50 h, from which the maximum was extracted, must be relatively large. Very often the actual number of observations is unknown, and individual observations are not recorded. Example 8.4 will give some indication of the *parent* sample size required.

EXAMPLE 8.3 Yield Stress of Steel

Tooling is being designed for a high-production part made from low-carbon steel. The primary operation is bending the workpiece in a brake press. The designer wishes to estimate the maximum bending force required in the production run of 10,000 pieces.

The bending force is directly related to the yield stress of the steel, for which the following Weibull distribution is available:

$$f(S_y) = 0.5539 \left(\frac{S_y - 37.4}{8.72} \right)^{3.83} \exp \left[- \left(\frac{S_y - 37.4}{8.72} \right)^{4.83} \right],$$

$$S_y \geqslant 37.4$$

$$= 0, \quad S_y < 37.4 \tag{8.3.16}$$

where S_y is in kips/in^2. The engineer decides to accept a probability of 0.95 that a specified maximum will not be exceeded, and use that value as the basis for choosing the brake press. Any pieces with a higher yield stress will be rejected. It is thus necessary to solve the following expression for S_s, the maximum specified value, using (8.3.14):

$$0.95 = \left[\int_{37.4}^{S_s} f(S_y) \, dS_y \right]^{10,000} \tag{8.3.17}$$

In the case of the Weibull distribution it is easier to work directly with the cumulative distribution function:

$$0.95 = [F_{S_y}(S_s)]^{10,000} \tag{8.3.18}$$

$$= \left\{ 1 - \exp \left[- \left(\frac{S_s - 37.4}{8.72} \right)^{4.83} \right] \right\}^{10,000} \tag{8.3.19}$$

This can be inverted to solve for S_s directly.

$$S_s = 52.03 \text{ kips/in}^2.$$

The engineer has some qualms about this quite high figure and plots the parent distribution, shown in Fig. 8.9. On seeing the density function, she decides that such a long extrapolation into the tail is unjustified.

The story is continued in Example 8.4.

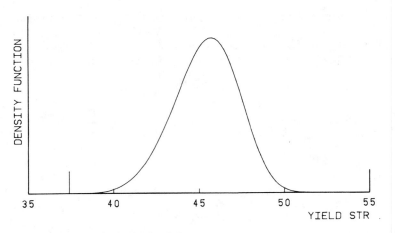

FIG. 8.9 Density function for yield stress from Example 8.3.

In applications of extreme value theory it is common for a sample of extreme values to be available. The parameters u and α can be estimated from these data by any of the methods described in Chapter 5. Thus (8.3.8) can be used to apply the method of moments. The maximum likelihood method leads to the following expressions, which can be solved for u and α (Gumbel, 1958):

$$\exp (\alpha u) \sum_{i=1}^{N} \exp (- \alpha x_{ei}) = N \tag{8.3.20}$$

$$\frac{\displaystyle\sum_{i=1}^{N} x_{ei} \exp(- \alpha x_{ei})}{\displaystyle\sum_{i=1}^{N} \exp (- \alpha x_{ei})} + \frac{1}{\alpha} = \mu_{e} \tag{8.3.21}$$

where

x_{ei} = ith sample extreme value

N = size of sample of extreme values

Mean rank-plotting is particularly suited for use with the extreme value distribution, since there is often considerable doubt that, on a priori grounds, the type 1 distribution is applicable. The plot can be linearized by transforming (8.3.2).

$$- \ln \{- \ln [F(x_e)] \} = y \qquad (8.3.22)$$

the value of y corresponding to the ith ordered sample value, x_{ei}, is obtained by replacing $F(x_e)$ by the mean rank.

$$y_i = - \ln \left[- \ln \left(\frac{i}{n + 1} \right) \right] \qquad (8.3.23)$$

Corresponding values of y_i and x_{ei} can now be plotted on ordinary graph paper and fitted by (8.3.3) to get estimates of u and α. This is done by MREX1.

EXAMPLE 8.4 Yield Stress of Steel (Cont.)

We return now to the probabilistic design problem in Example 8.3. The engineer, unhappy with her prediction of the maximum expected yield stress for the steel in a part being brake formed, discovered that the company's data bank had records for yield stress of the material in a similar part made sometime previously. Three hundred parts were made per day, and the maximum yield stress observed each day was recorded for a production run of 50 days. This seemed a likely situation for applying an asymptotic extreme value distribution; and since the parent distribution was Weibull, it would be type 1.

The engineer's first concern was whether the parent sample of 300 per day was large enough to justify an assumption that the asymptotic distribution was achieved. Because the parent distribution was "known," it was possible to estimate the asymptotic parameters using (8.3.9) and (8.3.10). These estimates would not, of course, be as good as those based directly on the extreme value data. But a plot of the corresponding density function could be compared with the 300th order statistic plot, to give an indication of asymptotic convergence. Fig. 8.10 shows, as a matter of interest, plots for parent sample sizes ranging from 10 to 1000.

The match for a sample size of 300, although generally good, was not as close as she would have liked in the critical upper tail area. However, the engineer proceeded to apply the mean-rank plotting method, using the following program, which includes the data for daily observed extreme values.* The results were plotted in Fig. 8.11.

---

*Although the parent Weibull distribution is based on real data, the extreme value data used here are hypothetical.

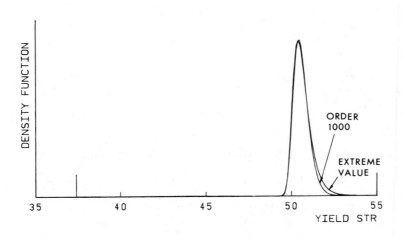

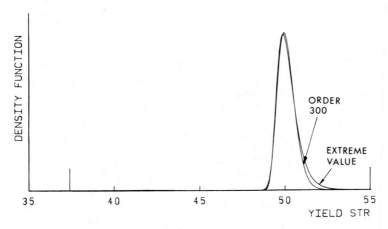

FIG. 8.10 Comparison of nth order and asymptotic density functions, for a Weibull parent distribution.

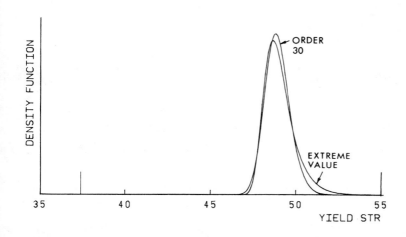

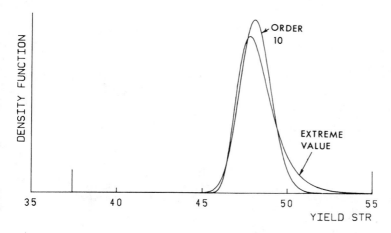

Fig. 8.10 (Continued)

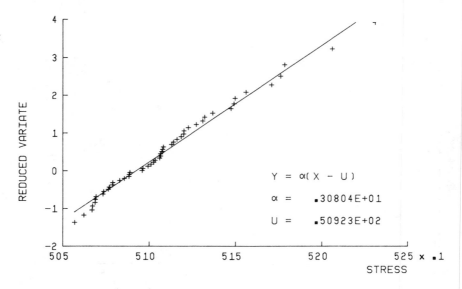

FIG. 8.11 Mean-rank plot for an extreme value distribution.

```
       DIMENSION XE(50),W(50)
       READ(5,4)XE
4      FORMAT(5E12.5)
       CALL MREX1(XE,50,6,6HSTRESS,1,UCHAR,ALPHA,W)
       CUMF=.95**.005
       Y=ALOG(-ALOG(CUMF))
       EXTSTR=Y/ALPHA+UCHAR
       WRITE(6,6)EXTSTR
6      FORMAT(* EXTREME VALUE OF STRESS= *,E12.5)
       STOP
       END
```

.50567E+02	.50621E+02	.50668E+02	.50670E+02	.50685E+02
.50687E+02	.50690E+02	.50730E+02	.50735E+02	.50762E+02
.50767E+02	.50785E+02	.50787E+02	.50827E+02	.50855E+02
.50882E+02	.50883E+02	.50887E+02	.50959E+02	.50962E+02
.50992E+02	.51010E+02	.51025E+02	.51034E+02	.51059E+02
.51064E+02	.51064E+02	.51073E+02	.51076E+02	.51081E+02
.51129E+02	.51140E+02	.51160E+02	.51184E+02	.51198E+02
.51200E+02	.51225E+02	.51272E+02	.51310E+02	.51320E+02
.51367E+02	.51474E+02	.51491E+02	.51498E+02	.51560E+02
.51709E+02	.51761E+02	.51785E+02	.52061E+02	.52310E+02

The good straight-line fit encouraged the engineer to accept the resulting prediction of the extreme value in 10,000 parts corresponding to 0.95 probability. Since the parameters from the mean-rank plot are related to a parent sample size of 300, the required relationship is

$$0.95 = [F_{s_e} (S_s)]^{10,000/300} \qquad (8.3.24)$$

The corresponding value of S_s is

$$S_s = 53,610 \text{ psi}$$

The engineer had one final concern. It seemed almost certain that many of the values from the parent samples of 300 each day would not be independent. Many parts would be cut from the same sheet, and perhaps many more from the same slab. Thus one theoretical requirement for asymptotic convergence would not have been met. However, there is some evidence (Gumbel, 1958) that full independence of all sample items is not essential for the type 1 extreme value distribution to apply. The engineer decided that the good straight-line fit, based on a fair-sized extreme value sample (50), was sufficient assurance that the model was satisfactory.

EXAMPLE 8.5 Ice Thickness on a Transmission Line

A designer wished to estimate the probability that an ice thickness of 3 in. accumulated on an electrical transmission cable will not be exceeded in 10 years. Maximum observed values occurring in 1 year are available for 22 previous years. She used subroutine MREX1 to generate a mean-rank plot for a type 1 extreme value distribution, shown in the following computer program and Fig. 8.12.

```
      PROGRAM TST(INPUT,OUTPUT,TAPE5=INPUT,TAPE6=OUTPUT)
C.... EX 8.5 - ICE THICKNESS
      DIMENSION X(22),W(22)
      DATA X/1.65,1.05,2.2,1.35,.075,.15,.27,.23,.72,.75,.71,.50,.35,
     1.48,.47,.37,1.32,1.25,1.13,.85,.95,1.05/
      CALL MREX1(X,22,9,9HICE THICK,1,UCHAR,ALPHA,W)
      WRITE(6,1)ALPHA,UCHAR
1     FORMAT(9H ALPHA = ,E12.5,/,9H UCHAR = ,E12.5)
      STOP
      END
```

```
      PROGRAM TST(INPUT,OUTPUT,TAPE5=INPUT,TAPE6=OUTPUT)
C.... PLOT OF EXTREME VALUE DENSITY FUNCTION FOR ICE THICK
      DIMENSION X(101),FX(101)
      BU=4.
      BL=-1.
      RANGE=5.
```

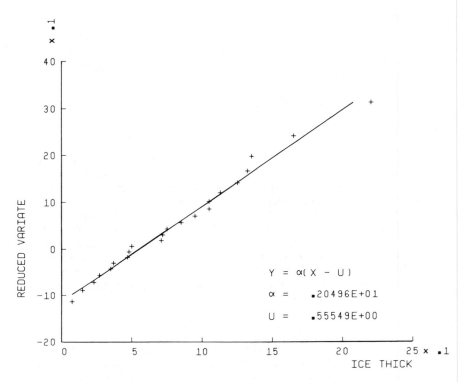

FIG. 8.12 Reduced variate type 1 extreme value plot for ice thickness.

```
      DO 1 I=1,101
      X(I)=RANGE*FLOAT(I-1)/100.+BL
1     FX(I)=DEXTR1(.555,2.05,X(I),0)
      CALL PLOTPL(1,X,FX,101,9HICE THICK,9,BU,BL)
      CALL PLOTPL(0,X,FX,101,9HICE THICK,9,BU,BL)
      STOP
      END
```

The linearity looks quite good. Her first reservation about its
validity was based on the likelihood that the parent sample size
would have been quite small. Most regions have relative few ice
storms in a year. However, she tentatively decided to go with it
because of the good linearity; and as a matter of routine plotted
the density function, shown in Fig. 8.13. She then realized that
the density function was unacceptable because of the large area
in the negative region of the variable. She therefore decided to
try a type 2 distribution, shown in the following section.

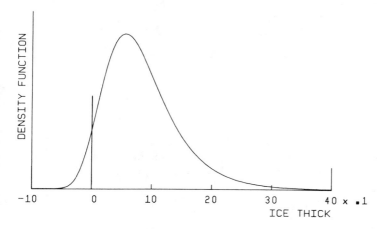

FIG. 8.13 Type 1 extreme value density function for ice thickness.

8.3.3 Type 2 Asymptotic Distribution

The type 2 extreme value function is asymptotic to the largest order statistic of a parent population having a form that is not generally defined, although it must satisfy certain criteria (Gumbel, 1958). It is thus not easy to relate this extreme value distribution to a parent distribution, and it is customarily selected on the basis of empirical fit to a data set. It has been used for maximum annual atmospheric vapor pressure, maximum annual flood, maximum atmospheric temperatures, and maximum wind speeds.

The largest-value type 2 distribution has the form (for $x > 0$)

$$F(x_e) = \exp\left[-\left(\frac{v}{x_e} \right)^k \right] \qquad (8.3.25)$$

$$f(x_e) = \frac{k}{v} \left(\frac{v}{x_e} \right)^{k+1} \exp\left[-\left(\frac{v}{x_e} \right)^k \right] \qquad (8.3.26)$$

where v and k are parameters and functions of the parent sample size. It has the property that moments do not exist from the kth and higher.

The mode is

$$\tilde{x}_e = \frac{vk}{k + 1} \qquad (8.3.27)$$

and the mean and variance are

$$\mu_e = v\Gamma\left(1 - \frac{1}{k}\right) \tag{8.3.28}$$

$$\sigma_e^2 = v^2\left[\Gamma\left(1 - \frac{2}{k}\right) - \Gamma^2\left(1 - \frac{1}{k}\right)\right] \tag{8.3.29}$$

Mean-rank plotting can be used after a transformation similar to that used for the Weibull function in Sec. 5.4. Operating on (8.3.25) gives

$$- \ln\left[- \ln F(x_e)\right] = k \ln x_e - k \ln v$$

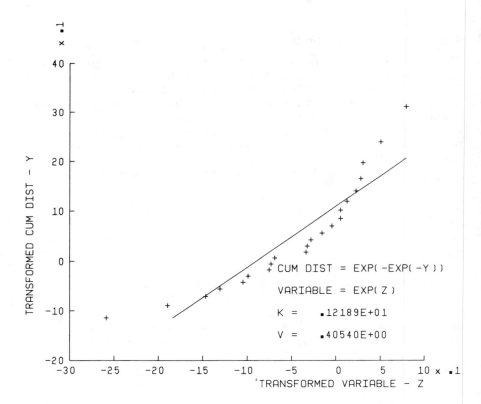

FIG. 8.14 Type 2 extreme value mean-rank plot for ice thickness.

We let

$$y = - \ln [- \ln F(x_e)] \qquad (8.3.30)$$

$$z = \ln x_e \qquad (8.3.31)$$

and have the following linear relationship for plotting:

$$y = kz - k \ln v \qquad (8.3.32)$$

Subroutine MREX2 provides a plot and calculation of k and v.

EXAMPLE 8.6   Ice Thickness on a Transmission Line (Cont.)

As noted in Example 8.5, the type 1 density function was unacceptable for this problem. The type 2 mean rank plot is shown in Fig. 8.14. The nonlinearity makes it clearly inadmissible. One final attempt is made in Example 8.7.

8.3.4 Type 3 Asymptotic Distribution

In the type 3 extreme value distribution, the function asymptotic to the smallest order statistic is the one of primary interest, and turns out to be the Weibull distribution, given in Chapter 3.
    The parent distribution must have the following properties:

1. It must have a finite minimum value of η (the Weibull location parameter).
2. The cumulative distribution function must have at least β derivatives (where β is the Weibull shape parameter), and the first $\beta - 1$ derivatives must vanish at x equal to η.

Justification for use of the distribution based on these criteria would commonly be rather difficult.

8.3.5 Other Extreme Value Distributions

In many applications of the classical extreme value distributions in the literature, the justification based on modeling is often rather tenuous or nonexistent. There would seem to be no reason why any theoretical distribution should not be a candidate, if the goodness of fit appears to justify it.

EXAMPLE 8.7   Ice Thickness on a Transmission Line (Cont.)

Examples 8.5 and 8.6 reached unsatisfactory conclusions insofar as achieving an acceptable estimate of the extreme value distribution for ice thickness on a transmission line. So the engineer

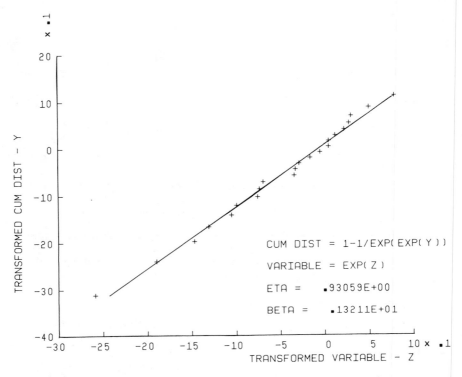

FIG. 8.15 Weibull function fit to ice thickness extreme values.

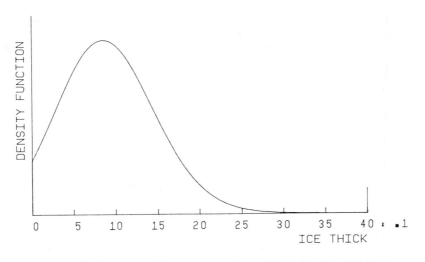

FIG. 8.16 Maximum entropy density function for ice thickness extreme values, using ranks. NL = 5, QPROB = 0.99, and NCYCLE = 2.

decided to try fitting a Weibull and a maximum entropy distribution. Both are assumed to have a lower bound of zero.

The mean-rank plot for the Weibull fit is shown in Fig. 8.15. Figure 8.16 shows the maximum entropy density function. Both appear to provide a good fit, so the Kolmogorov-Smirnov goodness-of-best-fit test was used, as follows.

```
C.... GOODNESS-OF-FIT TESTS FOR ICE THICKNESS
      DIMENSION X(22)
      EXTERNAL FW,FMEP
      DATA X/1.65,1.05,2.2,1.35,.075,.15,.27,.23,.72,.75,.71,.50,.35,
     1.48,.47,.37,1.32,1.25,1.13,.85,.95,1.05/
      FITW=FITT1(X,22,FW)
      FITM=FITT1(X,22,FMEP)
      WRITE(6,1)FITW,FITM
1     FORMAT(13H WEIBULL FIT=,E12.5,19H   MAX ENTROPY FIT=,E12.5)
      STOP
      END


      FUNCTION FW(X)
      FW=FWEIB(X,0.0,.93059,1.3211)
      RETURN
      END


      FUNCTION FMEP(X)
      DIMENSION AL(5)
      DATA AL/-1.4886,2.8280,-1.7579,.057132,.018481/
      FMEP=ENTRPF(AL,5,X)
      RETURN
      END


WEIBULL FIT= .91706E+00   MAX ENTROPY FIT= .93504E+00
```

The best choice is the Weibull function with a density function expressed by

$$f_e(x) = \frac{\beta}{\eta} \left(\frac{x}{\eta} \right)^{\beta-1} \exp\left[- \left(\frac{x}{\eta} \right)^{\beta} \right] \qquad (8.3.33)$$

Recalling Example 8.5, the problem was to estimate the probability that an ice thickness of 3 in. will not be exceeded in 10 years, whereas the density function is based on one year. Thus the required probability is

$$P(\text{thickness} < 3 \text{ in.}) = \left[\int_0^3 f_e(x)\ dx \right]^{10/22}$$

$$= [F_e(3)]^{10/22} = (0.9908)^{10} = 0.9117$$

8.4 SELECTION OF BOUNDS AND JUDGMENT OF OUTLIERS

It would seem intuitively preferable to adopt parent density functions having finite bounds rather than infinite tails. Such distributions include the beta, the maximum entropy, and the lower bounds of the Weibull and gamma. It was shown in Sec. 5.5.4 that the maximum entropy distribution was not generally too sensitive to the choice of bounds. On the other hand, selection of the Weibull lower bound is usually quite critical. It would be desirable to have some procedure to assist in the decision, and order statistics provide a possible tool.

If extreme value data are available, and an extreme value distribution has been generated with a known extreme value sample size N, it could be used to establish a bound based on some realistic risk of exceedance for a single trial, say 0.001. Equation (8.3.14) would be used as follows for an upper bound.

$$0.999 = F_x(x_b) = [F_{x_e}(x_b)]^{1/N} \qquad (8.4.1)$$

where the bound x_b is obtained by inverting (8.4.1).

If only a population sample is available, order statistics can be of some assistance. If we are concerned with the upper bound, we can use the nth order statistic (where n is the sample size), beginning with an arbitrarily assumed upper bound. The parent distribution is defined from the sample and this bound, and (8.2.5) used to define the nth order statistic distribution. The range of this order statistic can then be determined and used to indicate at least the amount of uncertainty associated with the bound.

EXAMPLE 8.8   Warehouse Floor Loads

Example 5.3 is used here to demonstrate the idea discussed above. The sample size is 220. The upper bound was initially set at the value of the largest observation, giving the following maximum entropy distribution.

$$f(x) = \exp\,(-\,6.9352 + 0.081214x - 0.98970 \times 10^{-3}\,x^2$$

$$+\,0.49905 \times 10^{-5}\,x^3 - 0.14505 \times 10^{-7}\,x^4),\quad 0 \leqslant x \leqslant 229.5$$

The highest order statistic distribution is generated by the following code.

```
C....  EXAMPLE 8.8 - DETERMINATION OF BOUND
       DIMENSION A(6),XD(220),SM(4),XP(1),CUM(1),X(200),FX(200),DIST(200)
       DIMENSION FORDER(200)
       COMMON/FD/AC(5),XMAX
```

```
        EXTERNAL FD,FCUM
        ORDMIN=150.0
        KSTART=2
        KDATA=0
        PROB=.01
        NSAMP=220
C.... READ SAMPLE DATA
        READ(5,1)XD
1       FORMAT(10F6.1)
C....CALCULATE MOMENTS
        CALL SMOM(XD,4,220,SM)
C.... FIND LOWER BOUND OF UPPER ORDER STATISTIC DISTRIBUTION
        CALL BOUNDS(XD,220,XMAX,XMIN)
        WRITE(6,10)XMAX
10      FORMAT(* XMAX=*,F6.1)
C.... GENERATE DISTRIBUTION
        CALL MEP1(4,SM,XMIN,XMAX,0,XP,KSTART,KDATA,A,CUM)
        CALL ENTPL(XMIN,XMAX,5,A,4,4HLOAD)
C.... OBTAIN NEW VALUE OF XMAX USING UPPER ORDER STATISTIC
        DO 5 I=1,5
5       AC(I)=A(I)
C.... DEFINE ORDER STATISTIC DENSITY FUNCTION
        ORDRGE=XMAX-ORDMIN
        DO 6 I=1,200
        X(I)=FLOAT(I-1)/199.*ORDRGE+ORDMIN
6       FX(I)=DORDER(FD,FCUM,NSAMP,NSAMP,X(I))
        DO 7 I=1,200
7       FORDER(I)=FX(I)/30.
        CALL PLOTPL(1,X,FORDER,200,5HLABEL,5,XMAX,ORDMIN)
C.... INVERT CUM DIST FUNCTION OF ORDER STATISTIC TO GET LOWER BOUND
        ORDRGE=XMAX-ORDMIN
        CALL FTOCUM(FX,DIST,ORDRGE,200)
        OBND=FINVRT(X,DIST,200,PROB)
        RANGE=XMAX-OBND
        WRITE(6,4)RANGE
4       FORMAT(* RANGE OF HIGHEST ORDER STATISTIC FOR SAMPLE IS*,F7.3)
        CALL PLOT(Y,Z,999)
        STOP
        END


        FUNCTION FD(X)
        COMMON/FD/AC(5),XMAX
        FD=ENTRPF(AC,5,X)
        RETURN
        END


        FUNCTION FCUM(X)
        DIMENSION XA(200)
        COMMON/FD/AC(5),XMAX
        FCUM=CDF(0.0,XMAX,X,AC,5)
        RETURN
        END
```

XMAX= 229.5

LAST VALUE OF DISTRIBUTION FUNCTION IS   1.00029387

RANGE OF HIGHEST ORDER STATISTIC FOR SAMPLE IS 19.083

The final density functions are plotted in Fig. 8.17.

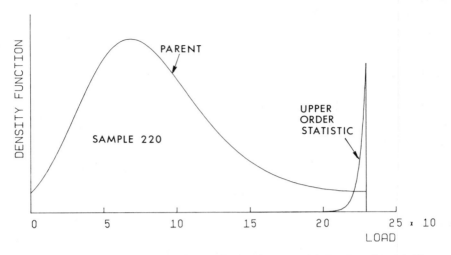

FIG. 8.17 Upper bound information using the highest order statistic.

The *outlier* problem is closely related to the problem of bounds. Sometimes a data sample will have one or more values at the extremes which are suspect, and a decision must be made whether or not to reject them. It is really not possible to make such a decision without using judgment related to knowledge of the physical basis for the random variable, but it can be helpful to use estimates of the probability of such an outlier legitimately occurring.

Many rules are given in the literature of statistics, but perhaps the most meaningful basis for the decision (augmenting judgment) is related to order statistics. The approach is to use the *distribution of distances*. We let w_{rs} be the random variable representing the distance or interval between the rth and sth order statistics. It can be shown (David, 1981) that the probability density for w_{rs} is given by

$$f_{rs}(w_{rs}) = \frac{n!}{(r-1)!\,(s-r-1)!\,(n-s)!}$$

$$\times \int_R [F_x(x)]^{r-1} f_x(x) [F_x(x+w_{rs}) - F_x(x)]^{s-r-1}$$

$$\times f_x(x+w_{rs})[1 - F_x(x+w_{rs})]^{n-s}\,ds, \quad r < s$$

$$(8.4.2)$$

Thus the outlier and the next highest observation yield the following density function for d_{n-1}, the distance between x_{n-1} and x_n.

$$f_{d_{n-1}}(d_{n-1}) = n(n-1) \int_R [F_x(x)]^{n-2} f_x(x) f_x(x + d_{n-1}) \, dx$$

(8.4.3)

The probability of the $(n-1)$ distance having a value equal to or greater than $(x_n - x_{n-1})$ will be a meaningful quantity, and is estimated by

$$P[d_{n-1} \geq (x_n - x_{n-1})] = \int_{x_n - x_{n-1}}^{\infty} f_{d_{n-1}}(d_{n-1}) \, dd_{n-1} \quad (8.4.4)$$

Thus if this probability is, say, 0.01 and the engineer is not entirely confident about the validity of the outlier observation, he might be inclined to reject it.

EXAMPLE 8.9 Outlier Data Point

In Example 5.4, which used the maximum entropy method to generate a probability density function for fracture toughness, there is some question about the largest data point. It had the appearance of a possible outlier, or a data point not really belonging to the population.

The parent density function was based on the following data: 69.5, 71.9, 72.6, 73.1, 73.3, 73.5, 74.1, 74.2, 75.3, 75.5, 75.7, 75.8, 76.1, 76.2, 76.2, 76.9, 77.0, 77.9, 78.1, 79.6, 79.7, 79.9, 80.1, 82.2, 83.7. 93.7. The outlier value was 93.70, the sample size was 26, and the upper distance was 10.0.

The required distance density function is given by (8.4.3); it is developed by the following coding and plotted in Fig. 8.18 for the data set with the outlier deleted.

```
      DIMENSION X(200),FX(200),ARG(200),FDIST(200),XDIST(200),A(6),
     1W(200),XD(26),SM(4),XSPEC(1),PSPEC(1)
C.... DEFINE DENSITY FUNCTION NUMERICALLY
      DATA(XD(I),I=1,25)/69.5,71.9,72.6,73.1,73.3,73.5,74.1,74.2,75.3,75
     1.5,75.7,75.8,76.1,76.2,76.2,76.9,77.0,77.9,78.1,79.6,79.7,79.9,80.
     11,82.2,83.7/
      CALL SMOM(XD,4,25,SM)
      INT=101
      NSAMP=25
      XMIN=67.0
      XMAX=83.70
      RANGE=16.70
      CALL MEP1(4,SM,XMIN,XMAX,0,XSPEC,4,0,A,PSPEC)
      DO 1 I=1,INT
```

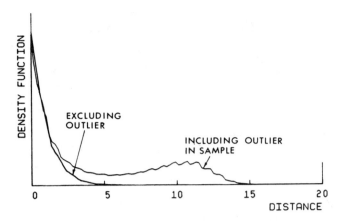

FIG. 8.18 Distance density function for an outlier.

```
        X(I)=XMIN+RANGE*FLOAT(I-1)/FLOAT(INT-1)
1       FX(I)=ENTRPF(A,5,X(I))
C.... DEFINE UPPER DISTANCE DENSITY FUNCTION
        DO 3 I=1,INT
        XDIST(I)=20.*FLOAT(I-1)/FLOAT(INT-1)
        DO 2 J=1,INT
        XPLUSD=X(J)+XDIST(I)
        IF(XPLUSD.GT.X(INT))GO TO 12
        IF(I.EQ.1.AND.J.EQ.1)GO TO 12
        TEMP1=FTABLE(X,FX,XPLUSD,INT)
        MINT=J
        IF(J.LT.5)MINT=5
        ARG(J)=CUM1(X,FX,INT,X(J),INT,W)**(NSAMP-2)*FX(J)*TEMP1
        GO TO 2
12      ARG(J)=0.0
2       CONTINUE
        AREA=FSIMP(ARG,RANGE,INT)
3       FDIST(I)=FLOAT(NSAMP*(NSAMP-1))*AREA
C.... CIIECK AREA OF DENSITY FUNCTION
        FAREA=FSIMP(FDIST,20.,INT)
        WRITE(6,9)FAREA
9       FORMAT(* AREA OF UPPER DISTANCE DENSITY FUNCTION IS *,E12.5)
        CALL SMOOTH(FDIST,W,INT)
        DO 5 I=1,INT
5       FDIST(I)=W(I)
C.... NORMALIZE AREA OF DENSITY FUNCTION
        CALL FNORM(FDIST,20.,INT)
        CALL PLOTPL(1,XDIST,FDIST,INT,8HDISTANCE,8,20.,0.0)
        CALL PLOTPL(0,XDIST,FDIST,INT,8HDISTANCE,8,20.,0.0)
C.... CALCULATE PROBABILITY THAT DISTANCE WOULD BE GREATER
C       THAN OBSERVED VALUE
        PROB=1.-CUM1(XDIST,FDIST,INT,10.0,INT,W)
        WRITE(6,4)PROB
4       FORMAT(* PROBABILITY THAT UPPER DISTANCE EXCEEDS 10.0 IS*,F8.5)
        STOP
        END
```

AREA OF UPPER DISTANCE DENSITY FUNCTION IS   .98470E+00

PROBABILITY THAT UPPER DISTANCE EXCEEDS 10.0 IS   .000

The procedure was repeated, including the outlier in the data, and gave the following result:

AREA OF UPPER DISTANCE DENSITY FUNCTION IS   .10279E+01

PROBABILITY THAT UPPER DISTANCE EXCEEDS 10.0 IS   .17781

The result would suggest that the engineer would need to have considerable confidence in the validity of the observed outlier, in order to accept it.

8.5 DISTRIBUTION OF EXCEEDANCES

In this topic we assume that we have observed a sample of size n (usually relatively small) of a continuous random variable x *with unknown distribution*. We wish to predict the probability that the ith ordered observation will be exceeded k times in N future trials (N will usually be large).

It can be shown (Thomas, 1948) the distribution of k is given by

$$f(k; n, N, k) = \frac{\binom{n}{i-1}(n - i + 1)\binom{N}{k}}{(N + n)\binom{N + n - 1}{k + n - i}} \qquad (8.5.1)$$

We will commonly be interested in the case when i equals n and k is 1.

$$f(1; n, N, n) = \frac{nN}{(N + n)\binom{N + n - 1}{2n - i}} \qquad (8.5.2)$$

A closely related concept is the *return period* (Gumbel, 1958; Thomas, 1948). It is a meaningful alternative way of representing risk associated with extreme value events. We let E represent such an event, say exceeding a specified near maximum amount of ore that is taken from a mine in 1 day. An exceedance of this specification would cause problems in processing or handling the ore above ground. The complementary event is $\overline{E}$. Then over a sequence of days, the event would occur randomly, indicated as follows:

$$\bar{E} \; \ldots$$

In most applications the interval between E's would be longer than indicated.

The *return period* is the random variable representing the interval between occurrences of the event, designated k. Extreme value theory can be used to determine the probability of the event E occurring, designated p. The random variable k has a geometric distribution.

$$f(k) = p(1 - p)^{k-1} \tag{8.5.3}$$

This random variable is not itself of great interest. We are not usually concerned with the probability of the specified event occurring in exactly k days. However, the mean value of k, or the mean return period (commonly just called the return period), is meaningful and has the value

$$\bar{k} = \frac{1}{p} \tag{8.5.4}$$

It represents, in our example, the average number of days until the event E occurs since the last time.

PROBLEMS

Problem 8.1 The acceleration of a point on the frame of a moving vehicle is continuously recorded during 10-hr trips between two cities. Let x be a random variable denoting the peak acceleration level (in g's) during the 10-hr period. Twenty observations were made and the data recorded are given below.

```
 9.42  12.80   7.23  18.30   9.18  14.10  12.30  10.40
13.30   8.37   8.24  14.90  13.70  17.80   8.07  12.60
11.00  10.10   8.76  10.20
```

A unit that is to be transported on this vehicle between the cities has a fragility level of 20 g's. Use extreme value theory to estimate the probability that the unit will be damaged. What will the probability of damage be if the unit makes a 40-hr trip subject to the same mode of random acceleration? What are the requirements in the modeling for these to be good estimates?

(This problem was adapted from *ASME Paper No. 75-DET-123.*)*

---

*Wirsching, P. H., and L. H. Jones (1975). On the Use of the Extreme Value Distribution in Reliability Analysis and Design, *Amer. Soc. Mech. Eng.* Paper No. 75-DET-123.

Problem 8.2 The maximum observed wind speed in mph in 1 year has been recorded as follows for 20 years:

67.0, 50.0, 50.0, 57.0, 59.5, 56.0, 48.0, 34.0,

40.0, 40.0, 40.5, 47.0, 47.5, 42.5, 42.5, 38.0,

36.0, 37.0, 37.5, 38.0

Try fitting types 1, type 2, and the maximum entropy distributions to the data. Select the best, justifying your choice. What is the return period for a 80-mph wind? If we are designing for an 70-mph wind on a structure, what is the probability of it not being exceeded in 1 year, 20 years, 50 years, and 100 years? Plot the 1-y, 20-y, 50-y, and 100-year density functions, together, and draw the 70-mph specification line.

Problem 8.3 Repeat the generation of Fig. 8.10 for some other distribution having an asymptotic tail. Determine how large the sample from which the extreme is observed need be to achieve asymptotic convergence.

Problem 8.4 A cable has n strands. The density function for the strength of a strand is f(t), where t is the breaking strength of a strand. We assume that α is constant, where α is defined by the relationship

$$p = \alpha e$$

where p is the load and e is the extension of a wire. Thus all strands have the same extension and load. The load on the cable is T. We wish to determine the probability that the cable will break under a given load T. The event of cable failure corresponds to the following set of events:*

$0 \leqslant t_1 \leqslant \dfrac{T}{n}$ (failure of weakest strand)

$t_1 \leqslant t_2 \leqslant \dfrac{T}{n-1}$ (failure of second weakest strand)

$\vdots$

$t_{n-1} \leqslant t_n \leqslant T$ (failure of strongest strand)

where t_i is the ith order statistic of a sample represented by the n

---

*H. E. Daniels (1945), The Statistical Theory of the Strength of Bundles of Threads, *Proc. R. Soc. Lond. A*, Vol. 183, p. 405.

cable strands. The cable is not considered to have failed until all strands have successively failed. At any stage in this process a strand may not fail and the cable is considered to then have not failed. Use the joint probability density function for n order statistics to develop an expression for the probability of failure of the cable.

Problem 8.5 The maxima of gust velocities, in ft/sec, for 485 traverses of thunderstorms, is taken from Gumbel and Carlson.*

Velocity	3	5	7	9	11	13	15	17	19	21
Frequency of observation	4	11	27	48	62	58	55	60	61	36

Velocity	23	25	27	29	31	33	35	37	39
Frequency of observation	17	18	8	7	6	3	1	2	1

Although the data are grouped, it is still possible to do a mean-rank plot by using the geometric mean of the first and last ranks in an interval.† The rank r'_k in the kth interval corresponding to $m + 1$ grouped observations ranging from r_i to r_{i+m} is thus

$$r'_k = \sqrt{r_i r_{i+m}}$$

This is taken as the rank value corresponding to the midpoint value of the kth interval. The mean ranks for the above data are, therefore,

$$2/486 \quad 8.66/486 \quad \cdots \quad 485/486$$

The rule above does not apply if the last interval has more than one observation. In this event the midpoint value is assigned to a mean rank corresponding to the first rank of the interval, and also to the last rank of the interval. Thus if the last interval has a midpoint value of x_p, and two observations corresponding to N and $N - 1$, the following pairs are plotted: x_p, $N/(N + 1)$ and x_p, $(N - 1)/(N + 1)$. Fit the data above to a type 1, type 2, and a maximum entropy distribution, and compare.

---

*E. J. Gumbel and P. G. Carlson (1954), Extreme Values in Aeronautics, *J. Aerosp. Sci.*, Vol. 21, p. 389.
†Ibid.

Problem 8.6 In Example 8.1 the beam streangth was w (ft lb) with density function given by (8.2.19).

$$f_w(w) = 6f_x(w)[1 - F_x(w)]^5$$

The maximum applied loading on the same beam in Example 8.2 was F (ft lb) with density function given by (8.2.22).

$$f_F(F) = 25[F_y(F)]^{24}f_y(F)$$

A margin of strength m can be defined given by

$$m = w - F$$

Use one method from Chapter 6 to obtain the probability of failure.

$f_x(x)$ = beta distribution, q = 3.0, r = 1.5,

$\ell = 3000$, u = 6500

$f_y(y)$ = Weibull distribution, $\beta = 2.96$,

$\eta = 2630$, $\ell = 2060$

Problem 8.7 A vehicle is used in a service in which the load carried for a daily run varies substantially from day to day. Records have been kept making it possible to observe the maximum load carried in any month and these are available for 30 months. The ranked data are as follows:

8,080	11,600	14,000	16,000	16,800	20,300
22,600	24,900	31,000	36,400	42,000	42,400
46,400	54,800	58,600	60,100	64,300	75,400
80,400	81,400	84,200	92,100	102,000	108,000
113,000	122,500	135,000	152,000	185,000	230,000

A new model of the vehicle is to be designed and we wish to predict the capacity to be used as a design specification if the probability of this capacity being exceeded in use during 1 month is to be less than 0.01. What is the probability of this load being exceeded in 1 year? in 10 years? Comment on the validity of this specification.

Problem 8.8 A steel cable has 18 strands. Each strand is assumed to have the same modulus of elasticity and share the load equally. The tensile strength of the strands is a random variable with known distribution. What is the probability of failure if we define failure of the cable as failure of any one strand?

Problem 8.9 A cooling system for a building is being designed and the engineer would like to use an extreme value approach to specify a maximum outside ambient design temperature. Records are available for daily maximum air temperatures for the past 40 years. Comment on the use of asymptotic or other extreme value distributions in this application. More specifically, is it a suitable application for extreme value theory, and how would you go about doing it?

Problem 8.10 One could argue that the use of mean-rank plotting with variable transformations to get a straight-line fit to a specified theoretical distribution is incorrect because the transformations of the mean rank give incorrect weight to the "errors" in the least-squared approximation. For example, when fitting the Weibull function, the transformation of the mean rank is

$$y_i = \ln \ln \frac{1}{1 - [i/(n + 1)]}$$

Suggest an alternative algorithm for obtaining the best-fitting values of the distribution parameters, but still using the mean-rank approximation to sample values of the cumulative distribution function.

Problem 8.11 Suggest an algorithm for deciding if a straight-line mean-rank plot "really is" a straight line. In other words, does the sample really come from a population represented by the proposed theoretical distribution?

REFERENCES

David, H. A. (1981), *Order Statistics*, 2nd ed., Wiley, New York.

Gumbel, E. J. (1958), *Statistics of Extremes*, Columbia University Press, New York.

Guttman, I., S. S. Wilks, and J. S. Hunter (1971), *Introductory Engineering Statistics*, 2nd ed., Wiley, New York.

Kendall, M., and A. Stuart (1977), *The Advanced Theory of Statistics*, Vol. 1: *Distribution Theory*, 4th ed., Macmillan, New York.

Thomas, H. A., Jr. (1948), Frequency of Minor Floods, *J. Boston Soc. Civil Eng.*, Vol. 35, p. 425.

Prediction of Failure Mode Probabilities

The concept of probabilistic failure modes and feasibility expressions •
The central role of random life • *Dependability and combined failure
modes* • *Fault trees*

9.1 INTRODUCTION

Our concern in this chapter is to estimate the probability of satisfactory performance of a device or system. This is an intricate topic: combinations of modes of failure can be very complex.

To clarify our approach it is necessary to define in a general way the various modes of failure. The term *feasibility*, or a *feasible design*, is commonly used in the terminology of design optimization, and can usefully be applied here.

In general, the performance of any device or system is controlled by a set of inequality and equality expressions, called *constraints*, that define the *feasible region* of a design in terms of the independent design variables. The hierarchy of design variables was outlined in Sec. 1.1. These constraint expressions can, by convention, be given the following general form:

$$\psi_i(x_1, x_2, \ldots, x_n) = 0, \quad i = 1, m \tag{9.1.1}$$

$$\phi_j(x_1, x_2, \ldots, x_n) \geqslant 0, \quad j = 1, p \tag{9.1.2}$$

where the x's are the functionally independent design variables. If we are optimizing, an optimizing function is used which expresses some criterion for "bestness" about the design. It can be given the general form

$$U(x_1, x_2, \ldots, x_n) = \text{maximum or minimum} \qquad (9.1.3)$$

This expression may be maximized or minimized, as required, by means of a numerical technique [see Siddall (1982)], subject to satisfying the constraint expressions. Equality constraints are relatively uncommon, and we shall temporarily defer their consideration.

Figure 9.1 illustrates how inequality constraints define a feasible region for a hypothetical two-variable design. These constraint functions commonly derive from expressions formulated in the following way:

$$y_i(x_1, x_2, \ldots, x_n) \geqslant S_i \qquad (9.1.4)$$

where y_i represents a design characteristic such as speed, capacity, cost, stress, deflection, and the like; and S_i represents a specification that must (or must not) be exceeded. For convenience this is then converted to the standard form.

$$\phi_i = y_i(x_1, x_2, \ldots, x_n) - S_i \geqslant 0 \qquad (9.1.5)$$

The expressions represented by $y_i(x_1, x_2, \ldots, x_n)$ are obtained from engineering modeling. *Each constraint thus represents a failure mode*, any one of which can cause the device or system to fail to perform satisfactorily.

If we are optimizing, and the independent design variables are considered determinate, we force their nominal values to fluctuate so as to find the optimal point, as shown in Fig. 9.1. If the x's are now considered random, we are faced with the *probabilistic or stochastic optimization problem*. This is a considerable complication of optimization and is explored in Chapter 13.

In this chapter we assume that the x's have been assigned nominal values, but are random variables. The range of the x's may or may not extend over the full feasible region. This is illustrated in Fig. 9.2.

In general, the probability of the device working satisfactorily is the probability of any particular embodiment of x_1 and x_2 falling in the feasible region. This can be expressed by using the constraint functions, as follows:

$$P(\text{no failure}) = P(\phi_1 \geqslant 0 \text{ and } \phi_2 \geqslant 0 \text{ and } \phi_3 \geqslant \text{and } \cdots \phi_m \geqslant 0)$$

$$(9.1.6)$$

The ϕ's are random variables whose distributions can be determined by probabilistic analysis.

We are now ready to return to the situation where we have equality constraints. The function $\phi_i(x_1, x_2, \ldots, x_n)$ is also a random

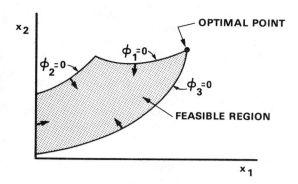

FIG. 9.1 Illustration of how inequality constraints define a feasible region for a design having two functionally independent variables.

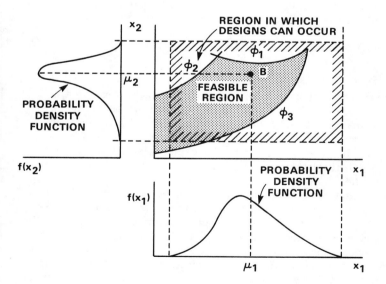

FIG. 9.2 Association of random, functionally independent design variables with a feasible design region.

variable whose probability distribution can be determined by probabilistic analysis. We thus cannot satisfy an equality constraint when the variables are random and independent, since there is zero probability that it will be satisfied. The best that can be done is to set the expected value of ψ_i equal to zero, as an equality constraint. We must accept the random deviation in ψ that will occur. This would happen if we are attempting to meet a specification, say that a spring is to have a specified deflection. However, equality constraints are not always of this type, and the problem is discussed further in Sec. 13.1.

There is another aspect of feasibility that requires clarification. Up to this point we have been talking about feasibility in operation, after the device has been built and is in the field. This can usefully be termed *function feasibility*, to distinguish it from *manufacturing feasibility*, which requires that the device or its components satisfy certain specifications as part of the manufacturing process. These can be related to performance characteristics, such as maximum horsepower of an engine; but more commonly are simply tolerances on independent design variables. Examples are resistance of a resistor, diameter of a shaft, and yield stress of a steel material. The probability of a device being rejected because of manufacturing infeasibility would be different from the probability of failure due to functional infeasibility.

The illustration in Figs. 9.1 and 9.2 is repeated in Fig. 9.3 to shown the effect of tolerances on x_1 and x_2. The probability of manufacturing feasibility is obtained from the density functions.

$$P(\text{manufacturing feasibility}) = P(x_{1\ell} < x_1 < x_{1u} \text{ and } x_{2\ell} < x_2 < x_{2u})$$

$$(9.1.7)$$

However, the density functions for x_1 and x_2 must be used in the truncated form shown in heavy outline in Fig. 9.3, when being used to estimate the probability of functional feasibility. They must be renormalized to compensate for the reduced range. The regions in the tails outside the tolerance ranges represent rejected components in the factory. Optimal tolerance assignment is a separate topic covered in Chapter 13.

There is a further conceptual problem to be resolved, related to life of a component or device. Some performance characteristics are time dependent, due to random occurrence of environmental effects, or due to deterioration effects such as wear, fatigue, corrosion, deposits inside heat exchanger tubes, and the like. In this situation it is necessary to use a life specification to control satisfactory performance. To achieve this, a relationship between the performance characteristic and life must be modeled. Three typical examples are indi-

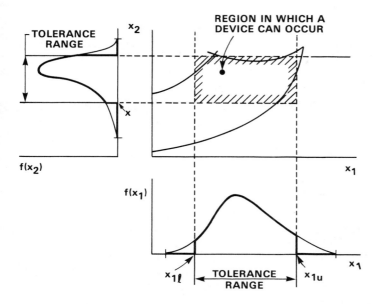

FIG. 9.3 Illustration of the effect of tolerance on the region in which a device can occur.

cated in Fig. 9.4: load in a structural member, power of an engine, and capacity of a heat exchanger. A special kind of relationship must be modeled between performance characteristics such as these and life:

$$\text{Life} = \zeta(y_i) \tag{9.1.8}$$

where y_i is a performance characteristic such as those shown in the figure, and $\zeta(\cdot)$ means that a nominal value of life (such as the mean) is related to y_i by the function. Life tends to be highly random, and there will be a different probability distribution for life corresponding to different values of the design characteristic.

The density function for life, corresponding to the specification value of the design characteristic, would be used to estimate the probability of a specified life being exceeded, as indicated on the curve for engine power. This is customarily called the *reliability* of the device, although more strictly, reliability is the probability of a device not failing due to any cause prior to the specified life.

Modeling the $\zeta(\cdot)$ relationships, with the corresponding density functions, is quite difficult to achieve; the loading on a member, for example may vary over the life. It is common practice to use experi-

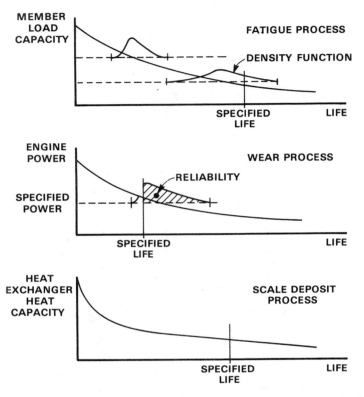

FIG. 9.4 Typical curves relating performance characteristic to life.

mental data to define life probabilities. This topic is explored in more detail in Chapter 11.

It is also common practice to ignore time effects if they are not too strong, and build into the specification an allowance for deterioration. Many of our examples in this book have, in effect, done this.

We are thus concerned in this chapter with identifying failure modes, and determining probabilities of failure modes, singly and in combination.

9.2 SINGLE FAILURE MODE PROBABILITIES

Up to this point in our study of probabilistic design we have seen how to determine the probability distributions for design characteristics which are functions of independent design variables with "known" dis-

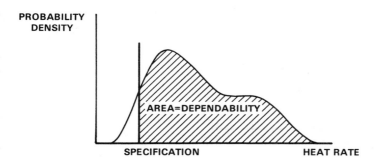

PROBABILITY DENSITY

AREA=DEPENDABILITY

SPECIFICATION   HEAT RATE

FIG. 9.5 Illustration of dependability.

tributions. We have also seen how probabilities of events associated with such design characteristics can be predicted.

The term *dependability* can usefully be defined as the probability that any design characteristic will satisfy a given specification. This is illustrated in Fig. 9.5 by a hypothetical density function for the heat transferred by a heat exchanger, a major design characteristic.

$$\text{Dependability} = D = 1 - F(q) \qquad (9.2.1)$$

where q is the heat rate.

The dependability concept can be broadened to the case where the specification is also random. We let y be a random variable representing a performance characteristic of a device or system, while S is the corresponding specification for the characteristic. S may also be a random variable, rather than a fixed quantity. Typical density functions are shown for these variables in Fig. 9.6. If we assume that the requirement is that $y \geqslant S$, the *dependability* is defined as the occurrence of this event, or *the probability that a specification will be satisfied*. Thus we can have strength dependability, life dependability, speed dependability, and so on. The *combined dependability* is the probability that all specifications are satisfied, or that the design is feasible.

Referring to Fig. 9.6, we can determine the dependability for $y \geqslant S$ by the following arguments, using infinitesimal events. The probability that S will have a value in the interval dS is $f_S(S)\,dS$. The probability that this will occur *and* y will exceed this value is

$$f_S(S)\,dS \int_S^\infty f_y(y)\,dy \qquad (9.2.2)$$

by the product law for independent events. We now argue that S can

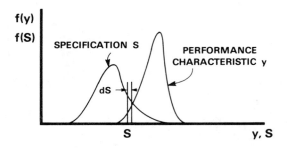

FIG. 9.6 Probability density functions for a performance characteristic and specification.

occur in any infinitesimal interval, and the summation law for mutually
exclusive events is represented by an integration of S.

$$D = \int_{-\infty}^{\infty} f_S(S) \int_S^{\infty} f_y(y) \, dy \, dS \qquad (9.2.3)$$

Or we could begin with dy to get

$$D = \int_{-\infty}^{\infty} f_y(y) \int_{-\infty}^{y} f_S(S) \, dS \, dy \qquad (9.2.4)$$

It is assumed that the variables are stochastically independent. Similar expressions could be written if the requirement was that $y \leqslant S$.

The existence of a specification that is a random variable is not,
in fact, an unusual one. It would occur in situations where many

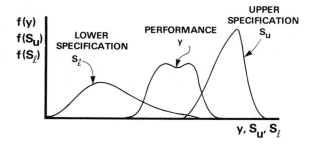

FIG. 9.7 Probability density functions with two-sided specifications.

users must be satisfied, and all users have different performance
requirements. Typical examples would be the capacity of a harvester
combine, the strength of a golf club, the capacity in water flow of an
irrigation pump, or the temperature of a domestic hot water tank.
The size of an automobile trunk would be an example where the performance characteristic is deterministic, but the specification random.
The quantity D is now the probability that the user will be satisfied,
and the density function for the specification can be thought of as an
indication of the probability that any customer will be satisfied by a
given performance characteristic.

It is also possible to have two-sided specifications. Figure 9.7
illustrates the probability densities for this situation. The dependability is given by

$$D = \int_{\infty}^{\infty} f_y(y) \int_y^{\infty} f_{S_u}(S_u) \, dS_u \int_{-\infty}^y f_{S_\ell}(S_\ell) \, dS_\ell \, dy \qquad (9.2.5)$$

It is apparent that we could use any of the methods of probabilistic analysis to determine the dependabilities given above. Consider,
for example, the situation represented by Fig. 9.6. We can define a
margin of performance,

$$m = y - S \qquad (9.2.6)$$

which is a random variable determined by probabilistic analysis. And
the dependability is

$$D = P(m \geqslant 0) \qquad (9.2.7)$$

The convolution integral defined by (9.2.3) may be easier to evaluate
numerically than by using the methods of Chapter 6.

EXAMPLE 9.1 Water Flow through a Pipeline

You are attempting to estimate the water flow through a pipeline
using the formula

$$\Delta p = f \frac{L}{D} \frac{V^2}{2g} \rho \quad \text{lb/ft}^3 \qquad (9.2.8)$$

where

　　　f = friction factor

　　　L = pipe length, ft

　　　D = internal pipe diameter, ft

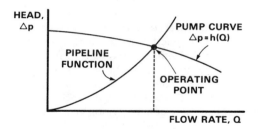

FIG. 9.8 Operating point for a pipeline system.

The pump being used can be represented by the following equation relating Δp to the flow rate Q:

$$\Delta p = h(Q) \tag{9.2.9}$$

The right side of this equation is an undefined function representing tabular data from the pump manufacturer. It represents in effect the characteristic curve for the pump being used. The interaction of (9.2.8) and (9.2.9) is shown in Fig. 9.8.

Quantities L, D, g, and ρ on the right side of (9.2.8) can be considered determinate. The friction factor f is a function of pipe roughness and Reynold's number. However, f is essentially independent of Re in flow regimes where Re exceeds 10^5, commonly used in water pipelines. There is usually considerable uncertainty associated with estimating f, and we shall assume that information is available permitting an estimate of its density function $f(f)$. The demand for water at the pipeline output is random, but with a known density function $f(d)$.

The solution begins by noting the relationship between velocity and flow rate:

$$V = \frac{4Q}{\pi D^2} \tag{9.2.10}$$

Substituting for V in (9.2.8) and combining with (9.2.9) gives

$$h(Q) = \frac{8LQ^2 \rho}{\pi^2 D^5 g} f \tag{9.2.11}$$

Or, in terms of f, this becomes

$$f = \frac{\pi^2 D^5 g}{8L\rho} \frac{Q^2}{h(Q)} \tag{9.2.12}$$

The density functions are related by (6.3.3).

$$f(Q) = \frac{f(f)}{|dQ/df|} \qquad (9.2.13)$$

The algorithm to obtain $f(Q)$ is as follows, assuming that $f(f)$ is defined numerically.

1. *Input*
   a. Equally spaced values of f, designated f_i, and corresponding values of $f(f)$, designated $f(f_i)$.
   b. Equally spaced values of Q and corresponding values of Δp, designated Q_j and Δp_j.
2. Determine unequally spaced values of f corresponding to Q_j, using (9.2.12), designated f_j. Interpolation will be necessary (FTABLE).
3. Determine corresponding values of $f(f)$ using input data and interpolation (FTABLE), designated $f(f_j)$.
4. Obtain a numerical estimate of df/dQ at station j, using the functional relationship defined by the tabular values in item 2. Use forward differences except for the last value, when backward differences are used. The derivative is designated $|df/dQ|_j$. Invert to get $|dQ/df|_j$.
5. Use (9.2.13) to get corresponding values of $f(Q_j)$, inserting data from items 3 and 4.

$$f(Q_j) = \frac{f(f_j)}{|dQ/df|_j} \qquad (9.2.14)$$

6. *Output.* Tabular values of Q and $f(Q)$ are available from items 1b and 5 and together define the density function.

Having $f(Q)$ defined numerically in terms of Q, we now could sketch the two known desnity functions $f(Q)$ and $f(d)$, as shown in Fig. 9.9. The required expression is

$$P(\text{demand will be met}) = \int_{\ell_Q}^{u_Q} \left[\int_{\ell_d}^{Q} f(d) \, dd \right] f(Q) \, dQ \qquad (9.2.15)$$

or

$$= \int_{\ell_d}^{u_d} \left[\int_{d}^{u_Q} f(Q) \, dQ \right] f(d) \, dd \qquad (9.2.16)$$

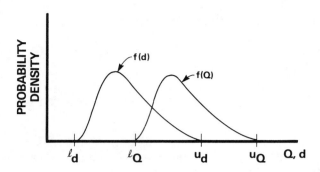

FIG. 9.9 Flow supply and demand density functions.

The algorithm to evaluate this expression is as follows:

1. Divide the total range of d and Q into equally spaced incre-
    ments corresponding to those already obtained for Q.
2. Obtain corresponding values of f(d) by interpolation
    (FTABLE).
3. DO on i.
4.   DO on j.
5.   Divide range of d from ℓ_d to Q_i into an even number of
        equal increments, d_j.
6.   Obtain values for f(d_j) by interpolation (FTABLE).
7.   END DO on j.
8.   Evaluate $\int_{\ell_d}^{Q_i} f(d)\, dd$ using the values above with
        numerical integration (FSIMP), designated A_i.
9.   Obtain values $A_i f(Q_i)$, designated B_i.
10. END DO on i.
11. Obtain the numerical integral for array B_i (using FSIMP),
    giving the required probability.

9.3 COMBINED FAILURE MODE PROBABILITIES

A device having a relatively large number of parts can have a large
number of modes of failure, which can be quite complex to identify.
The source of failure modes includes not only those of the device as a
whole, related to its performance characteristics, but also includes
failures of components; and each component may have different failure
modes.

The example of a forage harvester will help clarify this idea. A
forage harvester is an agricultural machine that cuts green hay,
chops it, and blows it into a wagon. We shall assume that it is trac-

tor drawn and powered by a power-takeoff from the tractor. Some typical modes are as follows:

1. Failure modes of whole device
    a. Production cost too high, dollars
    b. Capacity too low, lb hay/hr
    c. Power to drive too high, hp
    d. Draft to pull too high, lb
    e. Life to first failure too low, hr
2. Some typical component failures that cause failure of the device
    a. Failure of axle shaft
        (1) Fatigue
        (2) Yield
        (3) Corrosion
    b. Failure of gear in cutter drive
        (1) Bending fatigue
        (2) Wear fatigue
        (3) Scoring due to loss of lubricant
    c. Failure of knife
        (1) Fracture of knife section
        (2) Excessive wear of knife section

There could clearly be hundreds more component failure modes.

We may be able to predict or measure individual failure mode probabilities, and we thus need theories to predict combined dependabilities. Many of the component failure modes are life related, and can be subsumed under item 1e in our example.

Let us first look at failure modes of the whole device; and this could be generalized to systems. Each failure mode will have an individual dependability, D_i. Failure of the device will not occur if failure mode 1 does not occur *and* failure mode 2 does not occur *and* $\cdots$ *and* failure mode n does not occur. If the failure modes are stochastically independent, the product rule for combined events will apply.

$$D_c = D_1 D_2 \cdots D_n \qquad (9.3.1)$$

Two common situations can occur that will cause the dependabilities to be dependent.

1. The y's are functions of the same, or some of the same, functionally independent design variables or parameters.
2. The specifications are not independent. An example of this is the case where there are two different structural modes of failure caused by the same load system. Also, it would be a common situation for one user *generally* to require a higher level of performance than another user, and not just in one failure mode.

The foregoing type of combined failure modes, in which failure of a device will occur if any single failure mode occurs, is called *series combination of failure modes*. It is also possible to have combinations in which more than one failure mode of a particular type must occur before failure of the device occurs. Such a system is called a *redundant combination of failure modes* and is discussed in Sec. 9.5.

There are various ways in which nonindependent series combinations can be solved [when (9.3.1) does not apply]. One common type of solution occurs in *structural reliability*, discussed in Chapter 12. The most general and powerful solution technique is Monte Carlo simulation.

EXAMPLE 9.2 Laminated Wooden Beam

In Example 8.1 we considered the strength of a laminated wooden beam. We showed that the first order statistic density function for bending moment strength of an element represented the weakest element strength, given by

$$f_w(w) = 6f_x(w)[1 - F_x(w)]^5$$

where w is the failure bending moment of the weakest element, and $f_x(x)$ and $F_x(x)$ are probability functions for the parent distributions.

In Example 8.2 we applied a random sequence of 25 loads to the beam; the critical load corresponded to the 25th load, which corresponded to the 25th order statistic with density function

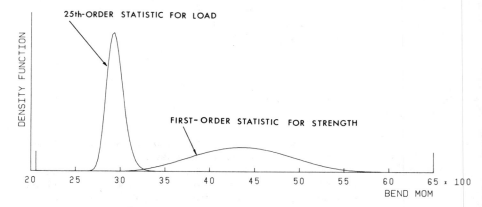

FIG. 9.10 Interacting random variables for strength of a laminated beam.

$$f_F(F) = 25[F_y(F)]^{24} f_y(F)$$

where $F_y(y)$ and $f_y(y)$ are the parent probability functions for the load distribution (expressed as a bending moment).

It is now clear that the criterion of failure is that $w > F$, illustrated in Fig. 9.10. The strength dependability comes from applying (9.2.3).

$$D = \int_{-\infty}^{\infty} f_F(F) \int_{F}^{\infty} f_w(w) \, dw \, dF \qquad (9.3.2)$$

The following computer program yields the result.

```
      DIMENSION BMF(101),ARGF(101),BMW(51),ARGW(51)
C.... GENERATE SET OF BENDING MOMENT LOAD VALUES
      DO 1 I=1,101
1     BMF(I)=2060.+FLOAT(I-1)*(6500.-2060.)/100.
C.... INTEGRATE
C
      DO 4 I=1,101
      ARGF(I)=FBMF(BMF(I))
      IF(ARGF(I).LT.1.E-6)GO TO 4
      RANGE=6500.-BMF(I)
      DO 3 J=1,51
      BMW(J)=BMF(I)+FLOAT(J-1)*RANGE/50.
3     ARGW(J)=FBMW(BMW(J))
      AREA=FSIMP(ARGW,RANGE,51)
      ARGF(I)=ARGF(I)*AREA
4     CONTINUE
      DEPEND=FSIMP(ARGF,4040.,101)
      WRITE(6,5)DEPEND
5     FORMAT(26H STRENGTH DEPENDABILITY IS,F8.5)
      STOP
      END
      FUNCTION FBMF(F)
C.... 25 TH ORDER STATISTIC DENSITY FUNCTION FOR LOAD
      DIMENSION BM(200),DENS(200),FCUM(200)
      DATA NCALL/0/
      IF(NCALL.EQ.1)GO TO 10
C.... OBTAIN DIST FUNCTION VALUES
      DO 1 I=1,200
      BM(I)=2060.+FLOAT(I-1)*(6500.-2060.)/199.
1     DENS(I)=FWEIB(BM(I),2060.,2630.,2.96)
      CALL FTOCUM(DENS,FCUM,4440.,200)
      NCALL=1
10    FBMF=25.*FWEIB(F,2060.,2630.,2.96)*FTABLE(BM,FCUM,F,200)**24
      RETURN
      END
      FUNCTION FBMW(W)
C.... FIRST ORDER STATISTIC DENSITY FUNCTION FOR STRENGTH
      DIMENSION BM(200),DENS(200),FCUM(200)
      DATA NCALL/0/
      IF(NCALL.EQ.1)GO TO 10
C.... OBTAIN DIST FUNCTION VALUES
      DO 1 I=1,200
      BM(I)=2060.+FLOAT(I-1)*(6500.-2060.)/199.
1     DENS(I)=FBETA(3000.,6500.,3.0,1.5,BM(I))
```

```
      CALL FTOCUM(DENS,FCUM,4440.,200)
      NCALL=1
10    FBMW=6.*FBETA(3000.,6500.,3.0,1.5,W)*(1.-FTABLE(BM,FCUM,W,200))**5
      RETURN
      END
```

STRENGTH DEPENDABILITY IS .90652

Example 9.2 leads us to an important concept in failure mode analysis. Returning to the general notation of Fig. 9.6, we are often concerned with a situation where there is only a small interaction between y and S, as indicated in Fig. 9.11. We may have very low probabilities of y < S, perhaps of the order of 10^{-5} or less. In this case accuracy of the tails is very important, precisely where accuracy is the lowest. However, we can focus on the tails, and improve the accuracy, if we use the largest extreme value density function for S, and the smallest extreme value function for y, rather than the parent distribution. This gives us the following probability.

P(no failure in a specified number of trials or a specified interval)

= P(smallest value of the design characteristic exceeds the largest value of the specification, in a specified number of trials or a specified interval)

However, this approach also may not be adequate when very high probabilities of no failure are being estimated. In this event it is necessary to work directly with sample data on actual tests or field trials, and observe the percent of failures. This experience can be projected to some extent on new similar design.

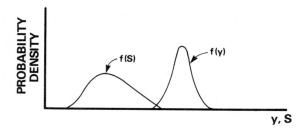

FIG. 9.11 High-dependability interaction.

9.4 SERIES COMBINATION OF FAILURE MODES

Failure modes are said to be in series combination if failure in any one mode will cause failure of the device. We have seen in Sec. 9.3 that the dependability of a device, having a series combination of failure modes, is given by the simple product of the dependabilities of the individual failure modes, if the events of failure in the individual modes are stochastically independent. It was also noted that techniques are available if these events are not independent.

Our concern here is primarily with time-dependent failure modes. These are commonly contributed by components of the device; and each component may have more than one. The probability of no failure for a specified life is commonly called the *reliability* of the device or component.

Consider, for example, the reliability of a large aircraft. It will, of course, have thousands of possible failure modes, but we shall only look at three, all at the subassembly level. Each subassembly reliability would be derived from failure modes of many components. The three modes considered here are fatigue failure of the wing structure due to gust loads, fatigue failure of the fuselage structure due to cyclical pressurization, and yielding of the landing gear due to landing impact. We presume that we know the reliability of each separate failure mode.

The standard argument in all texts on reliability theory for developing an expression for the combined reliability of failure modes in series goes as follows:

P(no failure of the aircraft) = P(no failure due to wing fatigue
and no failure due to fuselage fatigue
and no failure due to landing impact)

It is then argued that these three events are stochastically independent, because whatever causes that tend toward failure in one mode have no effect on, or are unrelated to, causes tending toward failure in the other modes. Thus the product rule applies, and in general

$$R_d = R_1 R_2 \cdots R_n \qquad (9.4.1)$$

where

R_d = device reliability

R_i = ith failure mode reliability

n = number of failure modes in series

It should always be possible to take the complementary viewpoint and get a consistent result. It turns out to be quite a lot more difficult in a series arrangement. To illustrate, we shall limit the system to two failure modes and define the following events.

A_i = failure in ith mode

$\overline{A}_i$ = no failure in ith mode

The probability of failure is

$$P(A_1 + A_2) = P(A_1 + \overline{A}_1 A_2) \tag{9.4.2}$$

The two events on the right side are mutually exclusive, giving

$$P(A_1 + A_2) = P(A_1) + P(A_1 A_2) \tag{9.4.3}$$

Equation (9.4.2) is not at all obvious but can be confirmed by careful study of a Venn diagram.

Consistency between complementary approaches can always be verified by using basic probabilities, discussed in Sec. 2.4. For the two-failure-mode case the known quantities are

$$P(\overline{A}_1\overline{A}_2) = R_1 R_2$$

$$P(A_1) = R_1$$

$$P(A_2) = R_2 \tag{9.4.4}$$

$$P(A_1 A_2) + P(A_1 \overline{A}_2) + P(\overline{A}_1 A_2) + P(\overline{A}_1 \overline{A}_2) = 1$$

The solution in terms of basic probabilities is

$$P(\overline{A}_1 A_2) = R_1 - R_1 R_2$$

$$P(A_1 \overline{A}_2) = R_2 - R_1 R_2$$

$$P(\overline{A}_1 \overline{A}_2) = R_1 R_2 \tag{9.4.5}$$

$$P(A_1 A_2) = 1 - R_1 - R_2 + R_1 R_2$$

Putting (9.4.3) in terms of basic probabilities gives

$$P(A_1 + \overline{A}_1 A_2) = 1 - R_1 R_2 \tag{9.4.6}$$

Thus we have confirmed that (9.4.3) is the complement of (9.4.1).

9.5 PARALLEL COMBINATIONS OF FAILURE MODES

9.5.1 Simple Parallel Redundancy

A combination of failure modes is said to be in simple parallel combination if all failure modes must occur before the device fails. An additional requirement is that all components with which the failure modes are associated are active, or in operation; and if one fails the load on the other is not increased. If the latter condition is not satisfied, then the reliability of the remaining components would change due to the increased load.

We can develop the formula by the following probabilistic arguments:

P(failure of the device) = P(failure in mode 1
              and failure in mode 2
              and $\cdots$ *and* failure in mode n)

If these events are stochastically independent, the product rule again applies and

$$R_d = 1 - (1 - R_1)(1 - R_2) \cdots (1 - R_n) \qquad (9.5.1)$$

It was more convenient in this case to develop the expression by beginning with the event of failure.

Examples of this kind of failure mode combinations are rather rare. Figure 9.12 shows two check valves, first *physically* in parallel and second in series. If the failure mode of each is failure to open, the reliability combination corresponds to the physical combination.

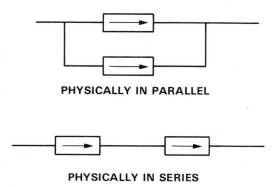

PHYSICALLY IN PARALLEL

PHYSICALLY IN SERIES

FIG. 9.12 Check valves in a fluid circuit.

However, if the failure mode is failure to close when the flow reverses, the pair that are physically in parallel are in series combination for reliability, and the pair physically in series are in parallel combination for reliability.

9.5.2 Standby Parallel Redundancy

Standby parallel redundancy is much more common. In this situation a component is held inactive until a parallel component fails. The standby unit is then "switched" into service. Switching may literally occur in electrical devices; and it may be manual or automatic. In mechanical devices the switching may be completely manual, such as changing a flat tire on an automobile. Or it may be automatic, as in the example of a standby diesel-powered electrical generator starting up in a hospital if the utility company power fails.

Although the inactive role of the standby unit reduces its risk of deterioration or risk of failure until the first unit fails, there is the added risk that the switching device will not work when called on.

Figure 9.13 is a logic diagram illustrating this type of combined failure modes. The reliability of the device now depends on the life density functions of the two components $f_1(t_1)$ and $f_2(t_2)$ and the probability that the switch will work, p_s. The event logic is as follows, for a combined life of t.

P(no failure of the device up to time t)
    = P[(no failure of component 1 up to time t)
        or (component 1 fails at time $\tau < t$ *and* the switch works
        and component 2 survives from time τ to t)]

The two "or" events are mutually exclusive. If we recall that the reliability function is simply the complement of the cumulative distribution function, it is clear that the probability of the first event is simply $R_1(t)$. We can now write

$$R_d(t) = R_1(t)$$

      + P(component 1 fails at time $\tau < t$ *and* the switch works *and* component 2 survives from time τ to t)

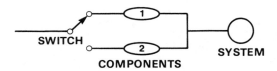

FIG. 9.13 Logic diagram to illustrate standby parallel redundancy.

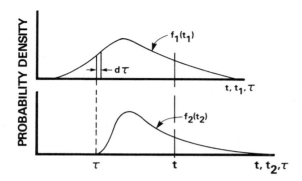

FIG. 9.14   Density functions for standby redundancy.

We shall assume that the three events in the second term are independent. The second probability will be easier to evaluate if reference is made to Fig. 9.14. The concept of the infinitesimal event is used to determine the second term. The result is

$$R_d(t) = R_1(t) + p_s \int_0^t f_1(\tau) \int_{t-\tau}^{\infty} f_2(t_2) \, dt_2 \, d\tau \qquad (9.5.2)$$

This could also be written as

$$R_d(t) = R_1(t) + p_s \int_0^t f_1(\tau) R_2(t - \tau) \, d\tau \qquad (9.5.3)$$

It is possible of course to have more than one standby unit. The relationship would be more complex, but the method would be the same.

9.5.3 Dependent Parallel Combination

This concept is based on components rather than on purely failure modes. A total of n components make up part of a device, and if one or more components fail the device will not necessarily fail. The remaining components must support the full load, using the term "load" in a general sense. Familiar examples are the multiengine aircraft and a steel cable. This is a complex reliability problem because the density function for life of surviving components changes each time a component fails and the load per component increases. So it is necessary to know how the density function changes with load. Even knowing this the calculation of the combined reliability function is quite difficult. Perhaps the most practical method is Monte Carlo simulation, done for a sequence of discrete times.

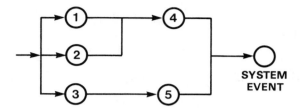

FIG. 9.15 Reliability block diagram.

9.6 COMPLEX SYSTEMS

9.6.1 Introduction

In systems having complex combinations of devices or components and failure modes it becomes difficult, if not impossible, to obtain a system reliability *function* by explicit analysis or formulas. It becomes necessary to do a sequence of calculations for successive times to failure, determining the probability of system survival for each specified life. The system reliability thus becomes defined numerically.

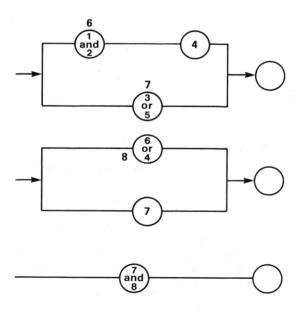

FIG. 9.16 Successive reduction of a failure block diagram.

Several methods are outlined below for systematically solving the system. However, the most straightforward way is by systematic reduction of a block diagram of the system. It is illustrated by the example shown in Fig. 9.15, which is a logic block diagram. The event of failure of a circled failure mode cuts the path to the system event. If no uncut path is left, the system has failed. The first stage could be to combine parallel events 1 and 2, and series events 3 and 5, shown in Fig. 9.16, which shows the subsequent reduction phases. The combined probabilities of the new events would be calculated for each phase. Even this approach is not, however, always applicable; see Problem 9.10.

9.6.2 Use of Basic Probabilities

This method will also be illustrated using the example shown in Fig. 9.15. We shall use the notation in which x_1 is the event of no failure of component 1, and so on. All basic probabilities are listed, and each one examined to see if it represents failure or no failure of the system. Failures are struck off.

$$
\begin{array}{lll}
x_1 x_2 x_3 x_4 x_5 & x_1 x_2 \bar{x}_3 x_4 \bar{x}_5 & x_1 \bar{x}_2 x_3 x_4 x_5 \\[4pt]
x_1 x_2 x_3 x_4 \bar{x}_5 & \cancel{x_1 x_2 \bar{x}_3 \bar{x}_4 x_5} & \cancel{x_1 \bar{x}_2 x_3 x_4 x_5} \\[4pt]
x_1 x_2 x_3 x_4 x_5 & \cancel{x_1 x_2 \bar{x}_3 \bar{x}_4 x_5} & x_1 \bar{x}_2 x_3 x_4 x_5 \\[4pt]
x_1 x_2 x_3 \bar{x}_4 x_5 & x_1 \bar{x}_2 x_3 x_4 x_5 & x_1 \bar{x}_2 x_3 x_4 \bar{x}_5 \\[4pt]
x_1 x_2 \bar{x}_3 x_4 x_5 & x_1 \bar{x}_2 x_3 x_4 x_5 & \cancel{x_1 \bar{x}_2 \bar{x}_3 \bar{x}_4 x_5} \\[4pt]
x_1 \bar{x}_2 x_3 \bar{x}_4 x_5 & x_1 \bar{x}_2 x_3 x_4 x_5 & \cancel{x_1 \bar{x}_2 x_3 \bar{x}_4 \bar{x}_5} \\[4pt]
\bar{x}_1 x_2 x_3 x_4 x_5 & x_1 x_2 \bar{x}_3 x_4 x_5 & x_1 x_2 x_3 x_4 x_5 \\[4pt]
\bar{x}_1 x_2 x_3 x_4 \bar{x}_5 & \cancel{x_1 x_2 \bar{x}_3 x_4 \bar{x}_5} & \cancel{x_1 \bar{x}_2 x_3 \bar{x}_4 \bar{x}_5} \\[4pt]
\bar{x}_1 x_2 x_3 \bar{x}_4 x_5 & \cancel{x_1 x_2 x_3 \bar{x}_4 \bar{x}_5} & \cancel{x_1 \bar{x}_2 x_3 \bar{x}_4 \bar{x}_5} \\[4pt]
\cancel{\bar{x}_1 x_2 \bar{x}_3 x_4 \bar{x}_5} & \bar{x}_1 x_2 x_3 x_4 x_5 & \cancel{\bar{x}_1 x_2 x_3 \bar{x}_4 \bar{x}_5} \\[4pt]
\cancel{x_1 x_2 x_3 \bar{x}_4 \bar{x}_5} & & \\[4pt]
\cancel{x_1 x_2 x_3 \bar{x}_4 \bar{x}_5} & &
\end{array}
$$

Because all basic events are mutually exclusive, the probability of no failure of the system is

$$R_d = P(x_1 x_2 x_3 x_4 x_5) + P(x_1 x_2 x_3 x_4 \bar{x}_5)$$

$$+ P(x_1 x_2 x_3 \bar{x}_4 x_5) + P(x_1 x_2 \bar{x}_3 x_4 x_5)$$

$$+ P(x_1 x_2 \bar{x}_3 x_4 \bar{x}_5) + P(x_1 \bar{x}_2 x_3 x_4 x_5)$$

$$+ P(x_1 \bar{x}_2 x_3 x_4 \bar{x}_5) + P(x_1 \bar{x}_2 x_3 \bar{x}_4 x_5)$$

$$+ P(x_1 \bar{x}_2 x_3 x_4 x_5) + P(x_1 \bar{x}_2 x_3 x_4 \bar{x}_5)$$

$$+ P(\bar{x}_1 x_2 x_3 x_4 x_5) + P(\bar{x}_1 x_2 x_3 x_4 \bar{x}_5)$$

$$+ P(\bar{x}_1 x_2 x_3 \bar{x}_4 x_5) + P(\bar{x}_1 x_2 \bar{x}_3 x_4 x_5)$$

$$+ P(\bar{x}_1 x_2 \bar{x}_3 x_4 \bar{x}_5) + P(\bar{x}_1 \bar{x}_2 x_3 x_4 x_5)$$

$$+ P(\bar{x}_1 \bar{x}_2 x_3 \bar{x}_4 x_5) \tag{9.6.1}$$

If the component reliabilities are independent, this reduces to

$$R_d = R_1 R_2 R_3 R_4 R_5 + R_1 R_2 R_3 R_4 (1 - R_5) + \cdots \tag{9.6.2}$$

This is probably the surest way of solving problems, but very lengthy.

9.6.3 Path Tracing

In this technique (Shooman, 1968) we determine all nonfailure paths by a systematic procedure. All elements in the block diagram are removed except one, and then all except two, and so on. In each configuration it is determined if there is a system no failure. Again using Fig. 9.15 to illustrate, we develop the following nonfailure paths:

One element: no successes
Two elements: $x_1 x_4$, $x_2 x_4$, $x_3 x_5$

We conclude from this that

$$R_d = P(x_1 x_4 + x_2 x_4 + x_3 x_5) \tag{9.6.3}$$

We need not examine larger groups of elements than two since (9.6.3) includes all possible modes of success. It may be expanded as follows:

$$R_d = P(x_1 x_4) + P(x_2 x_4) + P(x_3 x_5) - P(x_1 x_2 x_4)$$

$$- P(x_1 x_3 x_4 x_5) - P(x_2 x_3 x_4 x_5) + P(x_1 x_2 x_3 x_4 x_5)$$

$$= R_1 R_4 + R_2 R_4 + R_3 R_5 - R_1 R_2 R_4$$

$$- R_1 R_3 R_4 R_5 - R_2 R_3 R_4 R_5 + R_1 R_2 R_3 R_4 R_5 \qquad (9.6.4)$$

This method would appear to be shorter than the previous one using basic probabilities.

9.6.4 Decomposition

In this method (Shooman, 1968) a key element is selected which would seem likely to have a major role in successful basic events. This key element is designated x_k. The system reliability may be written

$$R_d = P(x_k) P(\text{system nonfailure} | x_k)$$

$$+ P(\bar{x}_k) P(\text{system nonfailure} | \bar{x}_k) \qquad (9.6.5)$$

This decomposition will commonly simplify the problem, particularly if x_k is well chosen. Returning again to Fig. 9.15, we select x_4 as the key event. Applying (9.6.5) gives

$$R_d = P(x_4) P(\text{system nonfailure} | x_4)$$

$$+ P(\bar{x}_4) P(\text{system nonfailure} | \bar{x}_4) \qquad (9.6.6)$$

Examination of Fig. 9.15 discloses the following:

$$R_d = P(x_4) P(x_1 + x_2 + x_3 x_5) + P(\bar{x}_4) P(x_3 x_5)$$

$$= R_1 R_4 + R_2 R_4 + R_3 R_5 - R_1 R_2 R_4 \qquad (9.6.7)$$

$$- R_1 R_3 R_4 R_5 - R_2 R_3 R_4 R_5 + R_1 R_2 R_3 R_4 R_5$$

In more complex systems, successive decompositions can be used.

9.6.5 Fault Tree Analysis

Fault tree analysis is a widely used technique for systematically defining the failure logic of a system (Barlow and Lambert, 1975; Dhillon and Singh, 1981; Henley and Kumamoto, 1981). It provides a kind of bookkeeping of failure events that helps prevent overlooking possible failure mode combinations that cause system failure. It also provides a convenient basis for estimation of failure probabilities.

Symbol	Name	Logic
	and	Output event occurs if all input events occur.
	or	Output event occurs if one or more input events occurs.
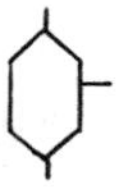	inhibit	Output event occurs when input occurs, if conditional event has occurred.
	priority and	Output event occurs if all input events occur in sequence from left to right.
	exclusive or	Output event occurs if one and only one input event occurs.
	M out of N	Output event occurs if M out of N input events occur.
N inputs		

FIG. 9.17 Boolean logic symbols.

     The basic tool of fault tree analysis is Boolean logic with its logic symbols (sometimes called gates). These symbols are itemized in Fig. 9.17. The logic symbols connect events, represented by the symbols in Fig. 9.18. In order to solve for system failure probability, each *event* must have a known probability associated with it.
     A full treatment of fault tree analysis is beyond the scope of this text. An example is given below, adapted from Amstader (1971).

     EXAMPLE 9.3 Heat Transfer Loop

     Figure 9.19 illustrates schematically the primary heat transfer loop in a design for a nuclear reactor energy conversion system. The heat transfer fluid, heated in the reactor, is a sodium-potassium eutectic mixture, circulated by a centrifugal pump-motor

Symbol	Meaning
 $\bigcirc$	Basic event
 $\diamondsuit$	Undeveloped event
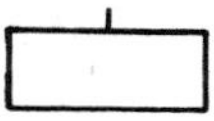 $\square$	Fault event
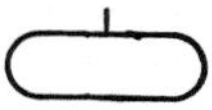 $\bigcirc\!\!\!\!\square$	Conditional event
 $\triangle$	Switch event
$\triangle$	Transfer symbol

FIG. 9.18 Event symbols.

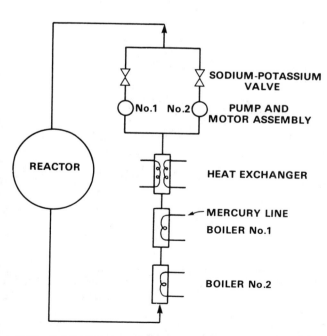

FIG. 9.19 Primary heat transfer loop in a nuclear reactor energy conversion system.

TABLE 9.1 Failure Mode Analysis of the Reactor Primary Loop

Failure mode	Effect
1 Pump-motor fails to operate	Switch to no. 2
2 Pump leaks externally	System fails
3 Heat exchanger fails internally	System fails: 0.20 prob.; no effect: 0.80 prob.
4 Heat exchanger fails externally	System fails
5 Boiler leaks externally	System fails
6 Boiler leaks internally	System fails
7 Boiler plugging on sodium-potassium side	System fails
8 Boiler plugging on mercury side	Switch to second turbine
9 Boiler performance degradation	Switch to second turbine
10 Erosion-corrosion of boiler on mercury side	Switch to second turbine
11 Structural failure of boiler	System fails
12 Valve leaks externally	System fails
13 Valve fails to close when pump-motor assembly fails	System fails
14 Valve fails to open when second pump-motor assembly required	System fails
15 Valve closes inadvertently	Switch to second pump-motor assembly
16 Valve at second pump-motor assembly opens inadvertently	Switch to second pump-motor assembly

assembly. One of these is on standby redundancy. Mercury is vaporized in the mercury boiler and operates a turbine. The second boiler is on standby redundancy on the mercury side, but in series on the sodium-potassium side. The heat exchanger is used during system startup.

A failure mode analysis is shown in Table 9.1. The corresponding logic diagram is shown in Fig. 9.20. The schematic symbols are used for clarity. This figure illustrates how the

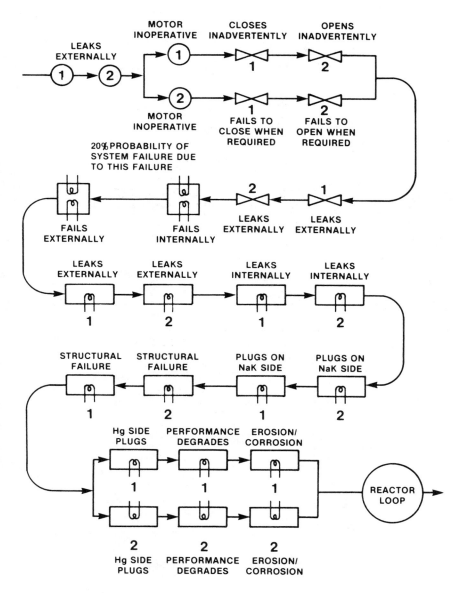

FIG. 9.20 Logic diagram for a heat transfer loop.

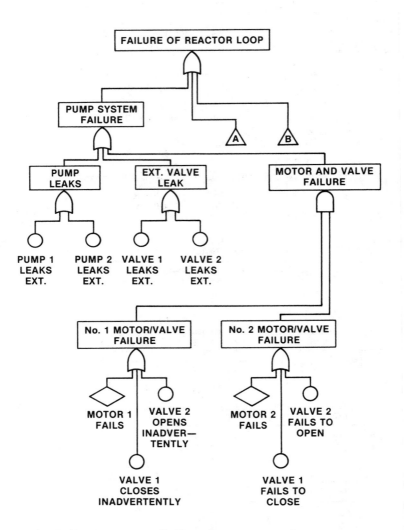

FIG. 9.21 Fault tree diagram for a reactor loop.

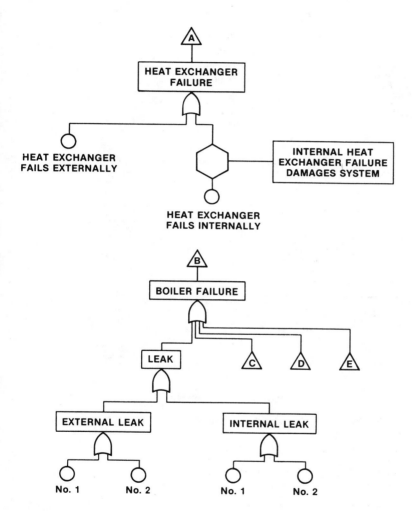

FIG. 9.21 (Continued)

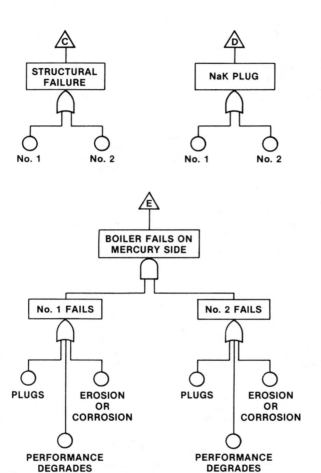

FIG. 9.21 (Continued)

same component may occur several times in a reliability logic diagram because of different modes of failure. For example, boiler 1 is present seven times. This example is quite simple from the point of view of combined reliabilities, being a straightforward combination of series and parallel componentns in effect, except for the minor complication associated with failure of the heat exchanger internally. The corresponding faul tree diagram is shown in Fig. 9.21.

Example 9.3 does not use all of the logic or event symbols. The switch event simply means that the event is preset to occur one way or the other, having in effect a probability of 1.

Although the fault tree is simply an alternative method of representing the failure logic of a system, it does conveniently lead to systematic techniques for calculating the probability of complex combined events. Both the construction and quantification of fault trees have been computerized (Barlow et al., 1975, p. 131; Henley and Kumamoto, 1981).

9.6.6 Monte Carlo Simulation of Failure Events

The Monte Carlo technique is a powerful and straightforward method for determining the combined probability of failure of a system having a large and complex combination of failure modes. The system must first be defined by either a logic diagram or a fault tree. The following algorithm is based on a fault tree.

1. The steps from items 1 to 8 are repeated for each trial.
2. Begin with the first basic event, determining its occurrence by comparing a uniformly distribution random number between 0 and 1 (FRAND) with the event probability.
3. If failure has been simulated, proceed up the fault tree to the first gate.
4. If the gate is an OR pass through it and continue upward until an AND gate is encountered, or the top event is reached. In the latter case, go to 8.
5. Each time an OR gate is passed, flag all other basic events going into the same gate. They can be ignored. Each time an AND gate is reached, flag that entrance as a failure.
6. When an AND gate is reached and flagged, drop back and continue scanning basic events that have not been discarded, following each branch up the tree as before.
7. The failure flags at the entrance to AND gates are monitored, and if all inputs are flagged this gate is passed.
8. The process stops when the top event is reached (a failure), or when all basic events have been scanned or discarded (a no failure).

9. Estimate the probability of system failure by calculating the ratio
    of simulated top event failures to the number of trials.

    A very large number of trials would be necessary to get an accu-
racy corresponding to the number of significant figures commonly
occurring in high reliability systems—possibly many millions of trials.
Reduced variance methods have been proposed and are described by
Henley and Kumamoto (1981) and Burt and Garmon (1970).

9.7 PREDICTING FAILURE MODES

One of the most difficult problems in design is predicting all possible
failure modes. It is probably safe to say that many more failures
occur due to unanticipated failure modes rather than due to poorly
modeled but anticipated modes.
    Collins et al. (1975) have proposed a promising technique for
assisting designers in the anticipation of failure modes. They define
a three-dimensional failure experience matrix, shown in Fig. 9.22.
A typical matrix element is shown, identified as follows:

Elemental mechanical function: force transmitting
Failure mode: fatigue
Corrective action: change of material

    The authors have developed extensive categories for each matrix
component. Using these it is possible to set up a data bank of observed failures for use in design, using the computer. When working

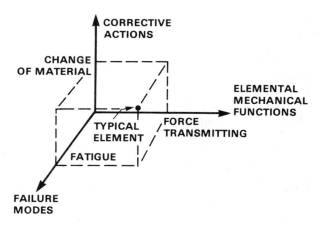

FIG. 9.22 Failure experience matrix, showing a typical matrix element. [From Collins et al. (1975).]

on a new design the designer could call up from the computer all failure modes and corrective actions associated with a given elemental
mechanical function related to a component being designed. This
would provide convenient historical evidence of failure modes and how
they have been successfully prevented. Details are provided in
Collins et al. (1975).

PROBLEMS

Problem 9.1 Suppose that we wish to predict the reliability of a device which has operated in two different environments. It operates
in environment 1 for the time period 0 to τ, and in environment 2 for
the period τ to t. The reliability function for environment 1 is $R_1(\tau)$,
and for environment 2 the reliability function is $R_2(t, \tau)$. Assume
that both are known and note that R_2 is a function of both τ and t.
a. Determine the expression for the reliability if τ is determinate.
b. Determine an expression for the reliability if τ is random with
    known density function $f(\tau)$.

Problem 9.2 A hydraulic system has two manually operated shutoff
valves in parallel as shown in Fig. 9.23. The system is put in operation daily and both are opened to ensure flow through the system.
When the system is shut down at the end of the day, both must be
closed or the system will fail. Let $R_c(t)$ be the probability that a
valve will not fail to close at time t, and $R_0(t)$ be the corresponding
probability that a valve will not fail to open at time t. What is the
reliability of this part of the system with regard to opening? With
regard to closing? What is the combined reliability for correct functional operation? How could it be improved using the same valves?
Suppose that the reliability for a human operator in actuating a valve,
either to open or close, is R_h. How would this affect the calculation?
Use the following numerical values:

$$R_c = 0.990, \quad R_0 = 0.995, \quad R_h = 0.999$$

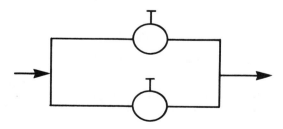

FIG. 9.23

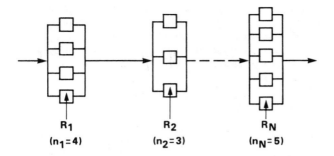

FIG. 9.24

Problem 9.3 a. Develop a general formula for a system of components in combined series and parallel combination, as shown in Fig. 9.24. Use the following notation:

R_d = reliability of assembly

R_i = reliability of the ith component in series

N = number of components in series

n_i = number of ith components in parallel

b. A device consists of five components in series with the following reliabilities:

R_1 = 0.9876

R_2 = 0.8725

R_3 = 0.9416

R_4 = 0.9370

R_5 = 0.9983

How could you use redundancy to achieve a device reliability of 0.9000?

Problem 9.4 A schematic reliability diagram is shown in Fig. 9.25 for a device having four types of components, all in series.

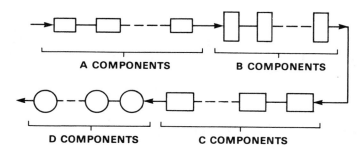

FIG. 9.25

Type	Number	Reliability	Cost/unit
A	16	0.99953	$ 1
B	8	0.99917	1
C	27	0.99992	8
D	32	0.99998	10

a. What is the reliability of the device?
b. If we wish to upgrade this to a device reliability of 0.99000, suggest how it might be done and give the additional cost.

Problem 9.5 Determine the system reliability for the system described by the reliability block diagram in Fig. 9.26.

$$R_1 = 0.972 \qquad R_3 = 0.843 \qquad R_5 = 0.937$$

$$R_2 = 0.916 \qquad R_4 = 0.996 \qquad R_6 = 0.905$$

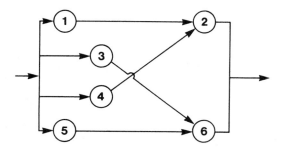

FIG. 9.26

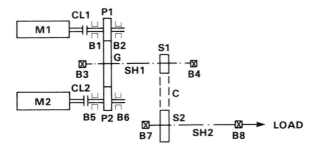

FIG. 9.27 Schematic diagram of a mechanical drive.

Problem 9.6 Figure 9.27 shows schematically a mechanical drive system in which the abreviated designations for components have the following meanings

M1,M2:	drive motors 1 and 2
B1,B2,...,B8:	bearings 1 to 8
P1,P2:	pinions 1 and 2
G:	gear
SH1,SH2:	shafts 1 and 2
S1,S2:	chain sprockets 1 and 2
C:	chain
CL1,CL2:	clutches 1 and 2

M2 is a standby drive which is clutched in if M1 fails. The probability of the operator failing to do this is p_h. For a given life, the probability of failure of all components is known. However, a pinion may fail in two ways—it may or may not jam if it fails and the probability either way is known. If it jams, the system fails; if it does not jam, the standby drive may be clutched in. Also, a clutch may fail in several ways as follows:

1. It fails while operating and remains in the engaged position.
2. It fails while operating and remains in the disengaged position.
3. It will not engage when the second motor is required from standby.
4. It will not disengage when the first motor failed and must be isolated from the system.

The clutches are only used to actuate standby operation and are not otherwise actuated during operation. Draw the logic diagram and the fault tree for the system and indicate how you would obtain the system reliability.

Problem 9.7 Use Monte Carlo simulation to determination the reliability of the nuclear reactor energy conversion system given in Example 9.3. Determine the reliability for a life of 1000 hr. Assume that all failure modes follow Weibull curve reliabilities given by

$$R(t) = e^{-(t/\eta)^\beta}$$

The following table gives values for β and η.

Failure mode	β	η
1	1.8	5,740
2	1.7	280,000
3	1.9	1,127,000
4	2.1	226,000
5	1.2	106,000
6	1.2	52,000
7	1.1	207,000
8	1.8	75,400
9	2.3	56,500
10	2.5	45,100
11	3.0	224,000
12	1.2	354,000
13	2.7	17,540
14	2.7	17,540
15	2.8	187,500
16	3.1	186,700

Where would you be inclined to attempt to improve the reliability with the given configuration, and how might the configuration be changed to improve reliability?

Problem 9.8 A device has four components and three out of the four must survive for the device to survive. The components have different reliabilities but they are independent. The life is specified, and the component reliability is independent of load.

a. Assume that only three out of the four are ever operating and if one fails the reserve one can be switched in. Sketch a diagram showing the necessary switching arrangment schematically and assume that the probability of any successful switch operation is known. How would you calculate the reliability of the device?

b. Assume now that component reliability does depend on load. How would you do part (a)?

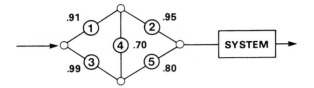

FIG. 9.28 Logic diagram for combined failure modes.

Problem 9.9 Determine the general expression for reliability of a device consisting of two identical units in standby combination. The density functions of the components are f(t) when in operation. They are subject to deterioration on standby with a density function of $f_s(t)$. The switching device is also subject to failure and deteriorates with time. Its probability of operating at any time t is p(t).

Problem 9.10 Find the probability of no failure of the system shown in the logic diagram of Fig. 9.28. Component dependabilities are shown. Can a fault tree be drawn for this system?

Problem 9.11 You are designing the shaft of the main rotor for an axial flow harvester combine. You wish to estimate the probability that the shaft will fail in service in any one season. The engineering modeling will be simplified so that the shaft is assumed to fail only in pure torsion, represented by the following:

$$P(\text{failure}) = P(S_y < \tau) = P(m < 0)$$

where

$$S_y = \text{yield stress, psi}$$

$$\tau = \text{applied stress, psi}$$

$$m = S_y - \tau = \text{margin of strength}$$

$$\tau = 16T / \pi d^3, \text{ psi}$$

$$T = \text{torque applied to shaft, in. lb}$$

$$d = \text{shaft diameter, in.}$$

It will also be simplifed so that only S_y and T are considered random variables, while d is treated as deterministic. Based on previous experience, it has been estimated that the yield stress can have its probability density function represented by a Weibull function.

$$F(S_y) = \frac{\beta}{\eta - \ell} \left(\frac{S_y - \ell}{\eta - \ell} \right)^{\beta-1} \exp\left[-\left(\frac{S_y - \ell}{\eta - \ell} \right)^{\beta} \right], \quad x \geq \ell$$

$$= 0, \quad x < \ell$$

The corresponding cumulative distribution function is

$$F(S_y) = 1 - \exp\left[-\left(\frac{S_y - \ell}{\eta - \ell} \right)^{\beta} \right]$$

The torque will be random due to random fluctuations in the density of the crop being harvested. This can vary from season to season, from one locale to another, and even quite markedly in one field. The accumulation of extensive field measurements have indicated that the maximum torque occurring in *any one season* is an *extreme value distribution*.

$$f(T) = \alpha \exp\left[-y - \exp(-y) \right]$$

$$F(T) = \exp\left[-\exp(-y) \right]$$

where α and u are parameters and y is a convenient intermediate variable related to T by

$$y = \alpha(T - u)$$

For this application the parameters have the following values, when stress is in units of 1000 psi and torque in units of 1000 in. lb:

$\ell = 48.7 \quad \alpha = 0.1202 \quad \beta = 2.72$

$\eta = 56.8 \quad u = 41.04$

The shaft diameter is 1.820 in.
For the assignment, do the following:

a. Plot the given density functions.
b. Run the problem for four values of NSAMP: 1000, 5000, 10,000, and 15,000.
c. Submit a report including your plots and a discussion of the effect of sample size and random number seed on the result. For the latter obtain results from several colleagues.

d. If 1000 machines of this design are built each year for 5 years, how many spare shafts for service repairs should be built each year for the next 10 years? Assume that a machine has a useful life of 8 years.

REFERENCES

Amstader, B. L. (1971), *Reliability Mathematics: Fundamentals, Practices, Procedures*, McGraw-Hill, New York.

Barlow, R. E., and H. E. Lambert (1975), Introduction to Fault Tree Analysis, in *Reliability and Fault Tree Analysis*, R. E. Barlow, J. B. Fussel, and N. D. Singpurwalla, eds., Society for Industrial and Applied Mathematics, Philadelphia.

Barlow, R. E., J. B. Fussell, and N. D. Singpurwalla, eds. (1975), *Reliability and Fault Tree Analysis*, Society for Industrial and Applied Mathematics, Philadelphia.

Burt, J. M., and M. B. Garman (1970), Monte Carlo Techniques for Stochastic Network Analysis, *Proc. of Fourth Conf. on Applications of Simulations, sponsored by ACM, AIIE, IEEE, SHARE, SEI, TIMS.*

Collins, J. A., B. T. Hagan, and H. M. Bratt (1975), The Failure-Experience Matrix—A Useful Design Tool, *ASME Paper No. 75-DET-122.*

Dhillon, B. S., and C. Singh (1981), *Engineering Reliability: New Techniques and Applications*, Wiley, New York.

Henley, E. J., and H. Kumamoto (1981), *Reliability Engineering and Risk Assessment*, Prentice-Hall, Englewood Cliffs, N.J.

Shooman, M. L. (1968), *Probabilistic Reliability: An Engineering Approach*, McGraw-Hill, New York.

Siddall, J. N. (1982), *Optimal Engineering Design: Principles and Applications*, Marcel Dekker, New York.

10

The Design Option Problem

The hierarchy of design variables • *The decision process in the design option problem* • *The inherent role of uncertainty* • *The key role of value theory*

10.1 THE DETERMINISTIC DESIGN OPTION PROBLEM

We are concerned in the design option problem with choosing the best of two or more competing designs, designs that are complete enough at least to permit good estimates for the design characteristics. The criterion for choice is that design having the highest total value.

We are concerned then with estimating the total value for a given design, using value concepts [see Siddall (1982)]. If we recall the hierarchy of design variables of Fig. 10.1 it is clear that the procedure is rather straightforward. Consider the specific example of the power lawn mower. We first consider the important design characteristics, and the values generated. Let us choose a few for illustration without necessarily arguing that it is an exhaustive list. These are shown in Table 10.1.

The designer relates these by value curves, subjectively drawn. The value scales are arbitrary and their relative levels of importance are codified in the value profile, shown in Fig. 10.2. Typical value curves are shown in Fig. 10.3.

The engineer now uses modeling or functional relationships to determine the specific capacity, noise level, and pushing force from the known independent variables of the design: dimensions, motor speed, friction coefficients, and the like. Hypothetical design points

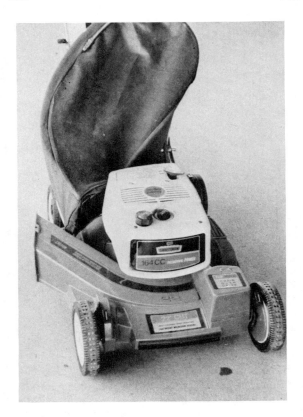

FIG. 10.1 Power lawn mower.

TABLE 10.1 Design Characteristics and Values
for a Lawn Mower

Design characteristics	Values
Capacity (ft^2/min)	Convenience
Noise level (dB)	Peace and quiet
Pushing force (lb)	Comfort

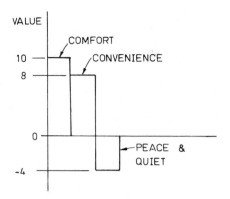

FIG. 10.2 Value profile for power lawn mower.

for these characteristics are shown on the value curves, with corresponding value levels, U_1, U_2, and U_3. The total value for this design is then

$$U = U_1 + U_2 + U_3 \qquad (10.1.1)$$

Note that there will tend to be a trade-off between capacity and pushing force, but this trade-off must be ignored when subjectively setting up the value curves. Trade-off relationships are important in design decision making, and are discussed in Siddall (1982).

There are also other important design characteristics which are assumed to be associated with *feasibility* rather than value. They must satisfy specifications set for the device. We have assumed, for example, that there is a maximum cost and a maximum weight specification that the design satisfies; otherwise, it would not be eligible for the decision process. It is not always easy to distinguish between specification-controlled and value-controlled design characteristics, and this point is discussed further in Siddall (1982).

There are additional important values which we have not considered: order, status, ritual, love of technological devices, and accumulation of material objects. These values are difficult to associate with specific design characteristics, and are generated by the external appearance or the overall design concept. Fortunately, not all of them are very sensitive to differences in configuration. However, we may be left with some that are, and the designer must intuitively integrate the contribution of these values with those given an analytical treatment. For example, one configuration of lawn mower being considered may have a radically new configuration—say a high-speed rotating wire which does the cutting. This could generate a high

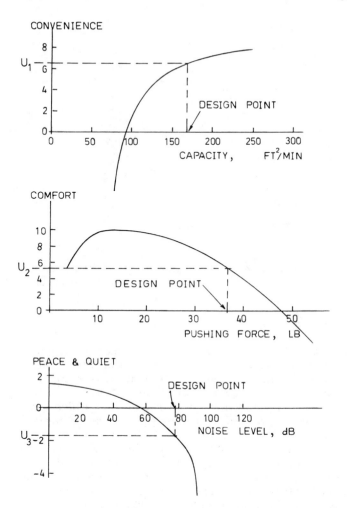

FIG. 10.3 Value curves for a lawn mower.

love of technology value which might well offset a lower level of total
values derived by analysis.

    A more conventional and simpler approach to this decision prob-
lem is to identify one dominant design characteristic as the decision
criterion, and assume that all other design characteristics are controlled by specifications. It is not then considered necessary to make
the value transformation, and a monotonic relationships is implicitly
assumed between the criterion quantity and value. This approach is

quite a valid one, particularly for a piece of capital equipment such as a heat exchanger, where the objective would be to minimize cost subject to specifications on amount of heat transferred, maximum length, and the like.

10.2 THE PROBABILISTIC DESIGN OPTION PROBLEM

The design option problem no longer has an obvious solution when the design characteristics become random variables, even for the apparently simple single criterion case. Let us begin with the assumption of only one design characteristic as the criterion. Figure 10.3 shows hypothetical probability density functions for this dependent variable, which could be cost, for two competing designs. We are happy about the lowest possible cost. One has more uncertainty than the other, but a lower mean value. Which should we choose? Design 1 looks attractive; it has the lower mean and a good probability of even lower values. But what about the rather high risk of even higher values than Design 2? Can the designer face this risk? It all depends on the shape of the value curve. One possible curve is shown on Fig. 10.4. The value curve can be thought of as a measure of aversion to risk. The designer would be very unhappy about the risk of getting the very low values of value on the right-hand end. It is clear then that *the decision should be based on the value probability distribution of each design.* The hypothesis used is that *the design with the maximum expected or mean value should be*

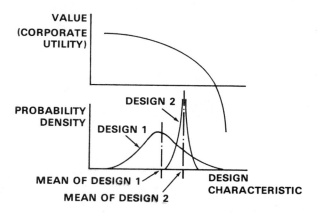

FIG. 10.4 Probability density functions and value curve for a common design characteristic for two competing designs.

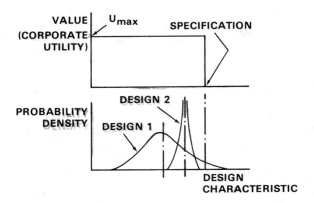

FIG. 10.5 Value curve converted to a specification.

selected. Regret, or aversion to risk, is automatically taken care of by this procedure.

Possibly this will be clarified somewhat if the value curve of Fig. 10.4 is converted to a specification, as in Fig. 10.5. Now the risk of design 1 violating the specification is the paramount consideration. The fact that it has a lower mean is irrelevant since both satisfy the specification and generate the same value. Design 2 clearly should now be chosen because it has a higher probability of satisfying the specification. This corresponds to the *dependability* of the design, discussed in Sec. 9.2. Design 2 also has the highest expected value. If D is the dependability and U_{max} is the value level for the satisfied specification, then for either design the expected value is proportional to D.

$$\bar{U} = 0(1 - D) + U_{max}D \qquad\qquad (10.2.1)$$

Returning to the more general case of Fig. 10.4, we need not establish the value density function in order to calculate the expected value. The following formula can be used, developed in Sec. 6.2.

$$U = \int_R U(y)f(y)\, dy \qquad\qquad (10.2.2)$$

where

$\quad$ $U(y)$ = value curve function
$\quad$ y = design characteristic
$\quad$ $f(y)$ = density function for the design characteristic

This can be evaluated by a relatively simple numerical integration on the computer. The density function $f(y)$ is obtained by probabilistic analysis.

Extension to multicriteria is quite simple in principle. If there are m value curves and corresponding design characteristics, the total expected value is

$$U = U_1 + U_2 + \cdots + U_m$$

$$= \sum_{i=1}^{m} \int_R U_i(y_i) f_i(y_i) \, dy_i \qquad (10.2.3)$$

The following example is hypothetical but credible.

EXAMPLE 10.1 Grain Unloader

A farm equipment manufacturer is considering two possible designs in a device to unload grain from the storage tank in a grain harvester to a truck. Design 1 is a slinger, a device similar in principle to a centrifugal pump. Design 2 is a chain conveyor.

Corporate utility curves for each of three important design characteristics are given in Figs. 10.6 to 10.8. These are percent damage to grain, power required, and number of parts. The scaling of the value ordinate indicates that damage is the most important, closely followed by the number of parts, and then power. The probability density curves for each characteristic and each design are superimposed on the value curves. Note that these are not scaled to give an area under the curve of 1. All the needed information is available to evaluate (10.2.3) for each design.

In this case U has three components:

$$U = U_1 + U_2 + U_3 \qquad (10.2.4)$$

so that by (10.2.3)

$$E(U) = E(U_1) + E(U_2) + E(U_3)$$

$$= \int f(y_1) U_1 \, dy_1 + \int f(y_2) U_2 \, dy_2 + \int f(y_3) U_3 \, dy_3$$

$$(10.2.5)$$

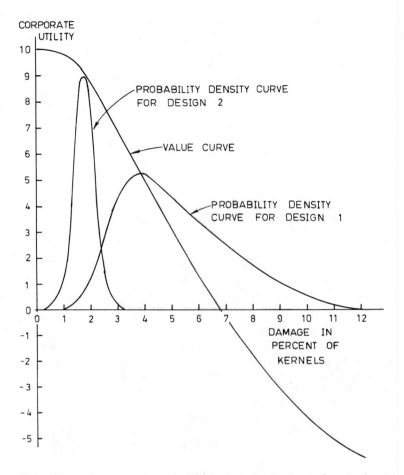

FIG. 10.6 Value and probability density curves for grain damage.

where

      y_1 = grain damage

      y_2 = power

      y_3 = number of parts

Algorithm

1. As given, the density functions are not normalized, so that the areas are 1. The range of each variable is divided into approximately 50 intervals. Subroutine FSIMP is used to

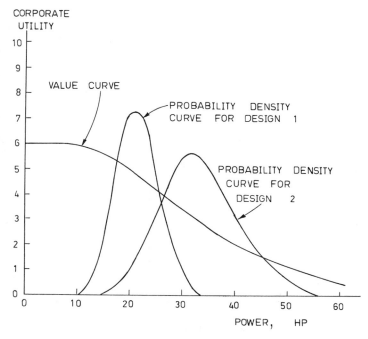

FIG. 10.7 Value and probability density curves for power.

determine the area of each curve, and the ordinate at each
station is divided by the area, in order to correct the curve.
2. The integrals in (10.2.5) are then evaluated numerically to
obtain expected values.

Program Listing

```
C       PROGRAM TO ESTIMATE MAXIMUM EXPECTED VALUE FOR ALTERNATE DESIGNS
C
C       FDENS(I)= ARRAYS FOR DATA DEFINING DENSITY FUNCTIONS
C       FDAM1(I)= ARRAY DEFINING DENSITY FUNCTION FOR DAMAGE IN DESIGN 1
C       FDAM2(I)= ARRAY DEFINING DENSITY FUNCTION FOR DAMAGE IN DESIGN 2
C       FPOW1(I)= ARRAY DEFINING DENSITY FUNCTION FOR POWER IN DESIGN 1
C       FPOW2(I)= ARRAY DEFINING DENSITY FUNCTION FOR POWER IN DESIGN 2
C       UDAM(I)= VALUE CURVE FOR DAMAGE
C       UPOW(I)= VALUE CURVE FOR POWER
C
        DIMENSION FDENS(61),FDAM1(61),FDAM2(61),FPOW1(61),FPOW2(61),
       1UDAM(61),UPOW(61),FUNC(61)
C
C       NORMALIZE DENSITY FUNCTIONS
C
        DO 1 I=1,4
        READ(5,10)(FDENS(J),J=1,61)
        READ(5,11)RANGE
```

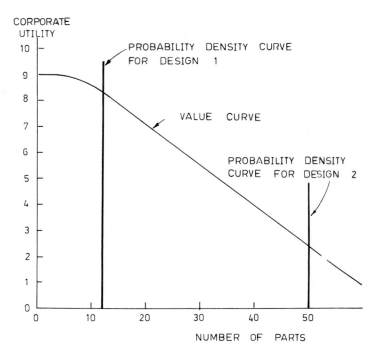

FIG. 10.8 Value and probability density curves for number of parts.

```
        AREA=FSIMP(FDENS,RANGE,61)
        DO 2 J=1,61
        GO TO(3,4,5,6)I
3       FDAM1(J)=FDENS(J)/AREA
        GO TO 2
4       FDAM2(J)=FDENS(J)/AREA
        GO TO 2
5       FPOW1(J)=FDENS(J)/AREA
        GO TO 2
6       FPOW2(J)=FDENS(J)/AREA
2       CONTINUE
1       CONTINUE
C
C       DETERMINE EXPECTED VALUES
C
        READ(5,10)(UDAM(J),J=1,61)
        READ(5,10)(UPOW(J),J=1,61)
        DO 20 I=1,6
        GO TO (30,31,32,33,34,35)I
30      DO 40 J=1,61
40      FUNC(J)=FDAM1(J)*UDAM(J)
        RANGE=12.
        E1=FSIMP(FUNC,RANGE,61)
        GO TO 20
31      DO 41 J=1,61
41      FUNC(J)=FDAM2(J)*UDAM(J)
C
```

```
          RANGE=12.
          E2=FSIMP(FUNC,RANGE,61)
          GO TO 20
   32     DO 42 J=1,61
          RANGE=60.
   42     FUNC(J)=FPOW1(J)*UPOW(J)
          E3=FSIMP(FUNC,RANGE,61)
   33     DO 43 J=1,61
   43     FUNC(J)=FPOW2(J)*UPOW(J)
          RANGE=60.
          E4=FSIMP(FUNC,RANGE,61)
          GO TO 20
   34     E5=8.17
          GO TO 20
   35     E6=2.20
   20     CONTINUE
          DES1=E1+E3+E5
          DES2=E2+E4+E6
   C
   C      OUTPUT
   C
          WRITE(6,100)
   100    FORMAT(/,1H ,15X,24HEXPECTED VALUES OF VALUE,//)
          WRITE(6,101)
   101    FORMAT(/,1H ,12X,6HDAMAGE,7X,5HPOWER,9X,5HPARTS,8X,5HTOTAL,//)
          WRITE(6,102)E1,E3,E5,DES1
   102    FORMAT(/,1H ,8HDESIGN 1,4X,F5.3,8X,F5.3,9X,F5.3,8X,F6.3)
          WRITE(6,103)E2,E4,E6,DES2
   103    FORMAT(/,1H ,8HDESIGN 2,4X,F5.3,8X,F5.3,9X,F5.3,8X,F6.3)
   10     FORMAT(16F5.2)
   11     FORMAT(F5.0)
          STOP
          END
```

```
                                  .1    .3    .5    .9  1.30 1.90 2.50 3.20 3.80 4.30
4.70 5.00 5.10 5.20 5.20 5.10 4.90 4.70 4.60 4.40 4.20 3.90 3.80 3.60 3.40 3.20
3.00 2.90 2.70 2.50 2.40 2.20 2.10 1.90 1.80 1.60 1.50 1.40 1.20 1.10 1.00  .90
 .80  .70  .60  .50  .40  .30  .30  .20  .10  .10
12.
                                  .1    .3    .7  1.20 2.80 6.00 8.40 9.00 7.60 4.80 2.60 1.10  .50  .20


12.
                                                          .1    .4    .8  1.3  2.3
3.60 4.60 5.90 6.90 7.20 7.30 7.20 6.80 6.00 4.80 3.80 3.00 2.40 1.90 1.20  .60
 .30  .10

60.

 .10  .30  .50  .70 1.00 1.40 1.80 2.20 2.70 3.10 3.60 4.10 4.70 5.10 5.50 5.60
5.60 5.50 5.30 5.00 4.70 4.30 4.00 3.60 3.20 2.90 2.50 2.10 1.90 1.60 1.40 1.20
1.00  .80  .70  .50  .40  .30  .20  .10
60.
10.0010.0010.0010.00 9.90 9.80 9.70 9.60 9.30 9.10 8.80 8.40 8.10 7.80 7.40 7.00
6.50 6.20 5.70 5.40 5.00 4.60 4.30 3.90 3.50 3.10 2.80 2.40 2.10 1.70 1.40 1.10
 .7   .4        -.3  -.6  -.9 -1.2 -1.4 -1.7 -2.0 -2.3 -2.5 -2.9 -3.0 -3.2 -3.4
-3.7 -3.9 -4.1 -4.3 -4.4 -4.6 -4.8 -5.0 -5.1 -5.3 -5.4 -5.5 -5.6
6.   6.   6.   6.   6.   6.   6.   6.   6.   6.   5.9  5.9  5.8  5.8  5.7  5.6
5.4  5.4  5.3  5.1  5.0  4.8  4.7  4.5  4.3  4.2  4.0  3.9  3.7  3.6  3.4  3.3
3.1  2.9  2.8  2.7  2.5  2.4  2.3  2.1  2.0  1.9  1.8  1.7  1.6  1.5  1.5  1.4
1.4  1.3  1.2  1.1  1.0  1.0   .9   .8   .8   .7   .6   .6   .5
/
```

Results

The results are shown below. The best choice is design 1.

Expected values of value

Design	Damage	Power	Number of Parts	Total
1	3.062	4.713	8.170	15.945
2	9.042	2.975	2.200	14.217

10.3 DISCUSSION

Is the design option decision-making procedure described in this chapter a practical technique? It is a fairly common practice for engineers, who are attempting to decide between two alternatives, to set up a matrix of design characteristics and give subjective rankings, weighted or unweighted, to grade the merits of each characteristic for both designs. The design having the highest total grade is chosen. An example is shown in Table 10.2 for the choice of an airport location in Mexico City. The conclusion was that alternative 6 was the best design. The methods being proposed here are a more rigorous evolvement from this simple exercise, codifying both value and risk judgments for an analytical treatment.

Essentially the same basic mathematical principles are used here for design option decision making as were developed in the classical statistical decision theory of von Neumann and Morgenstern (1944) and applied to management science (Schlaifer, 1969) and to systems

TABLE 10.2 Matrix of Rankings of Attributes for Airport Location Alternatives

Alternative	Attribute			
	Flexibility	Political effects	Externalities	Effectiveness
2	1	4	4	3
5A	2	3	3	2
5B	4	5	5	4
6	3	1	1	1
10	5	2	2	1

Source: From deNeufville and Marks (1974, p. 366).

design (deNeufville and Marks, 1974; Lifson, 1972). These sources have significant differences—classical economic utility is used rather than value concepts, and the standard gamble is used in generating utility functions.

Some mention should be made of a technique frequently referred to in the literature for making predictions. It is called the delphi method (Linstone and Turoff, 1975), and may have some application in the subjective estimation of value curves and probability curves. It is a form of structured communication in which a group of experts establish a consensus of the best possible prediction.

It is believed that the theory proposed here provides practical techniques for use in mechanical design in many applications, techniques that will make technology more responsive to society's true values.

PROBLEMS

Problem 10.1 The utility curve of a variable, X, is given in Fig. 10.9. The unnormalized probability density curves are also available for two alternative designs. Select the best design.

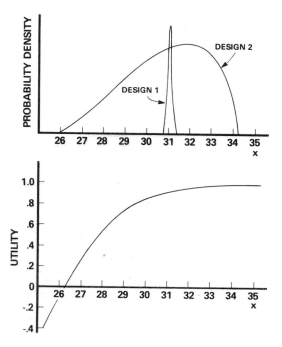

FIG. 10.9 Problem 10.1.

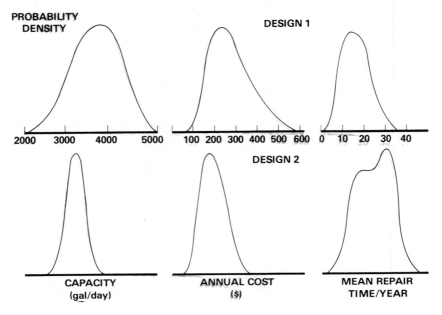

FIG. 10.10   Problem 10.2 density functions.

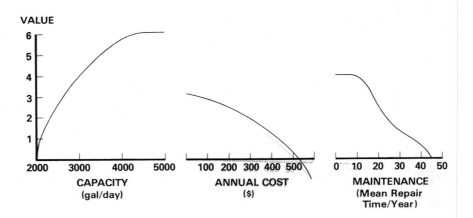

FIG. 10.11   Problem 10.2 value curves.

Problem 10.2 A farmer is trying to decide which of two irrigation systems to buy. His three most important design characteristics are capacity, annual cost, and mean repair time per year. The last is important aside from cost because it is a one-person operation and the farmer cannot spare much time for repair work. He has sketched his own value curves, and obtained subjective probability curves for the design characteristics, partly based on discussions with other farmers, and partly on manufacturer's data. Design 1 is new and innovative, and there is somewhat more uncertainty about its performance. The density functions are shown in Fig. 10.10 and the value curves in Fig. 10.11.

REFERENCES

deNeufville, R., and D. H. Marks, eds. (1974), *Systems Planning and Design*, Prentice-Hall, Englewood Cliffs, N. J.

Lifson, M. W. (1972), *Decision and Risk Analysis for Practicing Engineers*, Barnes & Noble, New York.

Linstone, H. A., and M. Turoff, eds. (1975), *The Delphi Method: Techniques and Applications*, Addison-Wesley, Reading, Mass.

Schlaifer, R. O. (1969), *Analysis of Decisions Under Uncertainty*, McGraw-Hill, New York.

Siddall, J. N. (1982), *Optimal Engineering Design: Principles and Applications*, Marcel Dekker, New York.

von Neumann, J., and O. Morgenstern (1944), *Theory of Games and Economic Behavior*, Princeton University Press, Princeton, N. J.

Reliability Theory

Historical basis for reliability • The hazard function as a conditional density function and as a failure rate • Some distributions commonly used • Setting up reliability programs • Maintainability and availability

11.1 INTRODUCTION

Reliability is, unfortunately, one of the modern engineering terms which has somewhat different meanings for different disciplines. Historically, engineers have called a device reliable if it would work when wanted. Aerospace and electrical engineers gave the term a more specific meaning when they found it necessary to cope analytically with problems in unreliable space vehicles. *Reliability was defined as the probability that a component or device or system will achieve a specified life without failure under a given loading.* The *life* of the device was thus recognized as being a random variable, and the reliability function is related to the cumulative distribution function by

$$R(t) = 1 - F(t) \tag{11.1.1}$$

where t is the time to failure.

Reliability theory, then, is simply a probabilistic design approach to the problem of the design characteristic life. Aerospace engineers found that the risks associated with design for life simply had to be codified in probabilistic terms.

The question then arises: Why just treat life in this way? Why not treat other design characteristics as random variables? It has been shown in earlier chapters that many characteristics can appropriately be treated as random variables. Life was the first because it tends to be the most random. The work of aerospace engineers was actually preceded by rolling element bearing manufacturers, who have long found it necessary to rate the lives of their bearings on a probabilistic basis because of the large spread in test results.

The term "reliability" was adopted by structural engineers to mean the probability of a structure not failing under a given load for a specified life. The random variable here is margin of strength. Other engineers have broadened the term to cover the probability of meeting any specification. This concept should be covered by the more general term *dependability* (see Sec. 9.2). With this terminology reliability becomes life dependability. In the following discussion this will be taken as the meaning of reliability if it is used without specification.

For the most part reliability distributions or functions are obtained by test or field experience, rather than by probabilistic analysis. Life is a difficult design characteristic to model.

Much of what is included in reliability books has already been covered in previous chapters. The discussion in Chapter 9 on combined failure modes and fault tree analysis is, for example, commonly considered part of reliability theory.

11.2 THE HAZARD FUNCTION

The hazard function is one of the standard probability functions. It has other names in the literature, such as intensity function and force of mortality. It is of particular importance in reliability work, where it is commonly called the *failure rate*. Its concept is not an easy one, and we shall approach it in two different ways, showing its relationship to the other probability functions, $f(t)$, $F(t)$, and $R(t)$.

11.2.1 The Hazard Function as a Conditional Density Function

The hazard function can be defined as a conditional density function. This means that it is analogous to the probability density function, $f(t)$, with the difference that the probability of an event determined from the hazard function is a *conditional probability*, and dependent on survival up to the time of the event. This is illustrated in Fig. 11.1. The event is the failure of a device between time t' and $t' + dt$. We designate the hazard function as $h(t)$.

P(failure between t' and $t' + dt$) = $f(t')$ dt

P(failure between t' and $t' + dt$ |no failure up to t') = $h(t')$ dt

P(survival up to t') = $R(t')$

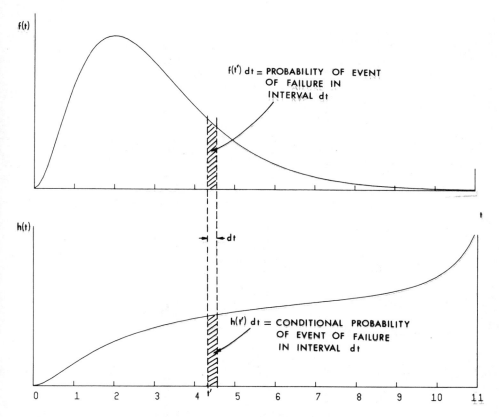

FIG. 11.1 Conditional infinitesimal event of failure.

Using the familiar probability law for conditional events, (2.3.8), gives

$$h(t') \, dt = \frac{f(t') \, dt}{R(t')}$$

We can eliminate dt and convert t' to t, since it is valid for all values of t.

$$h(t) = \frac{f(t)}{R(t)} \qquad\qquad (11.2.1)$$

Also, by definition, we can write

$$R(t) = \int_t^\infty f(t)\, dt$$

which gives a conversion relationship between h(t) and f(t).

$$h(t) = \frac{f(t)}{\int_t^\infty f(t)\, dt} \qquad (11.2.2)$$

Recalling the well-known relationship between the density function and cumulative distribution function (3.2.12), we have

$$f(t) = \frac{dF(t)}{dt} \qquad (11.2.3)$$

or in terms of the reliability function,

$$f(t) = \frac{d}{dt}\,[1 - R(t)] = -\frac{dR(t)}{dt} \qquad (11.2.4)$$

If we now substitute this into (11.2.1), we have

$$h(t) = -\frac{1}{R}\frac{dR}{dt} \qquad (11.2.5)$$

Integrating gives

$$-\int_0^t h(t)\, dt + \ln R \qquad (11.2.6)$$

or

$$R(t) = \exp\left[-\int_0^t h(t)\, dt\right] \qquad (11.2.7)$$

We thus have a conversion relationship between R(t) and h(t).
We can convert from h(t) to f(t) by using (11.2.4) and (11.2.5).

$$f(t) = R(t)h(t)$$

$$= h(t)\,\exp\left[-\int_0^t h(t)\, dt\right] \qquad (11.2.8)$$

It may be noted that we have taken the lower bound for life t, as zero. It could, in practice, be some finite positive value. Subroutines are available in the PROBVAR package in Appendix A for making these conversions.

11.2.2 The Hazard Function as a Failure Rate

The hazard function is commonly thought of as a failure rate in reliability work. More rigorously, from this viewpoint, it should be defined as the mean failure rate (or failures per unit time) per unit of *surviving* population.

Let us see how this arises from the previous definition. Returning to Fig. 11.1, we can imagine that we are counting the number of failures in a population of devices, in the interval dt. We let this be dN_f. We define the number of survivors up to this time as N_s, and the size of the original population as N_0. We can use frequency counts then to estimate probabilities.

P(failure in dt | no failure up to t)

$$= \text{fraction of survivors failed in dt} = \frac{dN_f}{N_s}$$

But we have previously shown that this probability is also h(t) dt. Equating the two and dividing by dt gives

$$h(t) = \frac{dN_f}{dt} \frac{1}{N_s} \tag{11.2.9}$$

which corresponds to our failure rate definition.

It is perhaps helpful to do the same thing with f(t).

$$f(t) \, dt = P(\text{failure in dt})$$

$$= \frac{dN_f}{N_0}$$

where N_0 is the original size of population. Dividing again by dt gives

$$f(t) = \frac{dN_f}{dt} \frac{1}{N_0} \tag{11.2.10}$$

Thus f(t) can be thought of as a failure rate per unit of *original* population. However, the hazard function failure rate is more mean-

ingful because our concern is normally with the fraction of current
population that is failing.

The cumulative hazard function is also of some interest. It is
defined in analogous fashion to the cumulative distribution function.

$$H(t) = \int_{-\infty}^{t} h(t) \, dt \tag{11.2.11}$$

It is clear from (11.2.7) that

$$R(t) = e^{-H(t)} \tag{11.2.12}$$

and therefore

$$H(t) = -\ln[R(t)] = \ln\left[\frac{1}{R(t)}\right] \tag{11.2.13}$$

The hazard function is, in some respects, more physically meaningful as a measure of reliability performance than is the reliability
function itself, and is commonly used in this role. If we are concerned with the probability of the event of failure between times t_1 and t_2,
then the area under the density function over this interval will give
us the absolute probability of any given device failing in that interval. The area under the hazard function from t_1 to t_2 will give us the
probability of any device in the population failing in that interval,
given that it has survived up to time t_1. The latter event is sometimes the one of most interest. For example, a farmer with a 3-year-
old implement may wish to know the risk of failure during the fourth
season.

11.3 PROBABILITY DISTRIBUTIONS USED IN RELIABILITY

The distributions most commonly used in reliability practice appear to
be the exponential and the Weibull—the former primarily for electronic
components and the latter for mechanical components. The justification for their use does not seem to be too well founded; and based
more on convenience and custom than strong experimental evidence.

The exponential distribution has a direct relationship to physical
modeling. If h(t) is constant, and designated λ, then (11.2.7) becomes

$$R(t) = e^{-\lambda t} \tag{11.3.1}$$

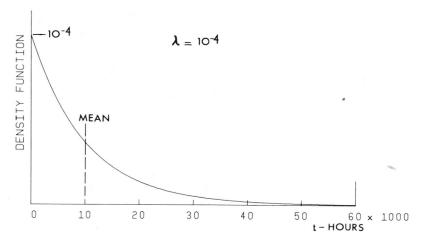

FIG. 11.2 Exponential density function.

The constant failure rate means that it represents a Poisson process, and a device is equally likely to fail at any time. We can anticipate that this is true by physical arguments for some components. An automobile tire is equally likely to pick up a nail at any time during its service, so the failure rate *for this mode of failure* would be constant.

The exponential function was the first one to receive common use, and many books on reliability tend to treat it as the *only* reliability function. The corresponding density function is

$$f(t) = \lambda e^{-\lambda t} \tag{11.3.2}$$

This distribution is shown in Fig. 11.2. It can be easily shown that the mean life is

$$\mu = \int_0^\infty t \lambda e^{-\lambda t} \, dt = \frac{1}{\lambda} \tag{11.3.3}$$

Thus the parameter λ is easily obtained from an observed sample of lives. The failure rate, being constant, is particularly meaningful, and this perhaps accounts for its dominant role in reliability work. Thus part of the jargon of the trade is *mean time to failure* (MTTF), which is simply the mean life. A component having an exponential distribution for life can be repaired, and the mean time to the next failure will be the same as the mean time to the first failure. The term *mean time between failures* (MTBF), is commonly used, but it only has meaning for an exponential distribution.

Weibull developed his distribution to represent fatigue strength of metals (Weibull, 1951). It was later applied to fatigue life (Freudenthal and Gumbel, 1953), and in recent years has been widely used to represent the life distribution of mechanical components. The expressions given in Sec. 3.4.7 are repeated here.

$$f(t) = \frac{\beta}{\eta - \ell} \left(\frac{x - \ell}{\eta - \ell} \right)^{\beta - 1} \exp \left[- \left(\frac{x - \ell}{\eta - \ell} \right)^{\beta} \right], \quad x \geq \ell$$

$$= 0, \quad x < \ell \qquad\qquad (11.3.4)$$

$$F(t) = 1 - \exp \left[- \left(\frac{x - \ell}{\eta - \ell} \right)^{\beta} \right] \qquad\qquad (11.3.5)$$

$$h(t) = \frac{\beta}{\eta - \ell} \left(\frac{x - \ell}{\eta - \ell} \right)^{\beta - 1} \qquad\qquad (11.3.6)$$

A typical Weibull distribution fit for life of a ball bearing is shown in Fig. 11.3. The normal distribution has been used to represent life to failure when wear is the failure mode, and the log normal has been used when fatigue is the failure mode.

There is no inherent reason why any of the distributions discussed in Chapter 3, or the maximum entropy distributions in Sec. 5.5, should not be candidates as well as the more commonly used functions mentioned above. Reliability engineers should be cautious about using conventional distributions only and not blindly use them without adequate evidence.

One difficulty in attempting to fit theoretical distributions to life data arises when a component is subject to different failure modes. A commonly presented curve in reliability work is shown in Fig. 11.4, the so-called bathtub curve. It represents the failure rate of a typical device as a function of time. The initial region has a high failure rate due to manufacturing errors, material defects, and the like. It can be considered a break-in period, and product warranties are often used to reduce the loss to the user for this period. The middle interval is the time in the life of the population when the dominant failure mode controls, such as fatigue in a mechanical component. The third region may be due to a failure mode such as wear or corrosion, which has a considerably higher lower bound, but a rapidly rising failure rate. It may be possible to eliminate the first region by

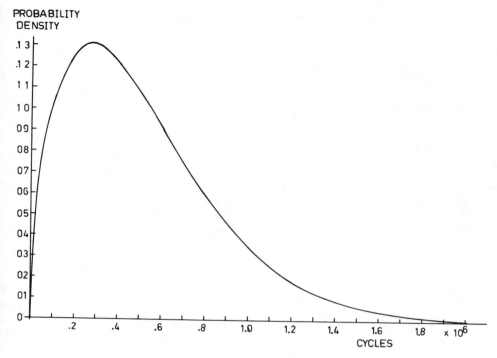

FIG. 11.3 Weibull function for a ball bearing.

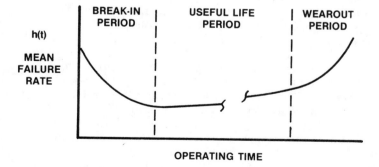

FIG. 11.4 Typical mean failure rate curve for combined failure modes.

having devices "run-in" at the factory, and the third region is sometimes eliminated by a field replacement policy.

11.4 RELIABILITY TESTING

Three kinds of *reliability testing* can be done. One type is essentially
an aspect of quality control and uses statistical acceptance testing
techniques (ARINC Research Corporation, 1964; Ireson, 1966; Kapur
and Lamberson, 1977).

The second type is more of a design and development tool, used
to determine if a new design has satisfactory reliability. The econo-
mics of engineering design and development often dictate that only
very small samples can be used—perhaps less than five and even only
one or two. In this situation the subjective techniques described in
Chapter 5 must be used. Accelerated testing can sometimes be used
to save time and costs.

The third type of testing is a technique of monotoring product
reliability to assure continuing product quality. Such a program requires highly organized and elaborate data gathering techniques that
report on operating failures in devices in service. Warranty data are
a common source of such information.

Techniques of processing the data obtained in the second and
third types of testing are similar. Fitting of theoretical distributions
to the data can be done using the methods of Chapter 5. Getting an
adequate sample size can be difficult and expensive due to the long
periods involved in life testing or monitoring. Fortunately, it is possible to use truncated data sets, so that we might begin by testing a
sample of 100 components, and then terminating the tests when 40
have failed. We thus have, in effect, the first 40 of an ordered sample of 100. These could be used in mean-rank plotting to estimate
parameters of a theoretical distribution, or they could be used to
generate a maximum entropy distribution. A good fit would be obtained to the lower portion of the curve, and the assumption made
that the upper portion can be successfully extrapolated.

Extreme value theory may be a useful technique in some situations. There may be some natural organization of data into N sets of
n trials, where the trials are tests or observed field failures. Or the
data can be organized in this way arbitrarily, as long as there is no
bias in the organization. Trials in any set need not be carried on, or
observed, past the first failure in the set. Trials in a set need not
even occur simultaneously, as long as we observe subsequent items
past the earliest failure time of previous trials in the set. We thus
obtain a sample of N extreme lowest values (out of n) with a minimum
number of observations or tests. It must be remembered, however,
that a priori modeling by means of the standard asymptotic extreme

value distributions requires that n be relatively large. Warranty failures can clearly be used as a source of this kind of truncated data, and are discussed further in Sec. 11.5.

We can fit an extreme value density function (not necessarily one of the asymptotic distributions) to the data, which represents the distribution of the least life to failure in a specified number of trials, n. The corresponding random variable is designated t_e. We can now estimate reliabilities as follows.

P(life will exceed a specified value in all of a group of n devices)

$$= \int_{t_s}^{\infty} f_{t_e} (t_e; n) = 1 - F_{t_e} (t_s; n) \tag{11.4.1}$$

This can be reduced to probabilities related to any single device.

P(life will exceed a specified value in any single device)

$$= 1 - [F_{t_e} (t_s; n)]^{1/n} \tag{11.4.2}$$

$$= 1 - F_t(t_s) = R(t_s) \tag{11.4.3}$$

Equation (11.4.2) arises from use of the familiar product rule for intersecting independent events. A probability density function, $f_t(t)$, could theoretically be generated from (11.4.3), but it represents an extrapolation from very local data. Its use should usually be restricted to the general region of the first order statistic.

The selection of bounds is an important problem in the estimation of life distributions. The use of distributions with zero lower bound and infinite upper bound tends to gloss over this problem, at the risk of improperly solving it. The use of order statistics to select bounds is discussed in Sec. 8.4. A technique for estimating the Weibull lower bound was given in Sec. 5.4.

11.5 RELIABILITY PROGRAMS AND DATA BASES

This section deals with the setting up of reliability programs for the purpose of assisting in the design of devices for reliability and the monitoring of the reliability of production devices. It does not include acceptance testing programs. The setting up of a reliability program includes the following aspects.

Definition of the Reliability Goals of the Company

The reliability goal of a company would generally be to maintain a satisfactory level of reliability for each of its products. However, it is desirable to have more specific goals for each product, such as the following. "The product reliability will be sufficient to provide a 90% probability that there will be no failure in the first year of service, and not less than 75% during each of the following four years". Such a specific goal gives a firm basis for design and testing, a basis for comparison with competitor's products, and a marketing feature that could be identified and understood by customers.

It may be necessary to be even more specific about the kind of failure. A classification suggested by Hollenback and Schmitt (1977) could be used.

1. *Full breakdown*: A failure that has caused a complete lack of operation or a complete lack of useful performance of the product. The product must be repaired to continue to operate.
2. *Partial breakdown*: A failure that could result in lower-level operation or performance and may cause complete loss of operation in the future. The machine is still operative but requires repair and downtime to restore 100% effectiveness.
3. *Nuisance failure*: A failure that causes loss of appearance or operator annoyance but does not significantly impair operation or performance.
4. *Non-time-related problems*: A problem that occurs repeatedly during operation that is not related to operating hours, but produces some degradation of performance.

Design of a Data Recovery System

The development of an adequate data recovery system is perhaps the most crucial and most difficult problem in setting up a successful reliability control problem.

As noted above, warranty information is an important data source on life to failure of components. It is necessary to have rather elaborate information gathering techniques on warranty failures, so that not only the component and time to failure and the cost is recorded, but also the particular *failure mode*. The warranty data have the advantage that they include the total population rather than a restricted sample. However, it is also important to obtain postwarranty data, and some kind of sampling scheme would be necessary—either from service points or direct from customers.

If data recovery is to be complete and accurate, it is necessary to devise simple to use techniques for recording the information, which is then rapidly transmitted to a central computer. Some procedures are described by ARINC Research Corporation (1964) and

Parescos (1977). Parescos gives the following principles for a successful failure reporting system.

1. *"Half the job concept"*: Personnel who are responsible for the repair of a failed equipment must be made aware that reporting (documenting) defects is of importance equal to repair of an item.
2. *"Prerecording statistical information"*: Analysis of most existing failure reporting systems reveals that a high percentage (up to 60%) of the reporter's time is spent recording statistical information. Wherever possible an optimum system has the statistical information prerecorded on the form during the printing process or by computer-printed headings following the printing process.
3. *"Eliminating code systems"*: The failure reporting form should be entirely in English language to eliminate the requirement of looking up or memorizing codes.
4. *"My card concept"*: The end user must be involved in the design of the card. It should not be designed by engineers alone. If the end user feels that the failure report form is his or hers, the program will succeed.
5. *"Human factored form design"*: The failure reporting form should be clear, simple, and highly visible so that it can be completed quickly and accurately.
6. *"Total system inputting"*: The failure report form should be more than a reporting media. It should be used to initiate work requests, order spare parts, specify personnel requirements, and so on. Let it be an optimum user system.
7. *"Rapid data inputting"*: Input the failure data rapidly using optical scanning and computer time-sharing techniques for total system use.
8. *"Feedback"*: Analyze the failure data as soon as possible. Report all evaluations, corrective actions, and results to all levels of management.

Design of Data Processing Software

In designing the software, one must begin with a set of specifications, which would include the following items.

1. *The general objectives of the software package*: These will follow directly from the general objectives for the whole reliability control system.
2. *The technical level of users related to programming skills and understanding of reliability concepts.*
3. *Input and output information to be used with the software system*: These will be discussed in more detail below.
4. *Input and output hardware to be used.*
5. *Reliability theory and data processing techniques to be used.*

6. *Memory storage available.*
7. *Documentation required.*

Manuals will be necessary that describe the programming functions, explain the system, give rules for use of the system, describe
the algorithms used, describe the way that the software is interfaced
with the computer operating software, and provide information for
future maintenance.

The input information will, of course, be related to the failure
report document. The following items could be included.

1. Identification of part, subassembly, and product
2. Environmental conditions: normal or pathological loading
3. Function or part
4. Operating severity of failure (see section above on the "Definition
    of the Reliability Goals of the Company")
5. Mode of failure
6. Time of failure
7. Time to repair

Output information is somewhat more difficult to identify. However, it is useful to categorize it from a functional point of view.

1. Information that identifies the normal reliability, maintainability,
    and availability functions for a part, subassembly, and product.
    This information will tell the engineering department if these
    functions meet target design specifications.
2. Information that warns the engineering and production depart-
    ments that a part has begun to develop a reliability outside of
    specifications.
3. Information that gives management a perception of the general
    level of product reliability and maintainability, and how well it is
    being maintained. Graphical displays would be highly desirable
    here, disclosing trends and target values for reliability levels and
    warranty costs.

Reliability Analysis Procedures

It is desirable to set up standard procedures for reliability
analysis of the data.

Feedback Procedures to Correct Reliability Failures

When the software system flags a reliability failure, or informal
field reports indicate a rapid breakdown in the reliability of a component, a formal procedure for failure analysis and correction must be
implemented.

Reliability Test Programs

Reliability test programs may be necessary during the design of new products, in order to predict whether or not the reliability goals will be met. Accelerated testing can be used for certain kinds of components, particularly those subjected to vibration and fatigue loading (Shooman, 1968). General testing procedures are also given in ARINC Research Corporation (1964) and Kapur and Lamberson (1977).

11.6 MAINTAINABILITY AND AVAILABILITY

Maintainability is related to the design variable *time to repair* a component or device. If this is considered a random variable, maintainability is the probability that the time to repair will be less than a specified time. It thus corresponds to the cumulative distribution function.

If the designer can define the random nature of the variables time to failure and time to repair, he or she is able to also define an important design characteristic, the *availability of a device*. This is defined as the average fraction of the specified life for which the device is available for use. This is, for the definition,

$$A = \frac{\mu_t}{\mu_t + \mu_r} \tag{11.6.1}$$

where μ_t is the mean life to failure and μ_r is the mean repair time. This will apply for repeated failures and repairs over the specified life if the repair returns the system to a new condition. The availability can also be thought of as the probability that the system will be available for use at any time over the specified life. These concepts can be used in developing an industrial replacement policy for maintenance (Goldberg, 1981; Jardine, 1979).

PROBLEMS

Problem 11.1 A sample of shafts were found to fail in fatigue with the following values of stress in 1000 psi: 57.3, 59.2, 62.5, 55.3, 61.4. It is believed that a three-parameter Weibull distribution should fit the data based on previous experience.
a. Determine the Weibull parameters.
b. Determine the mean.
c. Estimate the probability of failure if the actual stress is 50,000 psi.

Problem 11.2 A music wire material has the following reliability function based on fatigue failure at a stress level of 105,000 psi.

$$R(t) = \exp\left[-\left(\frac{x - x_0}{\theta}\right)^b \right], \quad x > x_0$$

where

$$b = 2.167$$

$$\theta = 50793$$

$$x_0 = 31692$$

$$R(t) = 1, \quad x \leqslant x_0$$

The variable x is the number of cycles to failure.

A device has six valve springs stressed to 105,000 psi. What is the reliability of the engine, based only on valve spring failures, for a life of 40,000 cycles? The failure of any one valve causes failure of the device. How could the reliability be increased?

Problem 11.3 Five gears were fatigue tested to failure and the following lives were obtained: 2.2×10^5, 0.51×10^5, 1.5×10^5, 3.0×10^5, 0.97×10^5. Assume that the distribution is Weibull, and estimate the Weibull function parameters, the mean life, and the reliability for a specified life of 0.25×10^5. Use either Weibull paper or a least-squares computer solution. Also plot failure rate versus life up to a life of 10^5 cycles.

REFERENCES

ARINC Research Corporation (1964), *Reliability Engineering*, Prentice-Hall, Englewood Cliffs, N.J.

Fruedenthal, A. M., and E. J. Gumbel (1953), On the Statistical Interpretation of Fatigue Tests, *Proc. R. Soc. Lond. A*, Vol. 216, p. 309.

Goldberg, H. (1981), *Extending the Limits of Reliability Theory*, Wiley, New York.

Hollenback, J. J., Jr., and G. L. Schmitt (1977), Combine Reliability Engineering, *Proc. International Grain and Forage Harvesting Conf.*, Sept. 25-29, American Society of Agricultural Engineers.

Ireson, W. G., ed. (1966), *Reliability Handbook*, McGraw-Hill, New York.

Jardine, A. K. S. (1979), Solving Industrial Replacement Problems, *Proc. 1979 Annual Reliability and Maintainability Symp.*

Kapur, K. C., and L. R. Lamberson (1977), *Reliability in Engineering Design*, Wiley, New York.

Parescos, E. J. (1977), A New Approach to the Establishment and Maintenance of Equipment Failure Rate Data Bases, in *Failure and Reliability*, American Society for Mechanical Engineers, pp. 263-268.

Shooman, M. L. (1968), *Probabilistic Reliability: An Engineering Approach*, McGraw-Hill, New York.

Weibull, W. (1951), A Statistical Distribution Function of Wide Applicability, *J. Appl. Mech.*, Vol. 18, p. 293.

12

Structural Reliability

The role of structural reliability in probabilistic design • *Analysis of single members and multimember structures*

12.1 INTRODUCTION

In the technical literature, structural reliability is generally defined as the probability that a structure will not fail under a given loading system, commonly random, and for a specified life. The predicted life is not usually treated as a random variable. Structural reliability is thus equivalent to *strength dependability* in the terminology of this text.

A particular feature of structures is that the failure modes are usually dependent. This is always true in discrete member structures because the individual loadings in each member are all a function of the external loading. For continuous structures it is usually true because the different failure modes are dependent on the same external loading system, or at least a related system.

One of the methods of probabilistic analysis must therefore be used that can accommodate dependent failure modes; and these include the Monte Carlo, generalized transformation of variables, and the independent variable cell methods. However, for *statically determinate discrete structures* a special technique can be used because of the special form of the problem, and it will be given in Sec. 12.3. It is also characteristic of discussions of structural reliability in the literature that member dimensions are considered deterministic, and only loadings and material properties are taken as random.

12.2 SINGLE MEMBERS

If we are concerned only with the strength dependability of a single
member, any of the methods can be used, and Example 6.2 is an illustration of this. However, the use of the infinitesimal event approach
is conceptually attractive, and because it is also the basis for the
method given in the next section for discrete structures, it will be
developed here with some ramification. It is the same method as that
explained in Sec. 9.2.

   We let the strength of a member (a load or a moment or a torque),
and L be the applied load. These are random variables with known
distributions, shown in Fig. 12.1. The probability of having a load
between L and L + dL is $f_L(L)$ dL. The probability of having this
load and a strength greater than this is

$$f_L(L) \, dL \int_L^\infty f_S(S) \, dS$$

since L can range from $-\infty$ to $+\infty$, we sum up for each infinitesimal dL.
This gives the probability of no failure.

$$R_S = \int_{-\infty}^\infty \int_L^\infty f_L(L) f_S(S) \, dS \, dL \qquad (12.2.1)$$

By noting that

$$\int_L^\infty f_S(S) \, dS = 1 - F_S(L)$$

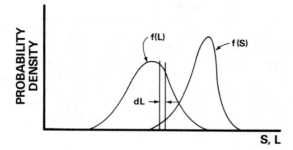

FIG. 12.1 Strength and load density functions for a single member.

we get an alternative form of (12.2.1),

$$R_S = \int_{-\infty}^{\infty} f_L(L)[1 - F_S(L)] \, dL \qquad (12.2.2)$$

We have assumed here that L and S are independent. If they are not, we must work with the joint density function $f_{S,L}(S, L)$. Equation (12.2.1) then has the form

$$R_S = \int_{-\infty}^{\infty} \int_{L}^{\infty} f_{S,L}(S, L) \, dS \, dL \qquad (12.2.3)$$

It is possible in a member for the resistance to positive and negative loads to be different, so that we must define new random variables S_p and S_n to represent resistance to positive and negative loads, respectively, if L spans both regions. Corresponding density curves are shown in Fig. 12.2. Now, failure of the member is defined as the event

$$(S_p \leqslant L, \, L > 0) + (S_n > L, \, L \leqslant 0)$$

These are mutually exclusive, so that (12.2.1) now has the form

$$R_S = 1 - \int_{-\infty}^{\infty} \int_{-\infty}^{L} f_L(L) f_{S_p}(S_p) \, dS_p \, dL$$

$$- \int_{-\infty}^{\infty} \int_{L}^{\infty} f_L(L) f_{S_n}(S_n) \, dS_n \, dL \qquad (12.2.4)$$

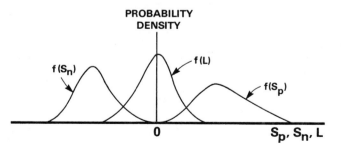

FIG. 12.2 Density functions where strength variable is different for positive and negative loads.

and (12.2.2) has the form

$$R_S = \int_{-\infty}^{\infty} f_L(L)[1 - F_{S_p}(L)] \, dL + \int_{-\infty}^{\infty} f_L(L)F_{S_n}(L) \, dL$$

(12.2.5)

The next consideration is a member subject to a repetition of loads L_1, L_2, ..., L_m. We shall assume that they are all one way, all independently random, and independent of S. It is more convenient to begin with the probability of having a strength between S and S + dS, which is $f_S(S)$ dS. The probability of having this strength *and* load L_1 less than this, *and* L_2 less than this, and so on, is

$$f_S(S) \, dS \int_{-\infty}^{S} f_{L_1}(L_1) \, dL_1 \int_{-\infty}^{S} f_{L_2}(L_2) \, dL_2 \cdots \int_{-\infty}^{S} f_{L_m}(L_m) \, dL_m$$

Since S can range from $-\infty$ to $+\infty$, we must sum these probabilities for each infinitesimal dS.

$$R_S = \int_{-\infty}^{\infty} \int_{-\infty}^{S} f_{L_1}(L_1) \, dL_1 \int_{-\infty}^{S} f_{L_2}(L_2) \, dL_2$$

$$\cdots \int_{-\infty}^{S} f_{L_m}(L_m) \, dL_m \, f_S(S) \, dS$$

(12.2.6)

An alternative form is

$$R_S = \int_{-\infty}^{\infty} F_{L_1}(S)F_{L_2}(S) \cdots F_{L_m}(S)f_S(S) \, dS$$

(12.2.7)

If there is cumulative damage, or fatigue, the loads and strength are not independent and (12.2.6) and (12.2.7) do not apply. It is also possible for L to be continuously varying and random. This requires use of the theory of random processes.

12.3 STATICALLY DETERMINATE DISCRETE STRUCTURES

In statically determinate structures the loadings in each member are independent of geometric variables and material property variables.

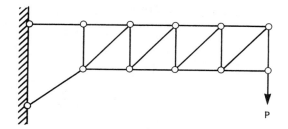

FIG. 12.3 Typical statically determinate structure.

They can therefore be defined as a constant proportion of the external load. This simplifies the probabilistic analysis, since each member can be treated separately.

It is not strictly true that member loadings are independent of the geometric dimensions defining the member connection locations. However, the variation in these will normally be small enough to have a negligible effect on member loadings.

We shall determine the expression for probability of no failure of a structure represented by Fig. 12.3. We define the following quantities:

  n = number of members

  P = external load

  a_i = fraction of P defining the load in the ith member

  L_i = load in the ith member

  S_i = strength of the ith member

It is possible to determine the a_i's by an equilibrium analysis. By definition

$$L_i = a_i P \qquad\qquad (12.3.1)$$

The probability of no failure is obtained in a manner similar to that for a single member. The probability of having a given load P in the interval dP, and having the corresponding load in each member less than the strength S_i, is given by

$$f_P(P) \ dP \ \prod_{i=1}^{n} \int_{a_i P}^{\infty} f_{S_i}(S_i) \ dS_i$$

The probability of no failure for any value of P is

$$R_S = \int_{-\infty}^{\infty} \left[\prod_{i=1}^{n} \int_{a_iP}^{\infty} f_{S_i}(S_i) \, dS_i \right] f_P(P) \, dP \tag{12.3.2}$$

An alternative form is

$$R_S = \int_{-\infty}^{\infty} \prod_{i=1}^{n} [1 - F_{S_i}(a_iP)] f_P(P) \, dP \tag{12.3.3}$$

This assumes that the strengths are stochastically independent. There will be a tendency for the strengths to be correlated due to the members being made from the same lot or run of material or fabrication. If they are perfectly correlated, they all tend in the same direction for any given structure. Since the loading in each member is always perfectly correlated, this means that the margin of strength of each member is perfectly correlated. It follows that the member having the lowest probability of no failure will always fail first. Thus, for perfect correlation, from (12.2.2)

$$R_S^{\,c} = \min_{i=1,n} \left\{ \int_{-\infty}^{\infty} [1 - F_{S_i}(a_iP)] f_P(P) \, dP \right\} \tag{12.3.4}$$

It seems clear that there will be fewer occasions of failure, if only one member can be critical, than if any member can be critical. Therefore, $R_S^{\,c}$ is always greater than R_S, and an assumption of independence is conservative.

It is of interest to note that if we try and invert our approach to (12.3.3), we run into trouble. Trying this, we say that

$$f_{S_i}(S_i) \, dS_i \int_{-\infty}^{S_i} f_{a_iP}(a_iP) \, d(a_iP)$$

is the probability that we have a given strength S_i in member i *and* the load on member i is less than S_i. Since

$$f_{a_iP}(a_iP) = \frac{f_P(P)}{|a_i|}$$

the expression reduces to

$$f_{S_i}(S_i) \, dS_i \int_{-\infty}^{S_i/a_i} f_p(P) \, dP$$

If we integrate over all values of S_i, we get R_i, the probability that the ith member will not fail.

$$R_i = \int_{-\infty}^{\infty} \int_{-\infty}^{S_i/a_i} f_{S_i}(S_i) f_P(P) \, dP \, dS_i \qquad (12.3.5)$$

The R_i's are not independent and thus (12.3.5) cannot be used except as an alternative in (12.3.4).

As we did for the single-member case, we next extend our analysis to repeated loads, P_1, P_2, ..., P_m. The probability of failing due to P_j is

$$1 - \int_{-\infty}^{\infty} f_{P_j}(P_j) \prod_{i=1}^{n} [1 - F_{S_i}(a_i P_j)] \, dP_j$$

The probability of failing due to P_1 or P_2 or P_3 and so on is, since they are mutually exclusive events,

$$\sum_{j=1}^{m} \left\{ 1 - \int_{-\infty}^{\infty} f_{P_j}(P_j) \prod_{i=1}^{n} [1 - F_{S_i}(a_i P_j)] \, dP_j \right\}$$

The system reliability is

$$R_S = 1 - \sum_{j=1}^{m} \left\{ 1 - \int_{-\infty}^{\infty} f_{P_j}(P_j) \prod_{i=1}^{n} [1 - F_{S_i}(a_i P_j)] \, dP_j \right\}$$

$$(12.3.6)$$

If all loads are identically distributed, then

$$R_S = 1 - m \left\{ 1 - \int_{-\infty}^{\infty} f_P(P) \prod_{i=1}^{n} [1 - F_{S_i}(a_i P)] \, dP \right\} \qquad (12.3.7)$$

We have assumed so far that we know the mode of failure of a member or structure. This may not be the case. A beam, for example, may fail in direct bending stress, shear stress, flange buckling, and web buckling. We let S_j be the strength of a single member in

the jth mode of r possible modes of failure. These would commonly
not be independent. Proceeding as usual we get the expression

$$R_S = \int_{-\infty}^{\infty} f_L(L) \int_L^{\infty} \int_L^{\infty} \int_L^{\infty} f_{S_1, S_2, \ldots, S_r}(S_1, S_2, \ldots, S_r)$$

$$\times \, dS_1 \, dS_2 \cdots dS_r \, dL \tag{12.3.8}$$

We may apply the same concept to discrete structures, where
S_{ij} is the jth mode of r_i failure modes in the ith member.

$$R_S = \int_{-\infty}^{\infty} f_P(P) \prod_{i=1}^{n} \left[\int_{a_i P}^{\infty} \int_{a_i P}^{\infty} \cdots \int_{a_i P}^{\infty} f_{S_{i1}, S_{i2}, \ldots, S_{ir}} \right.$$

$$\left. \times \, (S_{i1}, S_{i2}, \ldots, S_{ir}) \, dS_{i1} \, dS_{i2} \cdots dS_{ir} \right] dP \tag{12.3.9}$$

12.4 STATICALLY INDETERMINATE DISCRETE STRUCTURES

In general, statically indeterminate structures are analyzed as discussed in Sec. 12.1. The approach described above for determinate
discrete structures applies only if the variability of the cross-sectional and elastic properties is small, since the a_i's can then be calculated
as a function only of external loading.

PROBLEMS

Problem 12.1 You are asked to do the structural reliability analysis
for the design of the frame of a large earth-moving machine, shown in
Fig. 12.4. The load is evenly distributed in the bucket and has the
distribution given below, where L is the total load.

$$f(L) = \frac{1}{0.211 \times 10^6 \sqrt{2\pi}} \exp\left[-\frac{(L - 0.8 \times 10^6)^2}{2(0.211 \times 10^6)^2} \right]$$

A static analysis of the members has been made and is illustrated in
Fig. 12.5 for each member. The figure beside each member is the
fraction of L that it must carry. The strength of each member is defined by the following distributions.

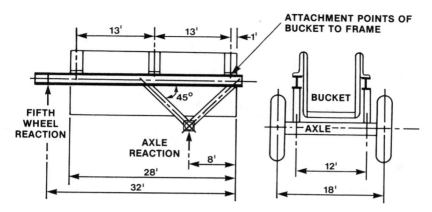

FIG. 12.4 Problem 12.1.

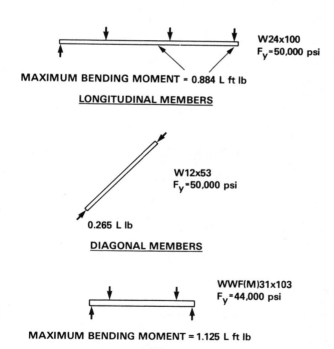

W24x100
F_y=50,000 psi

MAXIMUM BENDING MOMENT = 0.884 L ft lb

LONGITUDINAL MEMBERS

W12x53
F_y=50,000 psi

0.265 L lb

DIAGONAL MEMBERS

WWF(M)31x103
F_y=44,000 psi

MAXIMUM BENDING MOMENT = 1.125 L ft lb

AXLE

FIG. 12.5 Problem 12.1.

Longitudinal:

$$f(S) = \frac{4.46}{18.22 \times 10^4} \left(\frac{S - 1.059 \times 10^6}{18.22 \times 10^4} \right)^{3.46}$$

$$\times \exp \left[- \left(\frac{S - 1.059 \times 10^6}{18.22 \times 10^4} \right)^{4.46} \right]$$

Diagonal:

$$S = 411,000 \text{ lb}$$

Axle:

$$f(S) = \frac{4.87}{45.52 \times 10^4} \left(\frac{S - 1.570 \times 10^6}{45.52 \times 10^4} \right)^{3.87}$$

$$\times \exp \left[- \left(\frac{S - 1.570 \times 10^6}{45.52 \times 10^4} \right)^{4.87} \right]$$

a. Determine the structural reliability using a convolution integral.
b. Determine the structural reliability using Monte Carlo simulation or the generalized transformation of variables method or independent variable cell technique.
c. Determine where possible the probability that each member will be the first to fail.
d. Discuss how you could take care of other possible modes of failure for each member.

Problem 12.2 A structure corresponding to Fig. 12.6 is to be designed using square tubing. The mode of failure is assumed to be yielding for all members. The material is to be AISI 113T cold-drawn steel. Both the yield strength of the steel and the load L have probability distribution of the following form:

$$f(x) = \frac{b}{\theta} \left(\frac{x - x_0}{\theta} \right)^{b-1} \exp \left[- \left(\frac{x - x_0}{\theta} \right)^b \right]$$

The parameters have the following values for the yield stress and load:

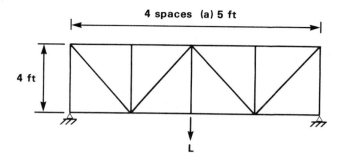

FIG. 12.6 Problem 12.2.

	$\bar{x}$	b	θ	x_0	σ
Yield stress (1000 lb)	98.1	1.41	6.48	92.2	4.24
Load (1000 lb)	50.3	3.99	11.92	39.5	3.04

The cross-sectional area of the members is assumed nonrandom. Also neglect the problem of discrete standard sizes.
a. Plot the density functions.
b. Design the truss based on the use of mean values and a safety factor of 2.
c. Design the truss using the methods in this chapter so that the reliability is 0.990.
d. Repeat part (c) using the Monte Carlo method.

Probabilistic Optimization

*An introduction to concepts in probabilistic optimization · Applications
to reliability assignment and tolerance assignment*

13.1 FORMULATION

The reader is referred to Siddall (1982) for the general theory of
deterministic optimization. When all quantities are deterministic, the
general formulation is as follows.

There is an optimization criterion function generalized as

$$U(x_1, x_2, \ldots, x_n) = \text{maximum or minimum} \tag{13.1.1}$$

where the x's are the independent design variables. There are, in
general, equality and inequality constraint functions representing
the failure modes of the device, and thus defining a feasible design.

$$\psi_i(x_1, x_2, \ldots, x_n) = 0, \quad i = 1, m \tag{13.1.2}$$

$$\phi_j(x_1, x_2, \ldots, x_n) \geq 0, \quad j = 1, p \tag{13.1.3}$$

In a stochastic treatment, some or all of the x's are treated as
random variables; and during the optimization process, their mean
values will be adjusted to achieve a "best" design. Furthermore, the
probability density function for an x may change it shape as the mean
is adjusted. In addition, there may be *random parameters*, such as a
coefficient of friction, that are not adjusted but, nevertheless, must
be considered as random. And, as well, some or all of the specifications may be sufficiently indeterminate so as to require that they be
treated as random variables, *but not as design variables.*

The general formulation now has the form

$$E[U(x_1, x_2, \ldots, x_n, a_1, a_2, \ldots, a_q)] = \text{maximum or minimum}$$

(13.1.4)

$$E[\psi_i(x_1, x_2, \ldots, x_n, a_1, a_2, \ldots, a_q, S_1, S_2, \ldots, S_r)] = 0,$$

$$i = 1, m \tag{13.1.5}$$

$$P[\phi_1(x_1, x_2, \ldots, x_n, a_1, a_2, \ldots, a_q, S_1, S_2, \ldots, S_r) \geq 0$$

$$and \ \phi_2(\ ) \geq 0 \ and \ \cdots \ and \ \phi_p(\ ) \geq 0] \geq D \tag{13.1.6}$$

where the a's are random parameters, the S's are random specifications, and D is the specified combined dependability of the device.

Several arbitrary assumptions are implied in the expressions above. The expected value of U is being used as the optimization criterion. There is no inherent reason why another characteristic measure should not be used instead, such as the mode or even a bound. A similar statement can be made about the ψ's. In executing the optimization, some characteristic measure must be used to represent a random x. The mean value would conform to traditional usage, but again, one could argue for the mode or median.

We have also increased the complexity of the solution enormously over a deterministic optimization. We must somehow determine how the probabilistic distribution of each random x varies as the mean varies. At every step in a numerical search technique, it is necessary to not only do an engineering modeling analysis, but also a probabilistic analysis to obtain the mean of U, the mean of any ψ's, and the joint probability of failure associated with the ϕ's.

It should be noted that we may also be interested in the density functions for the ψ's, at least for the optimum solution. For a given input, these cannot be controlled; however, their spread may be of considerable engineering interest. We may be similarly interested in the probability density function for U in the final solution. The indicated risk of getting values for U quite different from the mean could be unacceptable.

Equality constraints that are treated as indicated above would be of the type when the designer is predicting a performance characteristic and requiring it to meet a specification, say the deflection of a spring. Since the wire diameter, spring diameter, number of coils, and material properties must vary randomly, so must the resulting deflection. We can only aim to have the mean deflection of a population of springs satisfy the specification.

Some types of equality constraint functions cannot be treated in this manner. The engineering modeling governing the performance of a device or system may lead to a transcendental equation, represented generally as

$$f(x_1, x_2, \ldots, x_n, a_1, a_2, \ldots, a_q, S_1, S_2, \ldots, S_r) = 0$$

(13.1.7)

Normal procedure in an optimization formulation is to try to eliminate this equality by transforming (13.1.7) so that it can be represented as a function of one of the x's, and this x is eliminated by substitution.

It is not convenient to do this with a transcendental equation, and it could be simply made an equality constraint and carried into the optimization formulation. However, in stochastic optimization an equality constraint of this type must be satisfied by every embodiment of the design, it cannot be satisfied only by the means. Thus such expressions cannot be treated as equality constraints in stochastic optimization. One of the x's cannot be considered independently random, but instead strictly a function of the remaining random variables. We cannot avoid solving the transcendental equation (13.1.7).

13.2 EXECUTION

The formulation above has been structured in such a way so as to permit the use of deterministic optimization software. The probabilistic aspects are in the application subroutines and hidden from the library optimization subroutines.

Any of the methods for probabilistic analysis described in Chapter 6 can, in principle, be used. The fact that we do not require density functions for the U or the ψ's, except possibly at the conclusion, has some effect on the best choice. The software in PROBVAR can be used to make the programming relatively simple.

The moment transfer methods have the disadvantage that the combined dependability cannot be calculated without first calculating the maximum entropy density functions. Even these cannot be used if the ϕ's are not stochastically independent. The Monte Carlo method, the general transformation of variable method, and the independent variable cell method do not have this disadvantage.

The use of Monte Carlo strategy is quite straightforward and uses the following algorithm.

1. Define input: number of trials at each design point, optimization parameters.
2. Provide subroutines defining the optimization function, the constraint functions, and density functions.
3. Define starting values for means of the design variables (x's).

4.   Call the optimization subroutine.

5.   The optimization subroutine will use some strategy to find the optimum in step-by-step fashion. At each point, a set of mean values for the x's is defined. These in turn define the density functions to be sampled from.

6.   Loop on k to item 11 for each trial.

7.   Sample one value for each independent variable.

8.   Call the function subroutines to obtain a value for each function: U_k, ψ_{ik}, ϕ_{jk}.

9.   Obtain a successive summation to calculate means of U and ψ_i's.

10.   Check if all ϕ_{jk}'s are positive; if yes, *count one no failure*. Obtain successive summation of no failures.

11.   If k $\leqslant$ sample size, go to 6.

12.   Calculate current mean values of U, ψ_i's, and combined dependability of ϕ's using a frequency ratio.

13.   Return to optimization subroutine.

The generalized transformation of variable algorithm and the independent variable cell technique use a very similar cell approach. The method of calculating the mean of dependent variables used in the former can be easily adapted to the latter. If they are combined in this way, one call would give the mean of U and the ψ_i's, and the combined dependability of the ϕ_j's.

It will now be apparent why computer costs are high with either method. A full-scale probabilistic analysis is required at every optimization trial point; and at every Monte Carlo sample point, or every cell point in the cell method, a complete engineering modeling is required for each failure mode.

Optimization software can be found in Siddall (1982), and extracts from this may be integrated with extracts from PROBVAR in Appendix A in order to execute a probabilistic optimization.

13.3 OPTIMAL RELIABILITY ASSIGNMENT

It is fairly common practice, particularly in systems where high reliability is critical, for the specifications to include a required overall system specification. This is achieved by designing or selecting the components so that their combined reliability exceeds the specification. It would not be desirable to exceed the specification by too much, since high reliability is expensive to achieve.

The required combined reliability can be achieved in an infinite number of ways. If all components are in series, we could simply make all component reliabilities equal, and sufficiently high to achieve the desired specification. However, this may not be the optimum way to achieve the reliability goal. If cost is the primary criterion, equal reliabilities may be more expensive to achieve in some components than others. So the problem here is how to allocate reliability optimally to achieve a given overall requirement.

The formulation presented in the previous sections could theoretically be applied here. Reliability, or overall life dependability, would be incorporated as one of the probabilistic constraints in equation (13.1.6). However, this would require being able to model overall reliability as a function of component reliabilities, and component reliabilities as functions of the independent design variables. The latter is usually not practicable, as we have observed in Chapter 11.

Another approach is to treat component reliabilities as design variables in the optimization sense [see, e.g., Chatto (1975)]. Assuming that the criterion is cost, it may be possible to set up empirical cost functions relating cost to reliability for each component, represented generally as

$$c_i = g_i(R_i) \qquad\qquad (13.3.1)$$

The total cost is

$$U = \sum_{i=1}^{n} g_i(R_i) = \text{minimum} \qquad\qquad (13.3.2)$$

where there are n components. We also assume that there is some functional relationship between R_d, the device or system reliability, and the R_i's.

$$R_d = h(R_1, R_2, \ldots, R_n) \qquad\qquad (13.3.3)$$

If the specified overall reliability is R_s, we have the constraint function

$$\phi_1 = R_s - h(R_1, R_2, \ldots, R_n) \geqslant 0 \qquad\qquad (13.3.4)$$

If we are using parallel redundancy, with possibly more than one identical component in parallel, the number in parallel could also be treated as an integer design variable.

13.4 OPTIMAL TOLERANCE ASSIGNMENT

13.4.1 Introduction

The tolerance assignment problem is one part of the general problem of formulating a best possible design. There are several broad categories of decision making in design, but we are concerned here with one aspect of the optimization decision problem. This general problem will be outlined in detail below.

A tolerance can be defined generally as the limit imposed on the variability of some design variable or specification, from the point of view of producing the design. Although it is commonly associated with a machined dimension of a part, it should be generalized to be the bounds of any quantity. Examples are the yield point of a metal, the stiffness of a spring, and the horsepower of an engine. It is a recognition that a shaft cannot be made to an exact size, a steel cannot be made with an exact yield point, or an engine cannot be built with an exact maximum horsepower. The user must accept some tolerance on the nominal values. The user would usually prefer a tight tolerance, but the tighter the tolerance, the higher the cost.

We are concerned here then with the problem of the best possible trade-off between tolerance and cost. The problem should be stated more generally as one of choosing the tolerance that maximizes the overall value of the device.

The tolerance is now considered a design variable, and can be defined in terms of the nominal value of the base variable and equally spaced upper and lower deviations. The design variable thus has a value in the region

$$\bar{x}_i - \varepsilon_i < x_i < \bar{x}_i + \varepsilon_i \qquad\qquad (13.4.1)$$

where $\bar{x}_i$ is the nominal value, commonly the mean, and ε_i is the tolerance. The tolerance can also be defined simply by upper and lower bounds.

$$\ell_i < x_i < u_i \qquad\qquad (13.4.2)$$

However, this provides no convenient way to identify a nominal value for x_i, which is necessary in order to be able to optimize its nominal value.

The assignment of tolerances has traditionally been done wholly by judgment. The purpose of this study is to examine how it can be integrated into the overall optimization decision problem in an analytical way.

When tolerances are treated as design variables, they are generally assumed to generate changes in component cost. But they can also affect other design characteristics, such as life, accuracy, noise level, and so on. Such functional relationships are difficult to establish.

13.4.2 Probabilistic Aspects of Tolerance

Since the need for tolerance is due to the random variability of a design variable, the concept of tolerance is closely related to a probabilistic approach. A common convention, particularly for machined dimensions, is to assume a normal distribution and that the tolerance

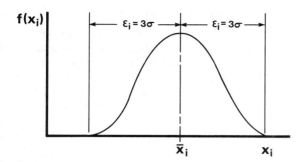

FIG. 13.1 Illustration of process tolerance.

ε_i is equal to 3σ, where σ is the standard deviation. This is illustrated in Fig. 13.1. However, this is the natural or process tolerance for
x_i, and may or may not correspond to the actual tolerance assigned by
the designer. The manufacturer may be able to adjust the process so
as to accommodate the specified tolerance. Otherwise, the tolerance
width may be more or less than the process range. If it is less, some
rejection must occur. The term *yield* is sometimes used to represent
the probability that a design variable will be within a specified tolerance. This is shown in Fig. 13.2.

Although machining processes often give near normal distributions to machined dimensions, this is not true of many other random
engineering quantities with tolerances. The normal distribution
should not be assumed without evidence.

Another aspect of tolerance is *process drift*. This is a recognition that the probability density function can change with time, particularly in the location of the mean.

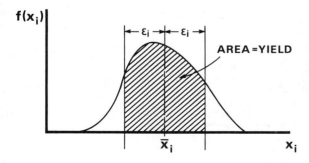

FIG. 13.2 Illustration of the concept of yield.

13.4.3 The Deterministic Tolerance Assignment Problem

We shall first assume that the tolerance width is fixed, and greater than or equal to the process bounds, for all variables having associated tolerances. The upper and lower tolerances will not necessarily be equal, but they will have a fixed ratio. The tolerances may actually vary with the magnitude of the nominal value, but this would be known. The general expressions would have the following form:

$$U(x_1', x_2', \ldots, x_n') = \text{maximum or minimum} \qquad (13.4.3)$$

$$\phi_i(\overline{x}_k) \geqslant 0, \quad i = 1, m$$
$$k = 1, 2^n \qquad (13.4.4)$$

$$\psi_j(\overline{x}_k) = 0, \quad j = 1, p$$
$$k = 1, 2^n \qquad (13.4.5)$$

where the x_i's are nominal values of the x_i's and $\overline{x}_k$ is a vector of x values defining the kth vertex of the region enclosed by the tolerances. At optimum at least one vertex will usually touch a constraint line, as shown in Fig. 13.3. The values of the x's are adjusted in the optimization search so that the tolerance region always stays inside the feasible region. This is sometimes called worst-case design in the literature.

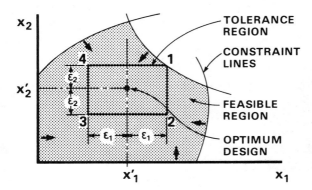

FIG. 13.3 Typical solution for the deterministic tolerance assignment problem.

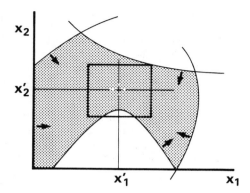

FIG. 13.4 Illustration of the difficulty in establishing a worst-case design.

It is possible, if the feasible region is nonconvex, for the approach described above to fail, as illustrated in Fig. 13.4. To ensure that this does not occur would require some kind of exploration for feasibility over the whole tolerance region.

We can extend the deterministic formulation to the case where the tolerance range is adjustable. The procedure is the same except that the tolerances are now included in the design variables; and the cost function will depend on the tolerances. Part of the cost associated with tolerances may be due to rejections of parts if the tolerance range is less than the process range, as shown in Fig. 13.5. This indicates that a fraction of the components defined by x_i, that are manufactured, will be rejected. The rejection rate will depend

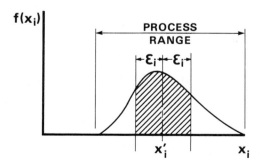

FIG. 13.5 Design variable with the tolerance range less than the process range.

on the tolerance range and the shape of the density function. Peters
(1970) has given a simple example of this formulation in the stacked
dimension problem, in which *only* the tolerances are treated as design variables, and the only cost is due to the rejections associated
with each component.

13.4.4 The Probabilistic Tolerance Assignment Problem

We now superimpose the tolerance problem onto the general stochastic
optimization decision problem. The tolerances are again design variables; and we permit the tolerance region to project outside the feasible region, as shown in Fig. 13.6, because there is a finite probability of infeasibility.

We are now optimizing the expected value of U, which is a function of the random variables x_i and the nonrandom tolerances. Figure
13.6 shows the design in the (x_1, x_2) plane, and a specific tolerance
region. In the general case the density function for x_i will be unsym-
metrical, and the tolerance bounds unequally spaced from the nominal
value of x_i. Not only this, but as the variable bounds move, they
redefine the mean value. In stochastic programming, we apply the
search strategy to the mean value of x_i. As it moves, the density
function moves with it, possibly changing its shape as it does so.

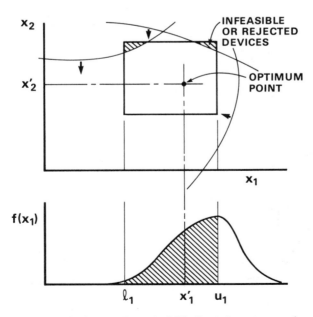

FIG. 13.6 General probabilistic tolerance assignment problem.

The combination of changing tolerance bounds modifying the density function and the location of the mean, and the search strategy simultaneously also doing so, considerably complicates the problem. For further details the reader is referred to Michael and Siddall (1981a,b).

PROBLEMS

Problem 13.1 A cargo floor panel is being designed for an aircraft, as shown in Fig. 13.7. The configuration is a sandwich panel, simply supported at the ends and free at the sides. The skin is aluminum alloy and the core is foamed polyurethane with adjustable density γ lb/in^3.

The span is 48 in. and the design requirement is support of a concentrated load of 8,000 lb at midpoint. The width b is 24 in. The total depth cannot exceed 3 in., and the skin thickness must not be less than 0.062 in. The stress in the skin is estimated by (Darvas, 1967)

$$\sigma_s = \frac{M}{bt_s(t_c + t_s)}$$

where M is the bending moment. The shear stress in the core is estimated by

$$\tau = \frac{V}{b(t_c + t_s)}$$

where V is the vertical shear. The failure modes are yielding of the skin and shear fracture of the core. The probability distribution for yield strength of the aluminum alloy is (Mischke, 1971)

$$f(S) = \frac{b}{\theta}\left(\frac{S - S_0}{\theta}\right)^{b-1} \exp\left[-\left(\frac{S - S_0}{\theta}\right)^b\right], \quad S \geq S_0$$

$$= 0, \quad S < S_0$$

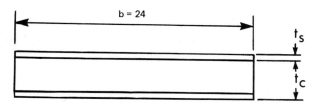

FIG. 13.7 Problem 13.1.

where

> S = yield stress in 1000 psi
>
> b = 1.42
>
> θ = 3.81
>
> S_0 = 47.4
>
> $\overline{S}$ = 50.8

The shear strength of the core is a function of its density

$$\tau = 14.9\gamma^{1.077} \quad \text{psi}$$

and the density has a normal distribution with standard deviation equal to 10% of its mean. The panel is to be as light as possible and the probability of failure must not exceed 0.005. Determine the optimum design.

Problem 13.2 The control system is being selected for a process that must have very high reliability. Any shutdown is extremely expensive. Six controllers are required at different stations in the plant. Sufficient redundant controllers at each station are to be provided so as to maximize the reliability for a maximum cost of $65,000. Alternative designs are available for each station. Reliability and cost data are given in the following table for all available designs.

Design alternative

Station	1 R	1 c	2 R	2 c	3 R	3 c	4 R	4 c
1	0.9983	2100	0.9967	1800	0.9835	1600	0.8476	1095
2	0.9992	3600	0.9906	2900	0.9037	2600	--	--
3	0.9846	1500	0.9637	1400	0.9630	1250	0.9580	1150
4	0.9996	952	0.9987	875	0.9965	825	0.9934	700
5	0.9998	5500	0.9994	5000	--	--	--	--
6	0.8692	1950	0.8543	1875	0.8317	1780	0.8176	1700

Design the system.

REFERENCES

Chatto, R. B. (1975), Mathematical Programming Applied to Reliability Analysis, *ASME Paper 75-DET-132.*

Darvas, R. M. (1967), Design Considerations in Sandwich Panel Construction, in *Cellular Plastics*, Publ. 1462, National Academy of Sciences—National Research Council, Washington, D.C., pp. 202-213.

Michael, W., and J. N. Siddall (1981a), The Optimal Tolerance Assignement with Less than Full Acceptance, in *Progress in Engineering Optimization—1981*, American Society for Mechanical Engineers, New York.

Michael, W., and J. N. Siddall (1981b), The Optimization Problem with Optimal Tolerance Assignment and Full Acceptance, *J. Mech. Des., ASME Trans.*, Vol. 103, No. 4 (Oct. 1981), pp. 842-848.

Mischke, C. R. (1971), Some Tentative Weibullian Descriptions of the Properties of Steels, Aluminums, and Titaniums, *ASME Publ. No. 71-Vibr-64.*

Peters, J. (1970), *Tolerancing the Components of an Assembly for Minimum Cost, J. Eng. Ind., ASME Trans.*, Ser. B, Vol. 92, No. 3 (Aug. 1970), pp. 677—682.

Siddall, J. N. (1982), *Optimal Engineering Design: Principles and Applications*, Marcel Dekker, New York.

appendix A

PROBVAR—A Software Package for Probabilistic Design

A.1 CLASSIFICATION AND DESCRIPTION OF SOFTWARE*

Family	Name	Description	Page
Analytical density functions (Sec. A.3)	ENTRPF	Maximum entropy function (see MEP1)	419
	FBETA	Beta function	402
	FEXP	Exponential function	402
	FGAMAC	Gamma function with noninteger parameter	403
	FGAMAI	Gamma function with integer parameter	403
	FGAUS	Normal function	403
	FRANK	Rank density function	403
	FWEIB	Weibull function	404
	POISS	Poisson function	404

---

*An alphabetical index of these programs, with page numbers, appears in Sec. A.14, p. 501.

TABLE (Continued)

Family	Name	Description	Page
Density function generation (Sec. A.4)	EVOL	Updates a density function by the evolutionary method	405
	FITT1	Kolmogorov-Smirnov goodness-of-fit test	406
	FITT2	Chi-square goodness-of-fit test	406
	HIST	Creation from data of a frequency histogram and a smoothed and normalized density function; plots histogram	407
	MEP1	Maximum entropy method using moments	411
	MEP2	Maximum entropy method using ranks	429
Miscellaneous functions (Sec. A.5)	GAMMA	Gamma function	436
Miscellaneous subprograms (Sec. A.6)	BFACT	Calculates factorial function NUM!(N!M!)	437
	BOUNDS	Determines upper and lower bounds in an array	437
	DERIV	Computes an array of derivative values from an array of equally spaced function values	437
	FACTO	Calculates a factorial	46
	FCUBIN	Table lookup with cubic interpolation	438
	FRAND	Generates uniformly distributed random numbers between zero and one	438
	FSIMP	Performs numerical integration	438
	FTABLE	Table lookup with linear interpolation	439
	SMOOTH	Smooths a set of data	440
	SORT	Sorts a sample from low to high	440
Moment calculations (Sec. A.7)	CMOM	Calculates central moments of a numerically defined density function	440
	CONVERT	Converts central moments to moments about the origin	441

TABLE (Continued)

Family	Name	Description	Page
	MOMENT	Calculates up to first four moments of a dependent variable from moments of independent variables by Taylor's series approximation	441
	SMOM	Calculates central moments from a sample of data	443
	SMOMG	Calculates central moments from a grouped sample of data	443
	TRANSM	Calculates all moments of a dependent variable from density functions of independent variables	444
Order statistics and extreme value distributions (Sec. A.8)	CUMRNK	Cumulative distribution function of a rank	446
	CUMXT1	Calculates local values of type 1 extreme value cumulative distribution function	447
	CUMXT2	Calculates local values of type 2 extreme value cumulative distribution function	447
	DEXTR1	Calculates local values of type 1 extreme value density function	447
	DEXTR2	Calculates local values of type 2 extreme value density function	448
	DORDER	Calculates local value of order statistic density function	449
	FRANK	Calculates local value of rank density function (see Sec. A.3)	403
Plotting density functions and histograms (Sec. A.9)	HIST	Plots a historgram (see Sec. A.4)	407
	MREX1	Type 1 extreme value plot using mean ranks	449
	MREX2	Type 2 extreme value plot using mean ranks	451
	MRWEIB	Weibull mean-rank plot	453
	PLOTPL	Plots a density function on a plotter	454
	PLOTPR	Plots a density function on the printer	456

TABLE (Continued)

Family	Name	Description	Page
Probabilistic analysis (Sec. A.10)	CARLO	Monte Carlo method	460
	CELLV	Cell method	471
	MANAL	Method using approximate moment transfers and maximum entropy distribution	476
	MOMINT	Method using moment transfers by integration	478
	TRANSF	Generalized transformation of variable technique	482
Operations on probability functions (Sec. A.11)	AMAX	Calculates maximum value of a density function and determines mode	490
	CUMTOF	Converts cumulative distribution function to density function	491
	CUMTOH	Converts cumulative distribution function to hazard function	491
	CUMTOR	Converts cumulative distribution function to reliability function	492
	CUM1	Determines value of cumulative distribution function at any given single point, when density function is defined numerically	492
	FINVRT	Finds the value of a random variable corresponding to a given value of the cumulative distribution function	493
	FNORM	Normalizes a density function	493
	FTOCUM	Converts a density function to a cumulative distribution function	493
	FTOH	Converts a density function to a hazard function	494
	FTOR	Converts a density function to a reliability function	495
	HTOCMH	Converts a hazard function to a cumulative hazard function	495

TABLE (Continued)

Family	Name	Description	Page
	HTOCUM	Converts a hazard function to a cumulative distribution function	495
	HTOF	Converts a hazard function to a density function	496
	HTOR	Converts a hazard function to a reliability function	496
	RTOCUM	Converts a reliability function to a cumulative distribution function	496
	RTOF	Converts a reliability function to a density function	497
	RTOH	Converts a reliability function to a hazard function	497
Probability laws (Sec. A.12)	BASIC	Generates a set of basic probabilities	499
	COMBO	Expresses all combinations of n events	500
	JOINT	Expresses any joint probability in terms of basic probabilities	499
	UNION	Expresses any combined probability in terms of basic probabilities	500
Sample analysis (Sec. A.13)	BOUNDS	Determines upper and lower bounds in a sample (see Sec. A.6)	437
	SMOM	Calculates central moments from a sample of data (see Sec. A.7)	443
	SORT	Sorts the sample from low to high (see Sec. A.6)	440

A.2 INTRODUCTION

This software system is designed to make probabilistic design rapid and convenient for the designer. Only relatively simple calling programs are required to use the subroutines. The simpler programs have the necessary documentation in the listing, whereas more complex ones have a separate user's manual provided.

The user can easily incorporate additional subroutines that are consistent with the PROBVAR package. It is strongly recommended

that graphics subroutines be added, using local facilities and software, that will make it easy and convenient to draw density functions
in every application. This experience is vital for developing judgmental skill in probabilistic design. Graphics subroutines are included here; and they may serve as a guide. They are, however, machine dependent. The machine-dependent plotting subroutines are
described in Sec. A.15.

A.3 ANALYTICAL DENSITY FUNCTIONS

Subprograms are included to provide local values for many of the commonly used analytical density functions.

```
          FUNCTION FBETA(XMIN,XMAX,Q,R,X)
C.... CALCULATES BETA PROBABILITY DENSITY FUNCTION
C         XMIN = LOWER BOUND
C         XMAX = UPPER BOUND
C         Q,R = PARAMETERS
C         X  = RANDOM VARIABLE
          IF(X.GT.XMAX+1.E-8.OR.X.LT.XMIN-1.E-8)GO TO 1
          IF(Q.LT.1..AND.X-XMIN.LT.1.E-6)GO TO 1
          IF(R.LT.1..AND.XMAX-X.LT.1.E-6)GO TO 1
          A=GAMMA(Q)
          B=GAMMA(R)
          C=Q+R
          D=GAMMA(C)
          DENOM=A*B/D*(XMAX-XMIN)**(C-1.)
          TEMP1=XMAX-X
          IF(ABS(Q-1.).LT.1.E-8)GO TO 2
          IF(ABS(TEMP1).LT.1.E-8)GO TO 1
          TEMP2=X-XMIN
          IF(ABS(R-1.).LT.1.E-8)GO TO 3
          IF(ABS(TEMP2).LT.1.E-8)GO TO 1
          GO TO 4
2         FBETA=TEMP1**(R-1.)/DENOM
          RETURN
3         FBETA=TEMP2**(Q-1.)/DENOM
          RETURN
4         FBETA=TEMP2**(Q-1.)*TEMP1**(R-1.)/DENOM
          RETURN
1         FBETA=0.0
          RETURN
          END




          FUNCTION FEXP(X,R)
C.... CALCULATES LOCAL VALUES OF EXPONENTIAL DENSITY FUNCTION
C.... INPUT
C         X = GIVEN VALUE OF RANDOM VARIABLE
C         R = PARAMETER VALUE
          IF(X.LT.-.000001)GO TO 1
          IF(X.LT..000001)GO TO 2
          FEXP=EXP(-R*X)*R
          RETURN
1         FEXP=0.0
          RETURN
          END
```

```
          FUNCTION FGAMAC(RATE,EVENTS,X)
C.... CALCULATES LOCAL VALUE OF GAMMA DENSITY FUNCTION WHEN PARAMETER
C         CORRESPONDING TO NUMBER OF EVENTS IS NON-INTEGER
C.... INPUT
C            RATE = RATE PARAMETER (LAMBDA)
C            EVENTS = NON-INTEGER EVENTS PARAMETER (K)
C            X = RANDOM VARIABLE
          IF(X.LE.0.0)GO TO 1
          FGAMAC=RATE*RATE**(EVENTS-1)*EXP(-RATE*X)/GAMMA(X)
          RETURN
1         FGAMAC=0.0
          RETURN
          STOP
          END




          FUNCTION FGAMAI(RATE,KEVENT,X)
C.... CALCULATES GAMMA DENSITY FUNCTION FOR INTEGER PARAMETER EQUALS
C         NUMBER OF EVENTS
C.... RATE = AVERAGE RATE OF OCCURENCE OF EVENTS
C         KEVENT = NUMBER OF EVENTS SPECIFIED
C         X = SPECIFIED INTERVAL
          IF(X.LT.-.000001)GO TO 1
          IF(X.LT..000001.AND.KEVENT.EQ.1)GO TO 2
          K=KEVENT-1
          IFACT=FACTO(K)
          FGAMAI=RATE*(RATE*X)**K*EXP(-RATE*X)/FLOAT(IFACT)
          RETURN
1         FGAMAI=0.0
          RETURN
2         FGAMAI=RATE
          RETURN
          END




          FUNCTION FGAUS(X,XMEAN,DEV)
C.... EVALUATES LOCAL VALUES OF NORMAL DENSITY FUNCTION
C.... INPUT
C            X = RANDOM VARIABLE
C            XMEAN = MEAN VALUE PARAMETER
C            DEV = STANDARD DEVIATION PARAMETER
          FGAUS=EXP(-(X-XMEAN)**2/(2.*DEV*DEV))/(DEV*SQRT(6.28318))
          RETURN
          END




          FUNCTION FRANK(RANK,I,N)
C.... COMPUTES LOCAL VALUES OF RANK DISTRIBUTION
C.... INPUT
C            RANK = VALUE OF RANK
C            I = ORDER
C            N = SAMPLE SIZE
          IF(RANK.LT.-1.E-10.OR.RANK.GT.1.+1.E-10)GO TO 8
          IF(I.EQ.1.AND.RANK.LT.1.E-10)GO TO 9
          IF(I.EQ.N.AND.RANK.GT.1.-1.E-10)GO TO 9
          P1=1
          DO 1 K=1,N
1         P1=P1*FLOAT(K)
          L=I-1
```

```
         P2=1.
         IF(I.EQ.1)GO TO 4
         DO 2 K=1,L
  2      P2=P2*FLOAT(K)
  4      L=N-I
         P3=1.
         IF(N.EQ.I)GO TO 7
         DO 5 K=1,L
  5      P3=P3*FLOAT(K)
  7      FRANK=P1/(P2*P3)*RANK**(I-1)*(1.-RANK)**L
         RETURN
  8      FRANK=0.
         RETURN
  9      FRANK=FLOAT(N)
         END




         FUNCTION FWEIB(X,XL,ETA,BETA)
C....  CALCULATES LOCAL VALUE OF WEIBULL DENSITY FUNCTION
C....  INPUT
C          X = RANDOM VARIABLE
C          XL = LOWER BOUND OF X
C          ETA = ETA PARAMETER (SCALE)
C          BETA = BETA PARAMETER (SHAPE)
         IF(X.LT.XL+.00001.AND.BETA.LT.1.)GO TO 1
         IF(X.LT.XL*1.00001.AND.BETA.LT.1.)GO TO 1
         IF(X.LT.XL)GO TO 1
         IF(X.LT.XL*1.00001.AND.ABS(BETA-1.).LT..00001)GO TO 2
         IF(X.LT.XL+.00001.AND.ABS(BETA-1.).LT..00001)GO TO 2
         TEMP1=X-XL
         TEMP2=ETA-XL
         IF(BETA.LT.1.)GO TO 3
         FWEIB=BETA/TEMP2*(TEMP1/TEMP2)**(BETA-1.)*EXP(-(TEMP1/TEMP2)**
        1BETA)
         RETURN
  1       FWEIB=0.0
         RETURN
  2      FWEIB=BETA/(ETA-XL)
         RETURN
  3      FWEIB=BETA/TEMP2*(TEMP2/TEMP1)**(1.-BETA)*EXP(-(TEMP1/TEMP2)**BETA
        1)
         END




         FUNCTION POISS (FREQ,X,INTVL)
         PROD=EXP(-FREQ*INTVL)
C....     CALCULATES POISSON PROBABILITY MASS FUNCTION
C          FREQ= MEAN FREQUENCY OF OCCURRENCE
C          X= NUMBER OF OCCURRENCES IN SPECIFIED INTERVAL
C          INTVL= SPECIFIED INTERVAL
C
         N=IFIX(X)
         IF (N.EQ.0) RETURN
         DO 1 I=1,N
         PROD=(PROD*INTVL*FREQ)/(X-FLOAT(I-1))
  1      CONTINUE
C
         POISS=PROD
         RETURN
         END
```

A.4 DENSITY FUNCTION GENERATION

Subroutine HIST can be used to provide the data defining a histogram from a given sample; and, if desired, plot it. It also determines the corresponding density function, based on smoothing the histogram midpoints, and using cubic interpolation to increase the number of points defining the density function to any desired number. There is also the option of normalizing the histogram so that its area equals 1, making it more directly comparable to the corresponding density function. The latter can be plotted on top of the histogram plot by means of PLOTPL with IFLAG set at 2. Subroutine HTPL, associated with HIST, can be used to plot a histogram when the interval and frequency data are provided.

Subroutine MEP1 generates a maximum entropy density function based on sample moments, whereas MEP2 is similar but based on sample ranks. These subroutines have separate documentation.

```
      SUBROUTINE EVOL(XOLD,FOLD,MINT,XDATA,NDATA,SCALE,XLOW,XHIGH,XNEW,
     1FNEW)
C.... DETERMINES EVOLUTIONARY PARAMETERS TO DEFINE A DENSITY FUNCTION
C     FOR A NEW RANDOM VARIABLE CONSIDERED SIMILAR TO AN OLD ONE THAT
C     IS KNOWN
C.... INPUT
C        XOLD(I) = ARRAY DEFINING EQUALLY SPACED VALUES OVER THE RANGE OF
C                  THE OLD VARIABLE, DIMENSION MINT
C        FOLD(I) = ARRAY DEFINING VALUES OF KNOWN DENSITY FUNCTION
C                  CORRESPONDING TO XOLD, DIMENSION MINT
C        MINT = SIZE OF ABOVE ARRAYS AND OF NEW DENSITY FUNCTION ARRAYS
C               - MUST BE LE 101
C        XDATA(I) = DATA FOR NEW RANDOM VARIABLE, DIMENSION NDATA
C        NDATA = NUMBER OF KNOWN DATA VALUES, MUST BE LE 100
C.... OUTPUT
C        SCALE = EVOLUTIONARY SCALE PARAMETER
C        XLOW = LOWER BOUND OF NEW VARAIABLE
C        XHIGH = UPPER BOUND OF NEW VARIABLE
C        XNEW(I) = ARRAY OF EQUALLY SPACE VALUES OF NEW VARIABLE,
C                  DIMENSION MINT
C        FNEW(I) = ARRAY OF VALUES DEFINING NEW DENSITY FUNCTION
C                  DIMENSION MINT
C
      DIMENSION XOLD(1),FOLD(1),XDATA(1),XNEW(1),FNEW(1)
      DIMENSION X(2),XSTRT(2),RMAX(2),RMIN(2),PHI(1),PSI(1),WORK(18)
      COMMON/DATA/NSAMP,XD(100),M,XO(101),FO(101),CUMOLD(101)
      COMMON/SEEK/IDATA,IPRINT,NSHOT,NTEST,MAXM,F,G,TOL,ZERO,R,REDUCE
      IDATA=0
      IPRINT=-1
C.... TRANSFER DATA TO COMMON
      NSAMP=NDATA
      M=MINT
      DO 1 I=1,NDATA
1     XD(I)=XDATA(I)
      DO 2 I=1,MINT
      XO(I)=XOLD(I)
2     FO(I)=FOLD(I)
C.... SORT SAMPLE DATA
      CALL SORT(XD,NDATA)
C.... FIND OLD CUMULATIVE DISTRIBUTION FUNCTION
      RANGE=XOLD(MINT)-XOLD(1)
```

```
      CALL FTOCUM(FOLD,CUMOLD,RANGE,MINT)
C.... DEFINE PARAMETERS FOR OPTIMIZATION SUBROUTINE - X(1)=SCALE, X(2)=
C     XLOW
      XSTRT(1)=1.0
      XSTRT(2)=XD(1)
      RMIN(1)=0.0
      RMAX(1)=5.
      RMIN(2)=0.5*XD(1)
      RMAX(2)=2.*XD(1)
C.... CALL OPTIMIZATION SUBROUTINE
      CALL SEEK(2,0,0,1,RMAX,RMIN,XSTRT,X,U,PHI,PSI,NVIOL,WORK)
C.... DEFINE OUTPUT
      SCALE=X(1)
      XNEW(1)=XLOW=X(2)
      FNEW(1)=FOLD(1)/SCALE
      DO 3 I=2,MINT
      XNEW(I)=XLOW+SCALE*(XOLD(I)-XOLD(1))
3     FNEW(I)=FOLD(I)/SCALE
      XHIGH=XNEW(MINT)
      RETURN
      END



      FUNCTION FITT1(X,NSAMP,F)
C.... KOLMOGOROV-SMIRNOV GOODNESS OF FIT TEST
C.... INPUT
C        X(I) = ARRAY OF SAMPLE VALUES, NEED NOT BE ORDERED, DIMENSION
C               NSAMP
C        NSAMP = SAMPLE SIZE
C        F(X) = EXTERNAL FUNCTION DEFINING THE THEORETICAL CUMULATIVE
C               DISTRIBUTION FUNCTION BEING TESTED - F(X) MUST BE
C               DECLARED EXTERNAL IN THE CALLING PROGRAM
C.... OUTPUT
C        FITT1 = TEST VALUE
      DIMENSION X(1)
      CALL SORT(X,NSAMP)
      TEST=0.0
      DO 1 I=1,NSAMP
      Q=FLOAT(I)/FLOAT(NSAMP)
      FITT1=ABS(F(X(I))-Q)
      IF(FITT1.GT.TEST)TEST=FITT1
1     CONTINUE
      FITT1=TEST
      RETURN
      END



      FUNCTION FITT2(X,NSAMP,NBARS,BU,BL,F,XFREQ,FREQ,XDENS,FDENS,W1,W2)
C.... CHI SQUARE GOODNESS OF FIT TEST
C.... INPUT
C        X(I) = ARRAY OF SAMPLE VALUES, NEED NOT BE ORDERED, DIMENSION
C               NSAMP
C        NSAMP = SAMPLE SIZE
C        NBARS = NUMBER OF INTERVALS TO BE USED
C        BU = UPPER BOUND
C        BL = LOWER BOUND
C        F(XX) = EXTERNAL FUNCTION DEFINING THE THEORETICAL DISTRIBUTION
C                BEING USED WHEN VARIABLE HAS A VALUE IN XX - F MUST BE
C                DECLARED EXTERNAL IN THE CALLING PROGRAM
C.... OUTPUT
C        FITT2 = TEST VALUE
C.... WORKING ARRAYS - DIMENSION AS FOLLOWS - XFREQ(NBARS+1),
C        FREQ(NBARS), XDENS(NBARS), FDENS(NBARS), W1(NBARS), W2(NBARS)
```

```
              DIMENSION X(1),XFREQ(1),FREQ(1),XDENS(1),FDENS(1),W1(1),W2(1)
              CALL SORT(X,NSAMP)
              FITT2=0.0
C.... OBTAIN HISTOGRAM OF FREQUENCIES
              CALL HIST(X,NSAMP,NBARS,BU,BL,1,0,1,1H ,NBARS,XFREQ,FREQ,XDENS,
             1FDENS,W1,W2)
C.... OBTAIN INTERVALS AND MID-POINTS
              DELX=(BU-BL)/FLOAT(NBARS)
              DO 1 I=1,NBARS
1             XFREQ(I)=XFREQ(I)+DELX/2.
C.... EVALUATE TEST VALUE
              DO 2 I=1,NBARS
              EI=F(XFREQ(I))
              IF(EI*(BU-BL)/2..LT.1.E-6)GO TO 3
2             FITT2=FITT2+(FREQ(I)-EI)**2/EI
              RETURN
3             WRITE(6,4)
4             FORMAT(59H CHI SQUARE TEST ABANDONED BECAUSE RANGE OF DATA IS OUTS
             1IDE/30H RANGE OF THEORETICAL FUNCTION)
              STOP
              END



              SUBROUTINE HIST (X,NSAMP,NBARS,BU,BL,NORM,IPLOT,N,LABEL,MINT,XFREQ
             1,FREQ,XDENS,FDENS,W1,W2)
C.... GENERATES A HISTOGRAM FROM A DATA SAMPLE; AND GENERATES A DENSITY
C     FUNCTION FROM THE HISTOGRAM
C.... INPUT
C         X(I) = ARRAY OF VALUES IN SAMPLE, DIMENSION NSAMP
C         NSAMP = SIZE OF SAMPLE
C         NBARS = NUMBER OF HISTOGRAM BARS - MUST BE 4 OR MORE
C         BU = UPPER BOUND
C         BL = LOWER BOUND
C         NORM = 0, FREQUENCIES ARE CALCULATED
C              = 1, FREQUENCIES ARE NORMALIZED TO MAKE HISTOGRAM AREA
C                   EQUAL ONE
C         IPLOT = 0, NO PLOT
C              = 1, PLOTS HISTOGRAM
C         MINT = NUMBER OF STATIONS DEFINING DENSITY FUNCTION
C         ICALL = 0, RAW DATA INPUT, DEFAULT VALUE
C              = 1, FREQUENCIES ARE INPUT - XFREQ AND FREQ ARE KNOWN,
C                   X,NSAMP NEED NOT BE DEFINED
C.... TO PLOT THE DENSITY FUNCTION, CALL PLOTPL IN MAIN PROGRAM. IF IT
C     IS TO BE SUPERIMPOSED ON THE HISTOGRAM, SET IFLAG=2
C         LABEL = NAME TO BE USED ON X AXIS, USE HOLLERITH CODE
C         N = NUMBER OF CHARACTERS IN LABEL
C         MINT = NUMBER OF VALUES TO DEFINE DENSITY FUNCTION
C.... OUTPUT
C         FREQ(I) = ARRAY OF FREQUENCIES, DIMENSION NBARS
C         XFREQ(I) = ARRAY OF X VALUES DEFINING INTERVALS, DIMENSION
C                    NBARS+1
C         XDENS(I) = ARRAY OF EQUALLY SPACED DISCRETE VALUES OF RANDOM
C                    VARIABLE DEFINING DENSITY FUNCTION, DIMENSION MINT
C         FDENS(I) = ARRAY OF DENSITY FUNCTION VALUES CORRESPONDING TO
C                    XDENS, DIMENSION MINT
C
C.... IN SETTING UP THE DENSITY FUNCTION A CUBIC IS FITTED TO FOUR
C     ADJACENT POINTS TO EACH BOUND AND USED TO EXTEND THE CURVE TO THE
C     BOUND. IF THE CURVE GOES NEGATIVE IT IS SET TO ZERO.
C.... IF A SPECIFIED BOUND FALLS INSIDE THE RANGE OF DATA IT IS IGNORED
C     AND THE EXTREME DATA VALUE IS USED. THUS A USER NOT WISHING TO
C     SPECIFY A BOUND CAN SET IT EQUAL TO ANY DATA VALUE.
C.... W1(I) = WORKING ARRAY, DIMENSION NBARS
C     W2(I) = WORKING ARRAY, DIMENSION NBARS
```

```
C
      DIMENSION X(1), FREQ(1), XFREQ(1), XDENS(1), FDENS(1), XMARK(20),
     1YMARK(20), W1(1), W2(1)
      COMMON /HIST/ ICALL
      DATA ICALL/0/
      IF (ICALL.EQ.1)XMAX=XFREQ(NBARS+1)
      IF (ICALL.EQ.1)XMIN=XFREQ(1)
      IF (ICALL.EQ.1) GO TO 7
C.... INITIALIZE
      DO 1 I=1,NBARS
      FREQ(I)=0.0
1     CONTINUE
C.... DETERMINE RANGE OF DATA AND SET UP INTERVALS
      CALL BOUNDS (X,NSAMP,XMAX,XMIN)
      XMAX=XMAX*1.0001
      RANGE=XMAX-XMIN
      DELX=RANGE/FLOAT(NBARS)
      XFREQ(1)=XMIN
      NS=NBARS+1
      DO 2 I=2,NS
      XFREQ(I)=XFREQ(I-1)+DELX
2     CONTINUE
C.... COUNT THE FREQUENCY OF OBSERVATIONS IN EACH INTERVAL
      DO 5 I=1,NSAMP
      DO 3 J=2,NS
      IF (X(I).GT.XFREQ(J)) GO TO 3
      KOUNT=J-1
      GO TO 4
3     CONTINUE
4     FREQ(KOUNT)=FREQ(KOUNT)+1.
5     CONTINUE
      IF (NORM.EQ.0) GO TO 7
      DO 6 I=1,NBARS
      FREQ(I)=FREQ(I)/(DELX*NSAMP)
6     CONTINUE
7     IF (IPLOT.EQ.0) GO TO 9
C.... PLOT HISTOGRAM
C
C.... DETERMINE BOUNDS ON FREQ
      FMIN=0.0
      FMAX=FREQ(1)
      DO 8 I=2,NBARS
      IF (FREQ(I).GT.FMAX) FMAX=FREQ(I)
8     CONTINUE
C.... DETERMINE SCALE OF FREQUENCY
      YSCALE=FMAX/3.5
C.... DETERMINE SCALE OF X
      TOP=BU
      BOT=BL
      IF (BU.LT.XMAX) TOP=XMAX
      IF (BL.GT.XMIN) BOT=XMIN
      RANGE=TOP-BOT
C     XSCALE = DATA UNITS PER INCH
      XSCALE=RANGE/6.5
C.... OBTAIN X AND Y INTERVALS
      CALL XAXIZ (XSCALE,BOT,TOP,XMARK,NMARK,RX,RK)
      CALL YAXIZ (YSCALE,FMIN,FMAX,YMARK,NMARKY,QX,QK)
C.... INITIATE PLOT
      V=-2.*XSCALE+XMARK(1)
      W=-2.*YSCALE
      XMIN=V
      YMIN=W
      XMAX=XMARK(NMARK)+XSCALE
      YMAX=FMAX+2.*YSCALE
```

```
             CALL PLTIN (XSCALE,YSCALE,V,W,XMIN,XMAX,YMIN,YMAX)
C.... PLOT X AXIS AND Y AXIS
             XAXIS=(XMARK(NMARK)-XMARK(1))/XSCALE+2.
             YAXIS=YMARK(NMARKY)/YSCALE+2.
C.... PLOT X AND Y AXES
             CALL XAXPL (XMARK,NMARK,XSCALE,RK,RX,XAXIS)
             CALL YAXPL (YMARK,NMARKY,YSCALE,QK,QX,YAXIS)
C.... LABEL X AXIS
             XLABEL=(XMARK(NMARK)-V)/XSCALE-FLOAT(N)*.15
             YLABEL=1.4
             CALL LETTER (N,.15,0.,XLABEL,YLABEL,LABEL)
C.... LABEL Y AXIS
             CALL LETTER (9,.15,90.,1.50,3.6,9HFREQUENCY)
C.... PLOT HISTOGRAM
             IF (IPLOT.NE.1) GO TO 9
             XAXIS=0.0
             CALL HTPL (XFREQ,FREQ,XAXIS,NBARS)
C.... OBTAIN DENSITY FUNCTION
C.... SMOOTH FREQUENCY DATA
9            CALL SMOOTH (FREQ,W1,NBARS)
             IF (ICALL.EQ.0) GO TO 10
             DELX=(XFREQ(NBARS+1)-XFREQ(1))/FLOAT(NBARS)
10           DO 11 I=1,NBARS
             W2(I)=XFREQ(I)+DELX/2.
11           CONTINUE
C.... SET UP XDENS
             IF (BL.GE.XFREQ(1)) GO TO 12
             XMIN=BL
             GO TO 13
12           XMIN=XFREQ(1)
13           IF (BU.LE.XFREQ(NBARS+1)) GO TO 14
             XMAX=BU
             GO TO 15
14           XMAX=XFREQ(NBARS+1)
15           XDENS(1)=XMIN
             RANGE=XMAX-XMIN
             DELD=RANGE/FLOAT(MINT-1)
             DO 16 I=2,MINT
             XDENS(I)=XDENS(I-1)+DELD
16           CONTINUE
C.... DETERMINE VALUES BETWEEN BL AND W2(1)
C        CUBIC INTERPOLATION TO GET FDENS(1)
             K=1
             FDENS(1)=FCUBIN(K,W2,W1,XDENS(1),NBARS)
             IF (FDENS(1).LT.0.0) FDENS(1)=0.0
C.... CUBIC ITERPOLATION FOR VALUES LESS THAN W2(1)
             FD10=(W1(1)-FDENS(1))/(W2(1)-XDENS(1))
             FD21=(W1(2)-W1(1))/DELX
             FD32=(W1(3)-W1(2))/DELX
             FD210=(FD21-FD10)/(W2(2)-XDENS(1))
             FD321=(FD32-FD21)/(2.*DELX)
             FD3210=(FD321-FD210)/(W2(3)-XDENS(1))
             I=2
17           IF (XDENS(I).GT.W2(1)) GO TO 18
             FDENS(I)=FDENS(1)+(XDENS(I)-XDENS(1))*FD10+(XDENS(I)-W2(1))*(XDENS
            1(I)-XDENS(1))*FD210+(XDENS(I)-W2(2))*(XDENS(I)-W2(1))*(XDENS(I)-XD
            2ENS(1))*FD3210
             IF (FDENS(I).LT.0.0) FDENS(I)=0.0
             I=I+1
             GO TO 17
C.... USE CUBIC INTERPOLATION BETWEEN W2(1) AND W2(NBARS)
18           DO 19 J=2,NBARS
             IF (XDENS(I).LE.W2(J)) GO TO 20
19           CONTINUE
```

```
20      K=J-2
        IF (K.LT.1) K=1
        IF (K.GT.NBARS-3) K=NBARS-3
        FDENS(I)=FCUBIN(K,W2,W1,XDENS(I),NBARS)
        IF (I.EQ.MINT) GO TO 22
        I=I+1
        IF (XDENS(I).LE.W2(NBARS)) GO TO 18
C.... USE CUBIC INTERPOLATION TO GET FDENS(MINT)
        K=NBARS-3
        FDENS(MINT)=FCUBIN(K,W2,W1,XDENS(MINT),NBARS)
        IF (FDENS(MINT).LT.0.0) FDENS(MINT)=0.0
C.... CUBIC INTERPOLATION FOR VALUES GT W2(NBARS)
        FD10=(W1(NBARS-1)-W1(NBARS-2))/DELX
        FD21=(W1(NBARS)-W1(NBARS-1))/DELX
        FD32=(FDENS(MINT)-W1(NBARS))/(XDENS(MINT)-W2(NBARS))
        FD210=(FD21-FD10)/(2.*DELX)
        FD321=(FD32-FD21)/(XDENS(MINT)-W2(NBARS-1))
        FD3210=(FD321-FD210)/(XDENS(MINT)-W2(NBARS-2))
21      FDENS(I)=W1(NBARS-2)+(XDENS(I)-W2(NBARS-2))*FD10+(XDENS(I)-W2(NBAR
       1S-1))*(XDENS(I)-W2(NBARS-2))*FD210+(XDENS(I)-W2(NBARS))*(XDENS(I)-
       2W2(NBARS-1))*(XDENS(I)-W2(NBARS-2))*FD3210
        IF (FDENS(I).LT.0.0) FDENS(I)=0.0
        IF (I.EQ.MINT) GO TO 22
        I=I+1
        GO TO 21
C.... NORMALIZE
22      CALL FNORM (FDENS,RANGE,MINT)
        RETURN
        END



        SUBROUTINE HTPL(XFREQ,FREQ,XAXIS,NBARS)
C.... PLOTS A HISTOGRAM AFTER AXES HAVE BEEN PLOTTED
C.... INPUT
C       XFREQ(I) = ARRAY OF VALUES DEFINING INTERVALS, DIMENSION NBARS+1
C       FREQ(I) = LOCATION OF TOP OF I TH BAR IN DATA UNITS, DIMENSION
C                 NBARS
C       XAXIS = LOCATION OF X AXIS IN DATA UNITS
C       NBARS = NUMBER OF HISTOGRAM BARS
C
        DIMENSION XFREQ(1), FREQ(1)
        CALL UNITTO (XFREQ(1),XAXIS,XP,YPO)
        CALL PLOT (XP,YPO,3)
        CALL UNITTO (XFREQ(1),FREQ(1),XP,YP)
        CALL PLOT (XP,YP,2)
        DO 1 I=1,NBARS
        CALL UNITTO (XFREQ(I+1),FREQ(I),XP,YP)
        CALL PLOT (XP,YP,2)
        IF (I.EQ.NBARS) GO TO 1
        CALL UNITTO (XFREQ(I+1),FREQ(I+1),XP,YP)
        CALL PLOT (XP,YP,2)
1       CONTINUE
        DO 2 I=1,NBARS
        CALL UNITTO (XFREQ(I+1),XAXIS,XP,YP)
        CALL PLOT (XP,YP,3)
        CALL UNITTO (XFREQ(I+1),FREQ(I),XP,YP)
        CALL PLOT (XP,YP,2)
2       CONTINUE
        RETURN
        END
```

Subroutine MEP1(N,CM,XMIN,XMAX,NXP,XP,KSTART,KDATA,AL, CUM)

Purpose

This subroutine provides the least biased estimate of the probability density function for the random variable x, where only the first n moments are used and the bounds are known. The density function y has the form

$$y = \exp (\lambda_1 + \lambda_2 x + \lambda_3 x^2 + \cdots + \lambda_{n+1} x^n) \tag{1}$$

The program gives the values of the λ's. It also provides values of the cumulative distribution function for given values of the random variable x.

How to Use

The calling program in its simplest form is set up as follows.

a. *DIMENSION statement*. Check through the list of input and output variables and include all subscripted variables, dimensioning as indicated.
b. *COMMON statement*. If nondefault values of the parameters are desired, a labeled COMMON statement must be added. See the section on programming information below.
c. *Define input data*. This may be done by simple arithmetic statements, DATA declarations, or READ statements. Each variable in the input list below must be included unless the default value is used.
d. Call subroutine MEP1.
e. Call the output subroutine. Library output subroutines can be used; see below.
f. Add STOP and END.

A sample calling program is given below.

```
DIMENSION TA(41),FTA(41),CMA(5),W(41),AL(7),XP(1),CUM(1)
COMMON/MEP1/KPRINT,TOL,MAXFN
KPRINT=0
TOL=1.E-4
MAXFN=200
DATA FTA/0.0,.034,.124,.230,.341,.468,.600,.744,.873,.954,
1.990,1.000,.990,.971,.949,.919,.887,.850,.811,.771,.728,
2.684,.640,.596,.550,.504,.461,.419,.375,.334,.292,.250,.211,
3.176,.141,.110,.080,.052,.030,.012,0.0/
       CALL FNORM(FTA,200.,41)
       DO 1 I=1,41
1      TA(I)=FLOAT(I-1)*5.
       CALL CMOM(TA,FTA,41,5,CMA,W)
       WRITE(6,100)CMA
100    FORMAT(5E12.5)
```

```
      CALL MEP1(5,CMA,0.0,200.,0,XP,1,0,AL,CUM)
      WRITE(6,101)(AL(I),I=1,6)
101   FORMAT(6E12.5)
      STOP
      END
```

Input Variables

N: number of first moments, should be less than or equal to 6.

CM(I): array containing the first N moments, dimensioned N. Moments above the first are central moments. MEP1 blows up if the mean is zero.

XMIN: lower bound of the variable.

XMAX: upper bound of the variable.

*TOL: The allowed relative error in the moment value; a solution exists where

$$R(I) = \frac{cc_i - cc_i'}{c_i} \leqslant TOL \tag{2}$$

where cc_i is the given moment and cc_i' is the predicted moment. The default value for TOL is 10^{-6}.

KDATA: = 1, all input data are printed out,
= 0, input data are not printed out.

*MAXFN: maximum number of iterations in the optimization subroutine (MPOPT). The default value is 50.

*KPRINT: prints results every KPRINT cycle, set = 0 for no intermediate output. If KPRINT $\neq$ 0, all intermediate results before optimization, the starting method name, and the starting values of the λ's, are printed out. In addition, the following are printed: cycle number; number of function evaluations (subroutine FUNCT); the norm gradient; total residuals ($\sum_i^n R_i^2$), where R_i is defined in (2); the values of the λ's; and the value of each individual R_i. The default value is zero.

KSTART: = 0, user-provided starting values placed in AL(I) (see the section on output variables),
= 1, normal assumption starting method,
= 2, uniform assumption starting method,
= 3, N-points assumption starting method,
= 4, step-by-step assumption starting method.
Set to 1 if no particular starting method is preferred. In this event the subroutine will try in sequence all four methods until one succeeds.

NXP: number of points for which it is required to calculate the cumulative distribution function. If none are required, set equal to zero.

XP(I): array containing the values of the independent variable for which the values of the cumulative distribution functions are to be calculated, dimensioned with the value of NXP. IF NXP is zero, dimension X(I) with 1.

Output Variables

AL(I): array containing the Lagrangian multipliers or λ's, dimensioned at (N + 2). Although there are only (N + 1) λ's, the (N + 2) subscript is used internally.

CUM(I): array containing the values of the cumulative distribution function of X, dimensioned with the value of NXP. If NXP is zero, dimension CUM(I) with 1.

Programming Information

MEP1 has full variable dimensioning. The calling program must provide dimensioning as given above. Input quantities with an asterisk have internally assigned default values. If the user wishes to change one of these it is necessary to add the following common statement:

COMMON /MEP1/KPRINT,TOL,MAXFN

Output subroutines are available for use with MEP1. Subroutine MEPOUT prints the values of the distribution parameters and values of the cumulative distribution function if they were requested.

CALL MEPOUT(AL,NL,NXP,XP,CUM)

Subroutine ENTPP provides a printer plot of the density function.

CALL ENTPP (XMIN,XMAX,NL,AL)

Subroutine ENTPL provides a plotter plot of the density function. The latter two subroutines are machine dependent.

CALL ENTPL (XMIN,XMAX,NL,AL,NLAB,LABEL)

where NL is the number of parameters, equal to N + 1, and LABEL is the random variable name with NLAB characters, coded in Hollerith form.

It is also possible to determine values of the cumulative distribution function directly, by calling function CDF(XMIN,XMAX,XP,AL,N). This can be done quite independently of MEP1. Argument information is given in the listing.

MEP1 has the feature that after a successful run the values of AL(I) will be available as starting values for a subsequent run from the same calling program, and KSTART will be reset at zero. This is useful when iterative runs are being used to determine sensitivity, or when different samples of the same variable are being tried.

If MEP1 does not work with the default values and KSTART equal to 1, it is often helpful to rerun with KPRINT set at 1. The detailed output will indicate if any of the starting methods are showing signs of converging. If one is, KSTART should be set at the corresponding flag value, and MAXFN set larger than the default value. It may also be necessary to use fewer moments or reduce th value of TOL.

Subroutines called by MEP1 are LINES, MPOPT, FUNCT, START, SIMPSON, MULTI, CONVERT, CDF, TRN1, TRN2, FACTO, and ENTRPF.

```
      SUBROUTINE MEP1(N,CM,XMIN,XMAX,NXP,XP,KSTART,KDATA,AL,CUM)
C.... EXECUTIVE PROGRAM FOR USING MAXIMUM ENTROPY METHOD CONSTRAINED BY
C     MOMENTS TO GENERATE A DENSITY FUNCTION
C.... CODING WAS ORIGINALLY DEVELOPED BY Y. DIAB
C
      DIMENSION AL(1), CM(1), ETA(4), XP(1),CUM(1),CC(8),ALS(10)
      COMMON /FAIL/ NFAIL
      COMMON /HELP/ S(41),XX(16,41),C(8),M
      COMMON/MEP1/KPRINT,TOL,MAXFN
      DATA KPRINT,TOL,MAXFN/0,1.E-6,50/
      IF (N.EQ.1) KSTART=2
C
C     WRITE THE INPUT DATA
C
      IF (KDATA.EQ.0) GO TO 1
      WRITE (6,24)
      WRITE (6,25) KDATA
      WRITE (6,26) KPRINT
      WRITE (6,28) N
      WRITE (6,29) XMAX
      WRITE (6,30) XMIN
      WRITE (6,31) (CM(I),I=1,4)
      IF (N.GT.4) WRITE (6,21) (CM(I),I=5,N)
      IF(ABS(CM(1)).LT.1.E-4)GO TO 48
      WRITE (6,32) TOL
      WRITE (6,33) NXP
1     CONTINUE
      NFAIL=0
      M=31
      X2MIN=0.0
      X2MAX=1.
C     SAVE CM
      DO 100 I=1,N
100   CC(I)=CM(I)
C
C     CALCULATE THE MOMENTS AT THE MODIFIED LIMITS
C
      CALL TRN1 (XMAX,XMIN,CC,X2MAX,X2MIN,N)
C
C     CALCULATE THE MOMENTS ABOUT THE ORIGIN FOR THE MODIFIED LIMITS,
C     STORE THEM IN COMMON IN C
C
      CALL CONVERT(CC,N)
C
C     GENERATE THE SIMPSON MULTIPLIERS AND STORE THEM IN HELP COMMON
C
      CALL SIMSON
C
C     GENERATE THE X,S POWER FOR SUBROUTINE FUNCT, STORE THEM IN HELP
C     COMMON ARRAY
C
      CALL MULTI (X2MAX,X2MIN,N)
C
C     DEFINE THE INPUT DATA FOR SUBROUTINE MPOPT
C
      ETA(1)=1.E-12
      ETA(2)=TOL
      ETA(3)=1.E-24
```

```
         ETA(4)=1.E-24
         MODE=1
         UMIN=0.0
C
C        WRITE THE INTERMEDIATE RESULTS YOU HAVE OBTAINED SO FAR
C
         IF (KPRINT.EQ.0) GO TO 2
         WRITE (6,34)
         WRITE (6,35) M
         WRITE (6,36) X2MAX,X2MIN
         WRITE (6,37) (CC(I),I=1,4)
         IF (N.GT.4) WRITE (6,22) (CC(I),I=5,N)
         WRITE (6,38) (C(I),I=1,4)
         IF (N.GT.4) WRITE (6,22) (C(I),I=5,N)
         WRITE (6,39) (ETA(I),I=1,4)
2        CONTINUE
C
C        FIND A STARTING POINT FOR SUBROUTINE MPOPT TO START THE OPTIMIZAT-
C        ION ALGORITHM
C
         IF(KSTART.EQ.0)GO TO 16
         IF(KSTART.EQ.4)WRITE(6,44)
         CALL START (X2MAX,X2MIN,AL,KSTART,CC,N,KPRINT,UMIN,MODE,MAXFN,ETA)
         IF (NFAIL.EQ.1) GO TO 9
C
C        PRINT THE STARTING VALUES
C
         IF (KPRINT.EQ.0) GO TO 7
         GO TO (3,4,5,6), KSTART
3        WRITE (6,40)
         WRITE (6,41) (AL(I),I=1,4)
         IF (N.GT.4) WRITE (6,22) (AL(I),I=5,N)
         GO TO 7
4        WRITE (6,42)
         WRITE (6,41) (AL(I),I=1,4)
         IF (N.GT.4) WRITE (6,22) (AL(I),I=5,N)
         GO TO 7
5        WRITE (6,43)
         WRITE (6,41) (AL(I),I=1,4)
         IF (N.GT.4) WRITE (6,22) (AL(I),I=5,N)
         GO TO 7
6        WRITE (6,41) (AL(I),I=1,4)
         IF (N.GT.4) WRITE (6,22) (AL(I),I=5,N)
         GO TO 7
16       CONTINUE
         RANGE=XMAX-XMIN
C.... CHANGE STARTING VALUES TO 0-1 DOMAIN FOR KSTART=0
C.... THIS ALGORITHM IS SIMILAR TO TRN2 BUT APPEARS TO GIVE BETTER
C        NUMERICAL RESULTS
         NPL=N+1
         IF(ABS(XMIN).LT.1.E-10)GO TO 19
         DO 17 I=2,NPL
         ALS(I)=0.0
         I1=I-1
         DO 18 J=I1,N
         ALS(I)=ALS(I)+FACTO(J)*XMIN**(J-I1)*RANGE**I1*AL(J+1)/FACTO(I1)
        1/FACTO(J-I1)
18       CONTINUE
17       CONTINUE
         GO TO 50
19       DO 20 I=2,NPL
20       ALS(I)=RANGE**(I-1)*AL(I)
```

```
C.... PUT AL(I) IN PROPER LOCATIONS
50    DO 51 I=1,N
51    AL(I)=ALS(I+1)
7     CONTINUE
      NFAIL=0
      IF (KPRINT.EQ.0) GO TO 8
      WRITE (6,45)
8     CONTINUE
      AL(N+1)=2.
      AL(N+2)=0.0
      CALL MPOPT (AL,N,ETA,UMIN,MAXFN,MODE,KPRINT)
      IF (NFAIL.EQ.0) GO TO 10
      IF (KSTART.EQ.4) GO TO 9
C
C     THE PROGRAM HAS FAILED SO FAR , TRY ANOTHER STARTING POINT AND TRY
C     AGAIN
C
      KSTART=KSTART+1
      IF (KSTART.EQ.4.AND.N.LE.2) GO TO 9
      GO TO 2
9     CONTINUE
      WRITE (6,46)
      CALL EXIT
10    CONTINUE
C
C     CALCULATE THE ZEROTH LAGRANGIAN MULTIPLIER
C
      SUM=0.0
      DO 12 I=1,M
      SZ=0.0
      DO 11 K=1,N
      SZ=SZ+AL(K)*XX(K,I)
11    CONTINUE
      SUM=SUM+S(I)*EXP(SZ)
12    CONTINUE
      NPL=N+1
      DO 13 I=1,N
      K=N+2-I
      AL(K)=AL(K-1)
13    CONTINUE
      DELTA=(X2MAX-X2MIN)/FLOAT(M-1)
      AL(1)=-ALOG(SUM*DELTA/3.)
      WRITE(6,101)UMIN
101   FORMAT(26H SUM OF RESIDUALS SQUARED=,E12.5)
      IF (KPRINT.EQ.0) GO TO 14
      WRITE (6,47) (AL(I),I=1,NPL)
14    CONTINUE
C.... RESET KSTART TO ZERO
      KSTART=0
C
C     CALCULATE THE LAGRANGIAN MULTIPLIERS FOR THE ORIGINAL LIMITS
C
      CALL TRN2 (XMAX,XMIN,AL,X2MAX,X2MIN,N)
C
C     CALCULATE THE CUMULATIVE DISTRIBUTION FUNCTION VALUE AT THE GIVEN
C     POINT
      IF(NXP.EQ.0)RETURN
      DO 15 I=1,NXP
      CUM(I)=CDF(XMIN,XMAX,XP(I),AL,NPL)
15    CONTINUE
      RETURN
C
21    FORMAT (57X,4E18.9,//)
22    FORMAT (57X,4E18.9,//)
```

```
24        FORMAT (1H1,//,20X,*INPUT DATA FOR SUBROUTINE    MEP1*,/,20X,33(*-*
         1),//)
25        FORMAT (* INPUT DATA IS PRINTED OUT FOR KDATA =1 ONLY . . .KDATA =
         1*,I18,/)
26        FORMAT (* INTERMEDIATE OUTPUT EVERY KPRINT(TH) CYCLE  . . KPRINT =
         1*,I18,/)
28        FORMAT (* NUMBER OF KNOWN FIRST MOMENTS . . . . . . . . . . .N=
         1*,I18,/)
29        FORMAT (* HIGHER LIMIT . . . . . . . . . . . . . . . . XMAX =
         1*,E18.9,/)
30        FORMAT (* LOWER LIMIT . . . . . . . . . . . . . . . . . XMIN =
         1*,E18.9,/)
31        FORMAT (* FIRST MOMENTS . . . . . . . . . . . . . . . CC(I) =
         1*,4E18.9,/)
32        FORMAT (* THE ALLOWED TOLERANCE IN LAGRANGIAN EQUATIONS . . .TOL =
         1*,E18.9,/)
33        FORMAT (* THE CUMULATIVE DISTRIBUTION REQUIRED AT NXP POINTS.NXP =
         1*,I18,/)
34        FORMAT (1H1,//,20X,*INTERMEDIATE RESULTS FOR SUBROUTINE   MEP*,/,2
         10X,41(*-*),//)
35        FORMAT (* NUMBER OF INTEGRATION STATION . . . . . . . . . . M =
         1*,I18,/)
36        FORMAT (* MODIFIED MAXIMUM AND MINIMUM LIMITS . . X2MAX , X2MIN =
         1*,2E18.9,/)
37        FORMAT (* MODIFIED MOMENTS ABOUT THE EXPECTED VALUE . . . .CC(I) =
         1*,4E18.9,/)
38        FORMAT (* MODIFIED MOMENTS ABOUT THE ORIGIN . . . . . . . C(I) =
         1*,4E18.9,/)
39        FORMAT (* SUBROUTINE MPOPT TOLERANCES . . . . . . . . . ETA(I) =
         1*,E18.9,/)
40        FORMAT (//,* NORMAL ASSUMPTION STARTING METHOD*/34(*-*),/)
41        FORMAT (* STARTING VALUES  . . . . . . . . . . . . . . AL(I) =
         1*,E18.9,/)
42        FORMAT (//,* UNIFORM ASSUMPTION STARTING METHOD*/35(*-*),/)
43        FORMAT (//,* N POINTS STARTING METHOD*/25(*-*),/)
44        FORMAT (//,* STEP BY STEP STARTING METHOD*/29(*-*),/)
45        FORMAT (//,* CYC NUMF   NORMGRAD       TOTAL*,24X,*VARIABLES*,40
         1X,*RESIDUALS*,/,* NO.*,22X,*RESIDUALS    X(1)        X(2)
         2    X(3)        X(4)       R(1)        R(2)        R(3)        R
         3(4)*,//)
46        FORMAT (* THE PROGRAM HAS FAILED*)
47        FORMAT (* THE MODIFIED LAGRANGIAN MULTIPLIERS ARE . . . . . . .
         1*,4E18.9/57X,4E18.9)
48        WRITE(6,49)
49        FORMAT(53H WARNING - MEAN IS NEARLY ZERO AND MEP1 WILL NOT WORK/12
         1H TRANSFORM X)
          STOP
          END



          FUNCTION CDF (XMIN,XMAX,XP,AL,N)
C         THIS FUNCTION SUBROUTINE IS TO CALCULATE THE CUMMULATIVE DISTRIBU-
C         TION FUNCTION AT A GIVEN POINT
C.... INPUT
C         XMIN = LOWER BOUND
C         XMAX = UPPER BOUND
C         XP = SPECIFIED POINT
C         AL(I) = ARRAY OF PARAMETERS, DIMENSION N
C         N = NUMBER OF PARAMETERS
          DIMENSION AL(1)
          IF (XP.LE.XMIN) GO TO 3
          IF (XP.GE.XMAX) GO TO 4
          RANGE=XMAX-XMIN
```

```
            RANGEN=XP-XMIN
            SS=RANGEN/RANGE*51.
            JSS=SS
            JSS=(JSS/2)*2+5
            AREA=0.0
            JSM1=JSS-1
            DELTA=RANGEN/FLOAT(JSM1)
            DO 1 I=2,JSM1,2
            X=XMIN+FLOAT(I-1)*DELTA
            AREA=AREA+4.*ENTRPF(AL,N,X)
      1     CONTINUE
            JSM1=JSM1-1
            DO 2 I=3,JSM1,2
            X=XMIN+FLOAT(I-1)*DELTA
            AREA=AREA+2.*ENTRPF(AL,N,X)
      2     CONTINUE
            AREA=AREA+ENTRPF(AL,N,XMIN)+ENTRPF(AL,N,XP)
            AREA=AREA*DELTA/3.
            CDF=AREA
            GO TO 5
      3     CDF=0.0
            GO TO 5
      4     CDF=1.
      5     CONTINUE
            RETURN
            END




            SUBROUTINE ENTPL(XMIN,XMAX,N,AL,NL,LABEL)
C.... PLOTS A MAXIMUM ENTROPY DENSITY FUNCTION ON THE PLOTTER
C.... INPUT
C         N = NUMBER OF PARAMETERS IN DENSITY FUNCTION
C         XMIN = LOWER BOUND
C         XMAX = UPPER BOUND
C         AL(I) = ARRAY OF PARAMETERS, DIMENSION N
C         NL = NUMBER OF CHARACTERS FOR LABEL OF RANDOM VARAIABLE
C         LABEL = LABEL FOR RANDOM VARIABLE AXIS
C                 USE HOLLERITH CODE
            DIMENSION AL(1),X(101),FX(101)
            M=101
            DELTA=(XMAX-XMIN)/FLOAT(M-1)
            DO 1 I=1,M
            X(I)=XMIN+FLOAT(I-1)*DELTA
      1     FX(I)=ENTRPF(AL,N,X(I))
            CALL PLOTPL(1,X,FX,M,LABEL,NL,XMAX,XMIN)
            RETURN
            END




            SUBROUTINE ENTPP(XMIN,XMAX,N,AL)
C.... PLOTS A MAXIMUM ENTROPY DENSITY FUNCTION ON THE PRINTER
C.... INPUT
C         N = NUMBER OF PARAMETERS IN DENSITY FUNCTION
C         XMIN = LOWER BOUND
C         XMAX = UPPER BOUND
C           AL(I) = ARRAY OF PARAMETERS, DIMENSION N
            DIMENSION AL(1)
            M=50
            DELTA=(XMAX-XMIN)/FLOAT(M-1)
            DO 1 I=1,M
            X=XMIN+FLOAT(I-1)*DELTA
            Y=ENTRPF(AL,N,X)
      1     CALL PLOTPT(X,Y,9)
```

```
        CALL OUTPLT
        RETURN
        END



        FUNCTION ENTRPF (AL,NPL,X)
C
C       FUNCTION TO EVALUATE THE ENTROPY DENSITY FUNCTION AT A GIVEN POINT
C.... INPUT
C          AL(I) = ARRAY CONTAINING PARAMETERS, DIMENSION NPL
C          NPL = NUMBER OF PARAMETERS
C          X = GIVEN VALUE
C
        DIMENSION AL(1)
        S=AL(1)
        DO 1 I=2,NPL
        S=S+AL(I)*X**(I-1)
1       CONTINUE
        ENTRPF=EXP(S)
        RETURN
        END



        SUBROUTINE FUNCT (N,AL,U,GRAD,RR)
C
C       THIS  SUBROUTINE IS USED TO CALCULATE THE OPTIMIZATION AND THE
C       GRADIENT AT ANY GIVEN POINT FOR SUBROUTINE POPT
C
        DIMENSION AL(1), GRAD(1), SUM(17), RR(1)
        COMMON /FAIL/ NFAIL
        COMMON /HELP/ S(41),XX(16,41),C(8),M
        N21=2*N+1
        ZERO=0.0
        DO 1 I=1,N21
        SUM(I)=0.0
1       CONTINUE
2       CONTINUE
        DO 4 I=1,M
        SZ=ZERO
        DO 3 K=1,N
        SZ=SZ+AL(K)*XX(K,I)
3       CONTINUE
        IF (SZ.GT.740.) GO TO 9
        SS=EXP(SZ)*S(I)
        SUM(1)=SUM(1)+SS
        DO 4 J=2,N21
        SUM(J)=SUM(J)+XX(J-1,I)*SS
4       CONTINUE
        DO 5 I=2,N21
        SUM(I)=SUM(I)/SUM(1)
5       CONTINUE
        U=0.0
        DO 6 I=1,N
        RR(I)=(SUM(I+1)-C(I))/C(I)
        U=U+RR(I)*RR(I)
6       CONTINUE
        DO 8 K=1,N
        GRAD(K)=0.0
        DO 7 J=1,N
        GRAD(K)=GRAD(K)+(SUM(J+K+1)-SUM(J+1)*SUM(K+1))*RR(J)/C(J)
7       CONTINUE
        GRAD(K)=GRAD(K)*2.
8       CONTINUE
```

```
        RETURN
9       CONTINUE
        AA=SZ-320.
        ZERO=ZERO-AA
        GO TO 2
C
        END



        SUBROUTINE LINES (FUNCT,X,H,AMBDA,N,F,G,NUMF,IER,EPS,EST,RR)
        COMMON /FAIL/ NFAIL
        DIMENSION H(1), X(1), G(1), RR(1)
        IER=0
        DY=0.
        HNRM=0.
        GNRM=0.
        DO 1 J=1,N
        HNRM=HNRM+ABS(H(J))
        GNRM=GNRM+ABS(G(J))
        DY=DY+H(J)*G(J)
1       CONTINUE
        IF (DY) 2,31,31
2       IF (HNRM/GNRM-EPS) 31,31,3
3       FY=F
        ALFA=2.*(EST-F)/DY
        IF (X(N+1).GT.0.) ALFA=X(N+1)*ALFA/2.
        AMBDA=1.
        IF (ALFA) 6,6,4
4       IF (ALFA-AMBDA) 5,6,6
5       AMBDA=ALFA
6       ALFA=0.
7       FX=FY
        DX=DY
        DO 8 I=1,N
        X(I)=X(I)+AMBDA*H(I)
8       CONTINUE
        CALL FUNCT (N,X,F,G,RR)
        IF (NFAIL.EQ.1) RETURN
        NUMF=NUMF+1
        IF (F.LT.FX) RETURN
        FY=F
        DY=0.
        DO 9 I=1,N
        DY=DY+G(I)*H(I)
9       CONTINUE
        IF (DY) 10,30,13
10      IF (FY-FX) 11,13,13
11      AMBDA=AMBDA+ALFA
        ALFA=AMBDA
        IF (HNRM*AMBDA-1.E10) 7,7,12
12      IER=2
        GO TO 31
13      T=0.
14      IF (AMBDA) 15,30,15
15      Z=3.*(FX-FY)/AMBDA+DX+DY
        ALFA=AMAX1(ABS(Z),ABS(DX),ABS(DY))
        DALFA=Z/ALFA
        DALFA=DALFA*DALFA-DX/ALFA*DY/ALFA
        IF (DALFA) 31,16,16
16      W=ALFA*SQRT(DALFA)
        ALFA=DY-DX+W+W
        IF (ALFA) 17,18,17
```

```
17      ALFA=(DY-Z+W)/ALFA
        GO TO 19
18      ALFA=(Z+DY-W)/(Z+DX+Z+DY)
19      ALFA=ALFA*AMBDA
        DO 20 I=1,N
        X(I)=X(I)+(T-ALFA)*H(I)
20      CONTINUE
        CALL FUNCT (N,X,F,G,RR)
        IF (NFAIL.EQ.1) RETURN
        NUMF=NUMF+1
        IF (F.LT.FX) GO TO 30
        IF (F-FX) 21,21,22
21      IF (F-FY) 30,30,22
22      DALFA=0.
        DO 23 I=1,N
        DALFA=DALFA+G(I)*H(I)
23      CONTINUE
        IF (DALFA) 24,27,27
24      IF (F-FX) 26,25,27
25      IF (DX-DALFA) 26,30,26
26      FX=F
        DX=DALFA
        T=ALFA
        AMBDA=ALFA
        GO TO 14
27      IF (FY-F) 29,28,29
28      IF (DY-DALFA) 29,30,29
29      FY=F
        DY=DALFA
        AMBDA=AMBDA-ALFA
        GO TO 13
30      AMBDA=AMBDA-ALFA
        RETURN
31      CONTINUE
        IF (DY.GE.0.) IER=-2
        IF (GNRM.LE.1.E-10) GO TO 32
        IF (HNRM/GNRM.LE.EPS) IER=-3
32      CONTINUE
        IF (DALFA.LT.0.) IER=-1
        NFAIL=1
        WRITE(6,33)
33      FORMAT(///,1X,*   THE PROGRAM HAS FAILED*)
        RETURN
C
C
        END




        SUBROUTINE MEPOUT(AL,N,NXP,XP,CUM)
C.... OUTPUT FOR MEP1 AND MEP2
        DIMENSION AL(1),XP(1),CUM(1)
        WRITE(6,1)(AL(I),I=1,N)
1       FORMAT(51H THE MAXIMUM ENTROPY DISTRIBUTION COEFFICIENTS ARE ,
       1/,(6E12.5))
        IF(NXP.EQ.0)RETURN
        WRITE(6,2)(XP(I),I=1,NXP)
2       FORMAT(/,44H CUM DIST FUNCTION VALUES ARE CALCULATED AT ,/,
       1(6E12.5))
        WRITE(6,3)(CUM(I),I=1,NXP)
3       FORMAT(/,30H CUM DIST FUNCTION VALUES ARE ,/,(6E14.5))
        RETURN
        END
```

```
      SUBROUTINE MPOPT (X,NDIM,ETA,EST,MAX,MODE,IPRINT)
      COMMON /FAIL/ NFAIL
      DIMENSION X(1), X1(10), X2(10), G1(10), G2(10), ALFA(10), H(10), P
     1(10,10), Y(10), PY(10), PE(10), ETA(1), BIGV(10), RR(8)
      EXTERNAL FUNCT
      KRST=0
      KTB=0
      IFLAG=0
      M=0
      N2=NDIM+1
      N1=NDIM+2
      NUMF=0
      IER=0
      DO 1 I=1,N1
      X1(I)=X(I)
1     CONTINUE
      CALL FUNCT (NDIM,X1,F1,G1,RR)
      NUMF=NUMF+1
      DO 2 I=1,NDIM
      X2(I)=X1(I)
      G2(I)=G1(I)
      H(I)=-G1(I)
2     CONTINUE
      F2=F1
      X2(N2)=X1(N2)
      X2(N1)=X1(N1)
3     CONTINUE
      KOUNT=0
      EPS=ETA(4)
      CALL LINES (FUNCT,X2,H,RO,NDIM,F2,G2,NUMF,IER,EPS,EST,RR)
      IF (NFAIL.EQ.1) RETURN
      IF (IER.NE.0) GO TO 30
      DO 4 I=1,N1
      BIGV(I)=X2(I)
      ALFA(I)=X2(I)
4     CONTINUE
      RO=-RO
      GG=0.
      DO 5 I=1,NDIM
      GG=GG+G2(I)*G2(I)
5     CONTINUE
      GG=SQRT(GG)
      IF (IPRINT.EQ.0) GO TO 7
      IF (MOD(KTB,IPRINT).NE.0) GO TO 6
      CALL OUTP (X2,F2,M,NDIM,GG,NUMF,RR)
6     KTB=KTB+1
7     DO 9 I=1,N1
      DO 8 J=1,N1
      P(I,J)=0.
8     CONTINUE
      P(I,I)=1.
9     CONTINUE
10    CONTINUE
      KOUNT=0
      KOUNT=KOUNT+1
11    DO 12 I=1,NDIM
      Y(I)=G2(I)
12    CONTINUE
      Y(N2)=F2
      Y(N1)=ETA(1)
      V=0.
      DO 13 I=1,NDIM
      V=V+X2(I)*G2(I)
13    CONTINUE
      YA=0.
```

```
          DO 14 I=1,N1
          YA=YA+Y(I)*ALFA(I)
14        CONTINUE
          VYA=V-YA
          BIGV(KOUNT)=V
          DO 15 I=1,N1
          PY(I)=0.
          PE(I)=P(I,KOUNT)
          DO 15 J=1,N1
15        PY(I)=PY(I)+P(J,I)*Y(J)
          EPY=PY(KOUNT)
          IF (ABS(EPY).LT.ETA(3)) GO TO 31
          PY(KOUNT)=PY(KOUNT)-1.
          DO 16 I=1,N1
          DO 16 J=1,N1
16        P(I,J)=P(I,J)-PE(I)*PY(J)/EPY
          DO 17 I=1,N1
          ALFA(I)=0.
          DO 17 J=1,N1
17        ALFA(I)=ALFA(I)+P(I,J)*BIGV(J)
          DEL=0.
          DO 18 I=1,NDIM
          DEL=DEL+G2(I)*(X2(I)-ALFA(I))
18        CONTINUE
          IF (ABS(DEL).GT.ETA(4)) GO TO 19
          IF (IFLAG.EQ.1) RETURN
          IFLAG=1
          GO TO 31
19        IFLAG=0
          DO 20 I=1,N1
          H(I)=X2(I)-ALFA(I)
          IF (DEL.GT.0) H(I)=-H(I)
20        CONTINUE
          DO 21 I=1,NDIM
          X1(I)=X2(I)
          G1(I)=G2(I)
21        CONTINUE
          F1=F2
          X1(N2)=X2(N2)
          X1(N1)=X2(N1)
          X2(N2)=ALFA(N2)
          X2(N1)=ALFA(N1)
          CALL LINES (FUNCT,X2,H,RO,NDIM,F2,G2,NUMF,IER,EPS,EST,RR)
          IF (NFAIL.EQ.1) RETURN
          IF (IER.NE.0) GO TO 30
          IF (DEL.GT.0) RO=-RO
          GG=0.
          DO 22 I=1,NDIM
          GG=GG+G2(I)*G2(I)
22        CONTINUE
          GG=SQRT(GG)
          KOUNT=KOUNT+1
          M=M+1
          IF (IPRINT.EQ.0) GO TO 23
          IF (MOD(KTB,IPRINT).NE.0) GO TO 23
          CALL OUTP (X2,F2,M,NDIM,GG,NUMF,RR)
23        CONTINUE
          KTB=KTB+1
          IF (MODE.EQ.2) GO TO 25
          IF (M.GT.MAX) GO TO 30
          NSOL=0
          DO 24 I=1,NDIM
          IF (ABS(RR(I)).GT.ETA(2)) NSOL=1
24        CONTINUE
          IF (NSOL.EQ.0) GO TO 26
```

```
        GO TO 29
25      IF ((GG.LT.ETA(1)).OR.(M.GT.MAX)) GO TO 26
        GO TO 29
26      CONTINUE
        IF (IPRINT.EQ.0) GO TO 27
        WRITE (6,33)
        CALL OUTP (X2,F2,M,NDIM,GG,NUMF,RR)
27      DO 28 I=1,NDIM
        X(I)=X2(I)
28      CONTINUE
        EST=F2
        NFAIL=0
        RETURN
29      CONTINUE
        IF (KOUNT.LE.N1) GO TO 11
        GO TO 10
30      PRINT 34, IER
        NFAIL=1
        RETURN
31      KRST=KRST+1
        IF (KRST.GT.10) NFAIL=1
        IF (NFAIL.EQ.1) RETURN
        DO 32 I=1,NDIM
        X1(I)=X2(I)
        G1(I)=G2(I)
        H(I)=-G1(I)
32      CONTINUE
        F1=F2
        X1(N2)=X(N2)
        X1(N1)=X(N1)
        X2(N2)=X(N2)
        X2(N1)=X(N1)
        GO TO 3
C
C
33      FORMAT (*  SOLUTION FOUND*)
34      FORMAT (///,1X,*  THE PROGRAM HAS FAILED---IER = *,I2)
        END




        SUBROUTINE MULTI (XMAX,XMIN,N)
C
C       THIS SUBROUTINE IS USED TO GENERATE THE X,S POWER FOR SUBROUTINE
C       FUNCT
C
        COMMON /HELP/ S(41),XX(16,41),C(8),M
        DELTA=(XMAX-XMIN)/FLOAT(M-1)
        DO 1 I=1,M
        XX(1,I)=XMIN+FLOAT(I-1)*DELTA
        NN=2*N
        DO 1 J=2,NN
        XX(J,I)=XX(J-1,I)*XX(1,I)
1       CONTINUE
        RETURN
        END




        SUBROUTINE OUTP (XNEW,FQ,KOUNT,N1,GG,NUMF,R)
        DIMENSION XNEW(1), R(1)
        WRITE (6,6) KOUNT,NUMF,GG,FQ,(XNEW(I),I=1,4),(R(I),I=1,4)
        IF (N1.LT.4) RETURN
        NN=N1-3
        GO TO (1,2,3,4,5), NN
```

```
1        RETURN
2        WRITE (6,7) XNEW(5),R(5)
         RETURN
3        WRITE (6,8) (XNEW(I),I=5,6),(R(I),I=5,6)
         RETURN
4        WRITE (6,9) (XNEW(I),I=5,7),(R(I),I=5,7)
         RETURN
5        WRITE (6,10) (XNEW(I),I=5,8),(R(I),I=5,8)
         RETURN
C
C
6        FORMAT (1X,I3,I4,6E14.5,4E11.3)
7        FORMAT (36X,E14.5,42X,E11.3)
8        FORMAT (36X,2E14.5,28X,2E11.3)
9        FORMAT (36X,3E14.5,14X,3E11.3)
10       FORMAT (36X,4E14.5,4E11.3)
         END



         SUBROUTINE SIMSON
C
C        THIS SUBROUTINE IS TO CALCULATE THE SIMPSON MULTIPLIERS
C
         COMMON /HELP/ S(41),XX(16,41),C(8),M
         S(1)=1.
         S(M)=1.
         N=M-1
         DO 1 I=2,N,2
         S(I)=4.
1        CONTINUE
         N=N-1
         DO 2 I=3,N,2
         S(I)=2.
2        CONTINUE
         RETURN
         END



         SUBROUTINE SOLVE (A,X,ID,N,NA)
         DIMENSION A(NA,1), X(1)
         D=0.
         DATA DIV/.693147181/
         DO 6 I=1,N
         AA=0.
         DO 1 J=1,N
         AB=ABS(A(J,I))
         IF (AB.LE.AA) GO TO 1
         K=J
         AA=AB
1        CONTINUE
         D=D+ALOG(AA)
         D=D+ALOG(AA)
         IF (I.EQ.N) GO TO 7
         IF (K.EQ.I) GO TO 3
         DO 2 J=I,N
         AB=A(I,J)
         A(I,J)=A(K,J)
         A(K,J)=AB
2        CONTINUE
         AB=X(I)
         X(I)=X(K)
         X(K)=AB
3        I1=I+1
```

```
        DO 5 J=I1,N
        AA=-A(J,I)/A(I,I)
        A(J,I)=0.
        DO 4 K=I1,N
        A(J,K)=A(J,K)+AA*A(I,K)
4       CONTINUE
        X(J)=X(J)+AA*X(I)
5       CONTINUE
6       CONTINUE
7       ID=D/DIV
        X(N)=X(N)/A(N,N)
        DO 9 II=2,N
        I=N+1-II
        I1=I+1
        AA=0.
        DO 8 J=I1,N
        AA=AA+A(I,J)*X(J)
8       CONTINUE
        X(I)=(X(I)-AA)/A(I,I)
9       CONTINUE
        RETURN
        END



        SUBROUTINE START (XMAX,XMIN,ALAMDA,KSTART,CC,NL,IPRINT,UMIN,MODE,M
       1AXFN,ETA)
C
C
C       THIS SUBROUTINE IS USED TO FIND A REASONABLE STARTING POINT FOR
C       SUBROUTINE   MPOPT
C
        COMMON /HELP/ S(41),XX(16,41),C(8),M
        DIMENSION R(11)
        DIMENSION CC(1), ETA(1)
        DIMENSION ALAMDA(1), X(10), Y(10), W(10,10)
        COMMON /FAIL/ NFAIL
        GO TO (3,1,5,26), KSTART
1       CONTINUE
        NFAIL=0
        DO 2 I=1,NL
        ALAMDA(I)=0.0
2       CONTINUE
        RETURN
3       CONTINUE
        NFAIL=0
        ALAMDA(1)=CC(1)/CC(2)
        ALAMDA(2)=-.5/CC(2)
        DO 4 I=3,NL
        ALAMDA(I)=0.0
4       CONTINUE
        RETURN
5       CONTINUE
        NFAIL=0
        NNN=NL/2
        NNN=NNN*2
        NP1=NL+1
        DELTA=(XMAX-XMIN)/FLOAT(NL)
        DO 6 I=1,NP1
        X(I)=XMIN+FLOAT(I-1)*DELTA
6       CONTINUE
        IF (NNN.NE.NL) GO TO 19
        W(1,1)=W(1,NP1)=1.
        DO 7 I=2,NL,2
        W(1,I)=4.
7       CONTINUE
```

```
          IF (NL.EQ.2) GO TO 9
          NM1=NL-1
          DO 8 I=3,NM1,2
          W(1,I)=2.
8         CONTINUE
9         CONTINUE
          DO 10 J=1,NP1
          DO 10 I=2,NP1
10        W(I,J)=W(I-1,J)*X(J)
          Y(1)=3./DELTA
          DO 11 I=1,NL
          Y(I+1)=C(I)*Y(1)
11        CONTINUE
          CALL SOLVE (W,Y,ID,NP1,10)
12        CONTINUE
          DO 13 I=1,NP1
          DO 13 J=1,NP1
13        W(I,J)=.0
          DO 14 I=1,NP1
          IF (Y(I).LE.0.0) Y(I)=.0002
14        CONTINUE
          DO 15 I=1,NP1
          Y(I)=ALOG(Y(I))
15        CONTINUE
          DO 16 I=1,NP1
          W(I,1)=1.
16        CONTINUE
          DO 17 I=2,NP1
          DO 17 J=1,NP1
17        W(J,I)=W(J,I-1)*X(J)
          CALL SOLVE (W,Y,ID,NP1,10)
          DO 18 I=1,NL
          ALAMDA(I)=Y(I+1)
18        CONTINUE
          RETURN
19        CONTINUE
          R(1)=3./8.
          R(4)=3./8.
          R(2)=R(3)=9./8.
          IF (NL.EQ.3) GO TO 22
          R(NL+1)=1./3.
          R(4)=R(4)+1./3.
          DO 20 I=5,NL,2
          R(I)=4./3.
20        CONTINUE
          IF (NL.EQ.5) GO TO 22
          NS=NL-1
          DO 21 I=6,NS,2
          R(I)=2./3.
21        CONTINUE
22        CONTINUE
          DO 23 I=1,NP1
          W(1,I)=R(I)
23        CONTINUE
          DO 24 J=1,NP1
          DO 24 I=2,NP1
24        W(I,J)=W(I-1,J)*X(J)
          Y(1)=1./DELTA
          DO 25 I=1,NL
          Y(I+1)=C(I)*Y(1)
25        CONTINUE
          CALL SOLVE (W,Y,ID,NP1,10)
          GO TO 12
26        CONTINUE
          N=2
```

```
      ALAMDA(2)=-.5/CC(2)
      ALAMDA(1)=CC(1)/CC(2)
      NFAIL=0
27    CONTINUE
      ALAMDA(N+1)=2.0
      ALAMDA(N+2)=0.0
      CALL MPOPT (ALAMDA,N,ETA,UMIN,MAXFN,MODE,IPRINT)
      IF (NFAIL.EQ.1) RETURN
      IF (N.EQ.NL) RETURN
      ALAMDA(N+1)=0.0
      N=N+1
      GO TO 27
      END



      SUBROUTINE TRN1 (X1MAX,X1MIN,C,X2MAX,X2MIN,NL)
C
C
C     THIS SUBROUTINE IS USED TO CALCULATE THE MOMENTS FOR THE MODIFIED
C     LIMITS
C
      DIMENSION C(1)
      SCL=(X1MAX-X1MIN)/(X2MAX-X2MIN)
      C(1)=C(1)/SCL-X1MIN/SCL+X2MIN
      IF (NL.EQ.1) RETURN
      DO 1 I=2,NL
      C(I)=C(I)/SCL**I
1     CONTINUE
      RETURN
      END



      SUBROUTINE TRN2 (X1MAX,X1MIN,X,X2MAX,X2MIN,N)
C
C
C     THIS SUBROUTINE IS USED TO CALCULATE THE LAGRANGIAN MULTIPLIERS
C     AT   THE ORIGINAL LIMITS
C
      DIMENSION X(1)
      S=(X1MAX-X1MIN)/(X2MAX-X2MIN)
      A=X2MIN-X1MIN/S
      X(1)=X(1)-ALOG(S)
      DO 1 I=1,N
      X(1)=X(1)+X(I+1)*A**I
1     CONTINUE
      IF (N.EQ.1) GO TO 6
      DO 5 J=2,N
      DO 3 I=J,N
      FAC=1.
      KK=I-J+2
      DO 2 K=KK,I
      FAC=FAC*FLOAT(K)
2     CONTINUE
      IF (X1MIN.EQ.0.0) GO TO 4
      X(J)=X(J)+FAC/FACTO(J-1)*A**(I-J+1)*X(I+1)
3     CONTINUE
4     X(J)=X(J)/S**(J-1)
5     CONTINUE
6     CONTINUE
      X(N+1)=X(N+1)/S**N
      RETURN
      END
```

Subroutine MEP2(NL,NSAMP,X,XMIN,XMAX,QPROB,MINT,NXP,XP, KSTART,KDATA,NCYCLE,RP,CUM,PHI,XD,W)

Purpose

This subroutine estimates the least biased density function for a continuous random variable using the ranks of the sample. The density function has the form

$$f(x) = \exp\,(r_1 + r_2 x + r_3 x^2 + \cdots + r_m x^{m-1})$$

and the program gives the values of the r's.

Any number of values of the cumulative distribution function are also available as an option.

How to Use

The calling program in its simplest form is set up as follows.

a. *DIMENSION statement.* Check through the list of input and output variables and include all arrays, dimensioning as indicated. Also include the working array PHI.
b. *COMMON statements.* A labeled COMMON statement for working arrays must be included; see below.
c. *Define input data.* This may be done by simple arithmetic statements, DATA declarations, or READ statements. Each variable in the input list below must be included unless the default value is used.
d. Call subroutine MEP2.
e. Call the output subroutine. Library output subroutines may be used; see below.
f. Add STOP and END.

A sample calling program is given below.

```
DIMENSION X(22),RP(6),CUM(1),PHI(44),W(68),XD(22),XP(1)
DATA X/1.65,1.05,2.2,1.35,.075,.15,.27,.23,.72,.75,.71,.50,.35,
1.48,.47,.37,1.32,1.25,1.13,.85,.95,1.05/
XP(1)=3.0
CALL MEP2(5,22,X,0.0,4.0,.99,51,1,XP,1,1,2,RP,CUM,PHI,XD,W)
CALL ENTPP(XMIN,XMAX,NL,RP)
CALL MEPOUT(RP,NL,NXP,XP,CUM)
CALL ENTPL(XMIN,XMAX,NL,RP,9,9HICE THICK)
STOP
END
```

Input Variables

NL: number of terms in the exponent of the function.
NSAMP: sample size.

X(I): array containing the sample elements, need not be sorted from low to high, dimensioned NSAMP.

XMIN: lower bound of the random variable—not necessarily equal to the smallest elements of the sample.

XMAX: upper bound of the random variable—not necessarily equal to the largest element of the sample.

QPROB: quantity defining the bounds on the rank distribution (see the section on programming information).

MINT: number of stations to be used for all numerical integrations done internally in the MEP2 package, maximum 101.

NXP: number of points for which the cumulative distribution function is required. Set at zero if none are required.

XP(I): array containing the values of the random variable, for which values of the cumulative distribution function are to be calculated; dimensioned NXP or 1 if NXP is zero.

KDATA: = 1, all input data are printed out.
        = 0, input data are not printed out.

KSTART: = 0, user provided starting values placed in RP(I); see the section on output variables.
        = 1, a normal start is used.
        = 2, a uniform start is used.
        = 3, a MEP1 feasible start is used. Set to 1 if no particular method is desired. In this event the subroutine will try in sequence all three methods until one succeeds.

NCYCLE: number of cycles used in shrinking rank bounds (see the section on programming information).

Output Variables

RP(I): array containing the density function parameters, dimensioned with N

CUM(I): array containing the values of the cumulative distribution function, dimensioned NXP, but not less than 1

Working Arrays

PH(I): dimensioned 2*NSAMP
XD(I): dimensioned NSAMP
W(I): dimensioned 5*NL + 2*NSAMP − 1

Programming Information

MEP2 has full variable dimensioning. The calling program must provide dimensioning as given above. If standard printed output of the output variables is desired, the calling program should include the statement

CALL MEPOUT(RP,NL,NXP,XP,CUM)

If a printer plot of the density function is desired, the calling program should include (note that this is a hardware-dependent feature)

CALL ENTPP(XMIN,XMAX,NL,RP)

If a plotter plot of the density function is desired, the calling program should include (note that this is a hardware-dependent feature)

CALL ENTPL(XMIN,XMAX,NL,RP,NLAB,LABEL)

MEP2 has the feature that after a successful run the values of RP(I) will be available as starting values for a subsequent run from the same calling program, and KSTART will be reset at 0. This is useful when iterative runs are being used to determine sensitivity to bounds; or when different samples of the same variable are being tried.

Experience with MEP2 has shown that it is usually necessary to start with wide bounds on the rank distributions, then use this solution as input to a second iteration with bounds reduced. A suggested initial value for QPROB is 0.99. The value of QPROB is reduced by 0.1 at each cycle. NCYCLE defines the number of cycles to be used. Two are commonly enough.

Subroutines called by MEP2, are ARO, CONST, RANKBD, START2, UREAL, FCUBIN, and FINVRT.

```
      SUBROUTINE MEP2(NL,NSAMP,X,XMIN,XMAX,QPROB,MINT,NXP,XP,KSTART,KDAT
     1A,NCYCLE,RP,CUM,PHI,XD,W)
C.... CALCULATES MAXIMUM ENTROPY DISTRIBUTION WITH RANKS
C.... INPUT
C          NL         = NUMBER OF PARAMETERS, INCLUDING FIRST
C          NSAMP      = SAMPLE SIZE
C          X(1)       = ORDERED SAMPLE VALUES, DIMENSION NSAMP
C          QPROB      = AREA OF RANK DENSITY FUNCTION ENCLOSING BOUNDS
C          KDATA      = 0, INPUT DATA NOT PRINTED
C                     = 1, PRINTED
C          XMIN       = ESTIMATED LOWER BOUND OF DISTRIBUTION
C          XMAX       = ESTIMATED UPPER BOUND OF DISTRIBUTION
C          MINT       = NUMBER OF INTEGRATION STATIONS USED - MAX 101
C          NXP        = NUMBER OF POINTS FOR WHICH VALUE OF CUMULATIVE
C                       DISTRIBUTION FUNCTION IS REQUIRED
C          XP(1)      = ARRAY OF VALUES OF RANDOM VARIABLE AT WHICH VALUES OF
C                       THE CUMULATIVE DISTRIBUTION FUNCTION ARE REQUIRED,
C                       DIMENSION NXP BUT NOT LESS THAN 1
C          KSTART     = FLAG FOR STARTING METHOD
C.... OUTPUT
C          RP(1)      = ARRAY OF PARAMETERS FOR DENSITY FUNCTION, DIMENSION
C                       NL
C          CUM(1)     = ARRAY OF VALUES OF CUMULATIVE DISTRIBUTION FUNCTION,
C                       DIMENSION NXP BUT NOT LESS THAN 1
C.... WORKING ARRAYS
C          PHI(1)     = DIMENSION 2*NSAMP
C          XD(1)      = DIMENSION NSAMP
C          W(1)       = DIMENSION 5*NL+2*NSAMP-1
      DIMENSION X(1),XP(1),RP(1),CUM(1)
      DIMENSION RMAX(8),RMIN(8),RSTRT(8),PHI(1),PSI(1),RX(8),XD(1),W(1)
      COMMON/FUNCT/DENS(101),ARG(101),RANK(101)
```

```
          COMMON/IN/NTERM,NS,QP,MI
          COMMON/MEP1/KPRINT,TOLER,MAXFN
          COMMON/SEEK/IDATA,IPRINT,NSHOT,NTEST,MAXM,F,G,TOL,ZERO,R,REDUCE
          COMMON/UREAL/NOPT
          DATA NCALL/1/
          NOPT=2
C....  SORT DATA
          CALL SORT(X,NSAMP)
C....  TRANSFER INPUT TO COMMON
          NTERM=NL
          NS=NSAMP
          QP=QPROB
          MI=MINT
C....  DEFINE PARAMETERS FOR OPTIMIZATION
          NCONS=NSAMP*2
          NEQUS=0
          MAXM=100
          TOL=1.E-3
          ZERO=1.E-3
          G=0.1
          NPENAL=1
          IPRINT=-1
          IDATA=0
C....  PRINT INPUT
          IF(KDATA.EQ.0)GO TO 200
          WRITE(6,100)
          WRITE(6,101)NSAMP
          WRITE(6,102)(X(I),I=1,NSAMP)
          WRITE(6,103)XMIN
          WRITE(6,104)XMAX
          WRITE(6,105)NTERM
          WRITE(6,106)KSTART
          WRITE(6,108)NXP
          IF(NXP.EQ.0)GO TO 200
          WRITE(6,109)(XP(I),I=1,NXP)
C....  INITIALIZE
200       NX=NTERM-1
C....  NX = UNKNOWN NUMBER OF PARAMETERS IN DENSITY FUNCTION
C....  TRANSFER SAMPLE TO 0-1 DOMAIN
          RANGE=XMAX-XMIN
          DO 1 I=1,NSAMP
1         XD(I)=(X(I)-XMIN)/RANGE
C....  CALCULATE BOUNDS ON RANK DISTRIBUTIONS
          DO 11 I=1,NSAMP
11        CALL RANKBD(I)
C....  PUT XD(I) INTO RANK(I) FOR TRANSFER TO CONST
          DO 15 I=1,NSAMP
15        RANK(I)=XD(I)
C....  FIND A START
          IF(KSTART.EQ.0)GO TO 3
2         CALL START2(XD,KSTART,RSTRT)
          GO TO 8
C....  THIS ALGORITHM IS SIMILAR TO TRN2 BUT APPEARS TO GIVE BETTER
C     NUMERICAL RESULTS
C....  CHANGE STARTING VALUES TO 0-1 DOMAIN FOR KSTART=0
C
3         IF(ABS(XMIN).LT.1.E-10)GO TO 13
          DO 4 I=2,NL
          RSTRT(I)=0.
          I1=I-1
          DO 12 J=I1,NX
          RSTRT(I)=RSTRT(I)+FACTO(J)*XMIN**(J-I1)*RANGE**I1*RP(J+1)/FACTO(I
          1)/FACTO(J-I1)
```

```
12        CONTINUE
4         CONTINUE
          GO TO 8
13        DO 14 I=2,NL
14        RSTRT(I)=RANGE**(I-1)*RP(I)
8         CONTINUE
C..... CHANGE RSTRT TO CORRESPOND TO RX
          DO 7 I=1,NX
7         RSTRT(I)=RSTRT(I+1)
C.... DEFINE BOUNDS FOR VARIABLES SUBMITTED TO SEEK   - RX(I),I=1,NX
22        DO 5 I=1,NX
          RMIN(I)=-5.*ABS(RSTRT(I))
          IF(ABS(RMIN(I)).LT.1.E-10)RMIN(I)=RMIN(1)
          RMAX(I)=ABS(RMIN(I))
          IF(RMAX(I).LT.1.E-10)RMAX(I)=RMAX(1)
5         CONTINUE
C.... CALL OPTIMIZATION SUBROUTINE
          CALL SEEK(NX,NCONS,NEQUS,NPENAL,RMAX,RMIN,RSTRT,RX,U,PHI,PSI,NVIOL
         1,W)
          CALL ANSWER(U,RX,PHI,PSI,NX,NCONS,NEQUS)
          IF(NVIOL.NE.0.AND.NCALL.EQ.1)GO TO 20
          IF(NCALL.GT.1.AND.NVIOL.NE.0)GO TO 24
C.... RERUN WITH SMALLER Q
          IF(NCALL.EQ.NCYCLE)GO TO 21
          NCALL=NCALL+1
          KSTART=0
          QPROB=QPROB-0.1
          DO 23 I=1,NSAMP
23        CALL RANKBD(I)
          GO TO 22
C.... SOLUTION FOUND, TRANSFER BACK TO ORIGINAL DOMAIN
21        DO 9 I=2,NL
9         RP(I)=RX(I-1)
          RP(1)=ARO(RX,NX)
          CALL TRN2(XMAX,XMIN,RP,1.,0.,NX)
          KSTART=0
C.... CALCULATE CUMULATIVE DISTRIBUTION FUNCTION VALUES
          IF (NXP.EQ.0)RETURN
          DO 10 I=1,NXP
10        CUM(I)=CDF(XMIN,XMAX,XP(I),RP,NL)
          RETURN
20        WRITE(6,113)KSTART
113       FORMAT(/,11H START NO. ,I1,11H HAS FAILED)
          KSTART=KSTART+1
          IF(KSTART.EQ.4)WRITE(6,110)
          IF(KSTART.EQ.4)STOP
          GO TO 2
24        WRITE(6,25)NCALL
25        FORMAT(70H OPTIMIZATION DID NOT SUCCEED IN HOLDING SEARCH FEASIBLE
         1 IN CYCLE NO. ,I3)
          STOP
100       FORMAT(///,24X,21HMEP2 DENSITY FUNCTION,/,24X,21H................
         1....,///)
101       FORMAT(34H SAMPLE SIZE ....................,I6,/)
102       FORMAT(34H SAMPLE DATA ...................,/,(5E13.5),/)
103       FORMAT(34H LOWER BOUND ....................,E13.5,/)
104       FORMAT(34H UPPER BOUND ....................,E13.5,/)
105       FORMAT(34H NUMBER OF DENS FUNCT PARAMETERS .,I6,/)
106       FORMAT(34H STARTING METHOD KSTART ..........,I6,/)
108       FORMAT(34H NUMBER OF CUM DIST FUNCT VALUES .,I6,/)
109       FORMAT(34H LOCATION OF CUM DIST FUNCT ......,E13.5,//)
110       FORMAT(25H CONVERGENCE NOT ACHIEVED,/
         122H CHECK FOR INPUT ERROR,/,
         243H TRY YOUR OWN STARTING VALUES WITH KSTART=0)
          END
```

```
         FUNCTION ARO(AL,N)
C....    CALCULATES FIRST PARAMETER OF MAX ENTROPY DIST IN TERMS OF
C        REMAINING PARAMETERS
         DIMENSION AL(1)
         COMMON/FUNCT/DENS(101),ARG(101),RANK(101)
         COMMON/IN/NL,NSAMP,QPROB,MINT
         DO 1 I=1,MINT
         X=FLOAT(I-1)/FLOAT(MINT-1)
         SUM=0.0
         DO 2 J=1,N
2        SUM=SUM+AL(J)*X**J
1        ARG(I)=EXP(SUM)
         AREA=FSIMP(ARG,1.,MINT)
         IF(ABS(AREA).LT.1.E-10)GO TO 3
         ARO=-ALOG(AREA)
         RETURN
3        ARO=-10.
         RETURN
         END




         SUBROUTINE CONST(RX,NCONS,PHI)
C....    CALCULATES CONSTRAINT FUNCTION VALUES
C....    INPUT
C           BL(I)    = LOWER RANK BOUND
C           BU(I)    = UPPER RANK BOUND
C           MINT     = NUMBER OF INTEGRATION STATIONS
C           NL       = NUMBER OF PARAMETERS IN MAX ENTROPY DIST
C
C....    AL(I) = ARRAY OF MAX ENTROPY DIST PARAMETERS, INCLUDING FIRST
C        BL(I) = LOWER BOUND OF RANK DISTRIBUTION
C        BU(I) = UPPER BOUND OF RANK DISTRIBUTION
C
         DIMENSION RX(1),PHI(1),AL(8)
         COMMON/FUNCT/DENS(101),ARG(101),XD(101)
         COMMON/BOUNDS/BL(101),BU(101)
         COMMON/IN/NL,NSAMP,QPROB,MINT
         R1=BU(1)-BL(1)
         NX=NL-1
         DO 3 J=2,NL
3        AL(J)=RX(J-1)
         AL(1)=ARO(RX,NX)
         DO 1 I=1,NSAMP
         DO 2 K=1,MINT
C....    CALCULATE LOCAL VALUES OF DENSITY FUNCT USING ENTRPF (FROM MEP1)
         X=FLOAT(K-1)/FLOAT(MINT-1)*XD(I)
2        DENS(K)=ENTRPF(AL,NL,X)
         TEMP=XD(I)
         CUMR=FSIMP(DENS,TEMP,MINT)
C....    WEIGHT PHI S TO ALLOW FOR DIFFERENT RANGES IN RANK DISTRIBUTION
         WATE=R1/(BU(I)-BL(I))
         PHI(I)=(CUMR-BL(I))*WATE
1        PHI(NSAMP+I)=(BU(I)-CUMR)*WATE
         RETURN
         END




         SUBROUTINE RANKBD(I)
C....    OBTAINS BOUNDS OF RANK DISTRIBUTION CORRESPONDING TO SPECIFIED
C        AREA OF DENSITY FUNCTION
C....    INPUT
```

```
C          QPROB = SPECIFIED AREA
C          NSAMP = SAMPLE SIZE
C          I     = ORDER NUMBER OF RANK IN SAMPLE
C.... OUTPUT
C          BL(I) = LOWER BOUND
C          BU(I) = UPPER BOUND
C
          COMMON/FUNCT/DENS(101),ARG(101),RANK(101)
          COMMON/BOUNDS/BL(101),BU(101)
          COMMON/IN/NL,NSAMP,QPROB,MINT
          DO 10 K=1,MINT
          RANK(K)=FLOAT(K-1)/FLOAT(MINT-1)
10        DENS(K)=FRANK(RANK(K),I,NSAMP)
C.... CONVERT TO DISTRIBUTION FUNCTION
          CALL FTOCUM(DENS,ARG,1.,MINT)
C.... GET INVERSE
          VALUE=(1.-QPROB)/2.
          BL(I)=FINVRT(RANK,ARG,MINT,VALUE)
          VALUE=1.-VALUE
          BU(I)=FINVRT(RANK,ARG,MINT,VALUE)
          RETURN
          END




          SUBROUTINE START2(XD,KSTART,RSTRT)
C.... SETS UP STARTING VALUES FOR MEP2 IN 0-1 DOMAIN
C.... INPUT
C          KSTART = CODE FOR STARTING METHOD
C          XD(I)  = ORDERED SAMPLE ARRAY ON 0-1 DOMAIN
C          NL     = NUMBER OF PARAMETERS IN MAX ENTROPY DIST, INCUDING
C                   FIRST
C.... OUTPUT
C          RSTRT(I) = ARRAY OF STARTING VALUES
          DIMENSION RSTRT(1),SM(8),AL(8),XP(1),YP(1),XD(1)
          COMMON/FUNCT/DENS(101),ARG(101),RANK(101)
          COMMON/IN/NL,NSAMP,QPROB,MINT
          IF(KSTART.GT.3)GO TO 10
C
          GO TO(1,3,5),KSTART
C.... NORMAL START
1         CALL SMOM(XD,2,NSAMP,SM)
          RSTRT(2)=SM(1)/SM(2)
          RSTRT(3)=-.5/SM(2)
          IF(NL.LT.4)RETURN
          DO 2 I=4,NL
2         RSTRT(I)=0.0
          RETURN
C.... UNIFORM START
3         DO 4 I=2,NL
4         RSTRT(I)=0.0
          RETURN
C.... MEP1 START
C     CALCULATE MOMENTS
5         NX=NL-1
          NS1=NSAMP-1
          SM(1)=1.
          DO 6 I=1,NSAMP
6         SM(1)=SM(1)+XD(I)
          SM(1)=SM(1)/FLOAT(NSAMP+1)
          DO 8 J=2,NX
          SM(J)=((XD(1)-SM(1))**(J+1)-(-SM(1))**(J+1))/(XD(1))+((1.-SM(1))**
         1(J+1)-(XD(NX)-SM(1))**(J+1))/(1.-XD(NX))
```

```
        DO 7 I=1,NS1
7       SM(J)=SM(J)+((XD(I+1)-SM(1))**(J+1)-(XD(I)-SM(1))**(J+1))/(XD(I+1)
        1-XD(I))
8       SM(J)=SM(J)/FLOAT((NSAMP+1)*(J+1))
        CALL MEP1(NX,SM,0.,1.,0,XP,1,0,AL,YP)
        DO 9 I=2,NL
9        RSTRT(I)=AL(I-1)
        RETURN
10      WRITE(6,11)
11      FORMAT(/,23H KSTART CANNOT EXCEED 3)
        CALL EXIT
        END



        SUBROUTINE UREAL(X,U)
        DIMENSION X(1)
        DIMENSION AL(8)
        COMMON/DATA/NDATA,XDATA(100),MINT,XOLD(101),FOLD(101),CUMOLD(101)
        COMMON/UREAL/NOPT
        COMMON/FUNCT/DENS(101),ARG(101),RANK(101)
        COMMON/IN/NL,NSAMP,QPROB,MINT2
        GO TO (100,101),NOPT
C.... THIS UREAL USED WITH EVOL
100     CONTINUE
        U=0.0
        SCALE=ABS(X(1))
        DO 1 I=1,NDATA
        AAA=XOLD(1)+(XDATA(I)-X(2))/SCALE
        CDF=FTABLE(XOLD,CUMOLD,AAA,MINT)
        ERROR=CDF-FLOAT(I)/FLOAT(NDATA+1)
1       U=U+ERROR*ERROR
        RETURN
101     CONTINUE
C.... THIS UREAL USED WITH MEP2
C.... CALCULATES VALUE OF ENTROPY FUNCTION
C.... INPUT
C        MINT = NUMBER OF INTEGRATION STATIONS
C        NL   = NUMBER OF PARAMETERS IN MAX ENTROPY DIST
C        AL(I)=ARRAY OF MAX ENTROPY DIST PARAMETERS, INCLUDING FIRST
C
        NX=NL-1
        AL(1)=ARO(X,NX)
        DO 3 I=1,MINT2
C.... CALCULATE LOCAL VALUE OF DENSITY FUNCTION USING ENTPRF (FROM MEP1)
        DO 2 J=2,NL
2       AL(J)=X(J-1)
        Y=FLOAT(I-1)/FLOAT(MINT2-1)
        DENS(I)=ENTRPF(AL,NL,Y)
3       ARG(I)=ALOG(DENS(I))*DENS(I)
        U=FSIMP(ARG,1.,MINT2)
        RETURN
        END
```

A.5 MISCELLANEOUS FUNCTIONS

```
        FUNCTION GAMMA(X)
C.... COMPUTES THE STANDARD GAMMA FUNCTION FOR A GIVEN ARGUMENT
C     THIS IS AN IMSL ROUTINE
```

A.6 MISCELLANEOUS SUBPROGRAMS

```
      FUNCTION BFACT(NUM,M,N)
C.... EVALUATES FACTORIAL FUNCTION
C     BFACT= FACT NUM/(FACT M * FACT N)
C     WHERE NUM GT M AND NUM GT N
      KMIN=MINO(M,N)
      KMAX=MAXO(M,N)+1
      BFACT=1
      DO 1 I=2,KMIN
1     BFACT=BFACT/FLOAT(I)
      DO 2 I=KMAX,NUM
2     BFACT=BFACT*FLOAT(I)
      RETURN
.     END
```

```
      SUBROUTINE BOUNDS(X,N,XMAX,XMIN)
C.... DETERMINES UPPER AND LOWER BOUNDS IN AN ARRAY
C       X(I) = INPUT ARRAY
C       N    = SIZE OF ARRAY
C       XMAX,XMIN = UPPER AND LOWER BOUNDS
      DIMENSION X(N)
      XMIN=XMAX=X(1)
      DO 1 I=1,N
      IF(X(I).LT.XMIN)XMIN=X(I)
      IF(X(I).GT.XMAX)XMAX=X(I)
1     CONTINUE
      RETURN
      END
```

```
      SUBROUTINE DERIV (FUNCT,DERIVF,RANGE,M)
C....    COMPUTES AN ARRAY OF DERIVATIVE VALUES GIVEN AN ARRAY OF
C        FUNCTION VALUES WHOSE ENTRIES CONSIST OF EQUIDISTANTLY SPACED
C        ARGUMENT VALUES
C        FUNCT(I)= FUNCTION ARRAY, DIMENSION M
C        DERIVF(I)= DERIVATIVE ARRAY, DIMENSION M
C        RANGE= RANGE OF VARIABLE
C        M= NUMBER OF STATIONS, MUST BE GE 3
      DIMENSION FUNCT(1), DERIVF(1)
      H=RANGE/FLOAT(M-1)
      HH=.5/H
      YY=FUNCT(M-2)
      B=FUNCT(2)+FUNCT(2)
      B=HH*(B+B-FUNCT(3)-3.*FUNCT(1))
      DO 1 I=3,M
      A=B
      B=HH*(FUNCT(I)-FUNCT(I-2))
      DERIVF(I-2)=A
1     CONTINUE
      A=FUNCT(M-1)+FUNCT(M-1)
      DERIVF(M)=HH*(FUNCT(M)*3.-A-A+YY)
      DERIVF(M-1)=B
      RETURN
      END
```

```
      FUNCTION FACTO (M)
C.... CALCULATES FACTORIAL OF M
```

```
          FACTO=1.
          IF (M.EQ.0) RETURN
          DO 1 I=1,M
          FACTO=FACTO*FLOAT(I)
1         CONTINUE
          RETURN
          END



          FUNCTION FCUBIN(I,X,FX,XVALUE,M)
C....  PROVIDES CUBIC INTERPOLATION USING NEWTONS DIVIDED DIFFERENCE
C      METHOD - SEE CARNAHAN, LUTHER AND WILKES, P. 12
C....  INPUT
C          I = SUBSCRIPT OF FIRST STATION IN FOUR BEING USED FOR INTERPOL-
C              ATION
C          X(I) = ARRAY OF TABULAR VALUES OF INDEPENDENT VARIABLE, DIMEN-
C                 M
C          FX(I) = ARRAY OF TABULAR VALUES OF DEPENDENT VARIABLE, DIMEN-
C                  SION M
C          XVALUE = GIVEN VALUE OF INDEPENDENT VARIABLE TO BE ENTERED INTO
C                   TABLE
C          M = SIZE OF ARRAYS
C....  OUTPUT
C          FCUBIN = INTERPOLATED VALUE OF FUNCTION
          DIMENSION X(1),FX(1)
          I1=I+1
          I2=I+2
          I3=I+3
          FD10=(FX(I1)-FX(I))/(X(I1)-X(I))
          FD21=(FX(I2)-FX(I1))/(X(I2)-X(I1))
          FD32=(FX(I3)-FX(I2))/(X(I3)-X(I2))
          FD210=(FD21-FD10)/(X(I2)-X(I))
          FD321=(FD32-FD21)/(X(I3)-X(I1))
          FD3210=(FD321-FD210)/(X(I3)-X(I))
          FCUBIN=FX(I)+(XVALUE-X(I))*FD10+(XVALUE-X(I1))*(XVALUE-X(I))*
         1FD210+(XVALUE-X(I2))*(XVALUE-X(I1))*(XVALUE-X(I))*FD3210
          RETURN
          END



          SUBROUTINE FRAND  (A,N,M)
C
C      FRAND IS A RANDOM NUMBER GENERATOR IN WHICH
C      A= STRING OF UNIFORMLY DISTRIBUTED RANDOM NUMBERS BETWEEN ZERO AND
C         ONE
C      N= NUMBER OF RANDOM NUMBERS IN THE STRING
C      M= NUMBER STARTING THE PSEUDO-RANDOM GENERATION. IT MAY BE SET AT
C         ANY INTEGER NUMBER. IF IT IS SET AT ZERO, GENERATION WILL
C         CONTINUE FROM THE LAST NUMBER OF THE PREVIOUS STRING
C
C      IF FRANDN IS CALLED INSIDE A DO LOOP, A DUMMY FIRST CALL SHOULD BE
C      USED WITH THE DESIRED VALUE FOR M
C
C....  ALMOST ANY LIBRARY RANDOM NUMBER GENERATOR CAN BE ADAPTED TO THIS
C      CALL
C....  SEE FOR EXAMPLE FRANDN IN SIDDALL,J.N.(1982), OPTIMAL ENGINEERING
C      DESIGN, PRINCIPLES AND APPLICATIONS, MARCEL DEKKER,N.Y.



          FUNCTION FSIMP(FUNC,RANGE,MINT)
C.....CALCULATES INTEGRAL BY SIMPSONS RULE WITH
```

```
C       MODIFICATION IF MINT IS EVEN
C.....INPUT
C           FUNC = ARRAY OF EQUALLY SPACED VALUES OF FUNCTION
C                  DIMENSION MINT
C           RANGE = RANGE OF INTEGRATION
C           MINT = NUMBER OF STATIONS
C.....OUTPUT
C           FSIMP = AREA
        DIMENSION FUNC(1)
C.....CHECK MINT FOR ODD OR EVEN
        XX=RANGE/(3.*FLOAT(MINT-1))
        M=MINT/2*2
        IF(M.EQ.MINT) GO TO 3
C.....ODD
        AREA=FUNC(1)+FUNC(M)
        MM=MINT-1
        DO 1 I=2,MM,2
1       AREA=AREA+4.*FUNC(I)
        MM=MM-1
        DO 2 I=3,MM,2
2       AREA=AREA+2.*FUNC(I)
        FSIMP=XX*AREA
        RETURN
C.....EVEN
C.....USE SIMPSONS RULE FOR ALL BUT LAST 3 INTERVALS
3       M=MINT-3
        AREA=FUNC(1)+FUNC(M)
        MM=M-1
        DO 4 I=2,MM,2
4       AREA=AREA+4.*FUNC(I)
        MM=MM-1
        DO 5 I=3,MM,2
5       AREA=AREA+2.*FUNC(I)
        FSIMP=XX*AREA
C.....USE NEWTONS 3/8 RULE FOR LAST THREE INTERVALS
        FSIMP=FSIMP+9./8.*XX*(FUNC(MINT-3)+3.*(FUNC(MINT-2)+FUNC(MINT-1))
1       +FUNC(MINT))
        RETURN
        END




        FUNCTION FTABLE(VAR,FUNC,XX,M)
C....  DOES LINEAR INTERPLOLATION IN A TABLE OF VALUES WHICH NEED NOT BE
C          EQUALLY SPACED
C....      VAR = ARRAY OF VALUES OF INDEPENDENT VARIABLE, DIMENSION M
C          FUNC = ARRAY OF VALUES OF FUNCTION , DIMENSION M
C          XX = GIVEN VALUES OF INDEPENDENT VARIABLE
C          M = NUMBER OF VALUES IN TABLE
C....  FTABLE = REQUIRED VALUE OF FUNCTION
C....  NOTE THAT LOGIC IS REQUIRED IN CALLING PROGRAM TO WARN USER OF
C          VALUES OF XX OUTSIDE RANGE OF TABLE
        DIMENSION VAR(1),FUNC(1)
        NEND=M-1
        DO 10 I=1,NEND
        INT=I
        IF(XX.GE.VAR(I).AND.XX.LE.VAR(I+1))GO TO 11
10      CONTINUE
11      FTABLE=FUNC(INT)+(XX-VAR(INT))*(FUNC(INT+1)-FUNC(INT))/(VAR(INT+1)
1       -VAR(INT))
        RETURN
        END
```

```
      SUBROUTINE SMOOTH (FUNC,SFUNC,M)
C....     TO COMPUTE A SET OF SMOOTHED FUNCTION VALUES GIVEN A
C         SET OF FUNCTION VALUES HAVING EQUALLY SPACED ARGUMENTS.
C         M MUST EXCEED OR EQUAL 5.
C         FUNC(I)= ARRAY OF GIVEN FUNCTION VALUES
C         SFUNC(I)= ARRAY OF SMOOTHED FUNCTION VALUES
C         M= NUMBER OF STATIONS
C
      DIMENSION FUNC(1), SFUNC(1)
      B=FUNC(1)
      C=FUNC(2)
      DO 2 I=5,M
      A=B
      B=C
      C=FUNC(I-2)
      D=C-B-FUNC(I-1)
      D=D+D+C
      D=D+D+A+FUNC(I)
      IF (I-5) 1,1,2
1     SFUNC(1)=A-.014285714286*D
      SFUNC(2)=B+.057142857143*D
2     SFUNC(I-2)=C-.0857142857*D
      SFUNC(M-1)=FUNC(M-1)+.057142857143*D
      SFUNC(M)=FUNC(M)-.014285714286*D
      RETURN
      END



      SUBROUTINE SORT(Y,N)
      DIMENSION Y(1)
C     Y IS THE ARRAY TO BE SORTED, I.E. AT COMPLETION Y(1)
C     WILL BE THE SMALLEST VALUE AND  Y(N) WILL BE THE LARGEST
      N1=N-1
      DO 1 I=1,N1
      J=I+1
      DO 2 K=J,N
      IF (Y(I).LT.Y(K)) GO TO 2
      TEMP=Y(I)
      Y(1)=Y(K)
      Y(K)=TEMP
2     CONTINUE
1     CONTINUE
      RETURN
      END
```

A.7 MOMENT CALCULATIONS

```
      SUBROUTINE CMOM(X,DENSI,M,NMOM,CMOMT,W)
C....     CALCULATES CENTRAL MOMENTS OF A DENSITY FUNCTION DEFINED BY A
C         SET OF DISCRETE VALUES
C....  INPUT
C         X(I) = RANDOM VARIABLE ARRAY, DIMENSION M
C         DENSI(I)= DENSITY FUNCTION ARRAY, DIMENSION M
C         XL, XU= LOWER AND UPPER BOUNDS OF VARIABLE
C         M = NUMBER OF STATIONS DEFINING VARIABLE, MUST BE ODD AND GT 6
C         NMOM= NUMBER OF MOMENTS CALCULATED
C ....  OUTPUT
C         CMOMT(I)= ARRAY OF CENTRAL MOMENTS, DIMENSION NMOM
C         W(I) = WORKING ARRAY DIMENSION M
```

```
C
      DIMENSION X(1),DENSI(1),CMOMT(1),W(1)
C.... CHECK AREA OF DENSITY FUNCTION
      RANGE=X(M)-X(1)
      AREA=FSIMP(DENSI,RANGE,M)
      IF(ABS(AREA-1.).GT.1.E-4)WRITE(6,10)AREA
C.... OBTAIN FIRST MOMENT
      DO 1 I=1,M
1     W(I)=X(I)*DENSI(I)
      CMOMT(1)=FSIMP(W,RANGE,M)
      IF(NMOM.EQ.1)RETURN
C.... OBTAIN HIGHER MOMENTS
      DO 3 J=2,NMOM
      DO 2 I=1,M
2     W(I)=(X(I)-CMOMT(1))**J*DENSI(I)
3     CMOMT(J)=FSIMP(W,RANGE,M)
      RETURN
10    FORMAT(//,39H WARNING - AREA OF DENSITY FUNCTION IS ,E12.5,//)
      END



      SUBROUTINE CONVERT (CM,NL)
C
C     THIS SUBROUTINE IS TO CALCULATE THE MOMENTS ABOUT THE ORIGIN
C
      COMMON /HELP/ S(41),XX(16,41),C(8),M
      DIMENSION CM(1)
      C(1)=CM(1)
      IF (NL.EQ.1) RETURN
      DO 2 J=2,NL
      C(J)=CM(J)-C(1)**J*(-1.)**J
      N=J-1
      DO 1 K=1,N
      C(J)=C(J)-(-1.)**K*FACTO(J)/(FACTO(K)*FACTO(J-K))*C(1)**(K)*C(J-K)
1     CONTINUE
2     CONTINUE
      RETURN
      END
```

Subroutine MOMENT (CM,N,CC,M)

Purpose

This subroutine calculates the first four moments of a random variable y, which is some general function of a set of independent random variables.

$$y = g(x_1, x_2, \ldots, x_n)$$

The first four moments of the x's, and the partial derivatives

$$\frac{\partial y}{\partial x_i} \quad \text{and} \quad \frac{\partial^2 y}{\partial x_i^2}$$

must be provided.

Method

The relationships are derived by taking a Taylor's series expansion about the mean of y; and then taking successive expected values to develop the moment relationships.

Input Variables

$CM(I,J)$: array giving the Jth moment of the Ith independent random variable, dimensioned N, M

M: number of moments of y required; cannot exceed four

N: number of independent random variables

Output Variables

$CC(J)$: array of moments of y, dimensioned with M

User-Provided Subroutine

The user must provide subroutine DERV, having the following form.

```
SUBROUTINE DERV(FUN,DE1,DE2,N,X)
DIMENSION DE1(1),DE2(1),X(1)
```

FUN = function of N independent random variables $X(I)$, defining

$$y = g(x_1, x_2, \ldots, x_n)$$

$DE1(1)$ = coding to define $\partial y / \partial x_1$
$DE1(2)$ = coding to define $\partial y / \partial x_2$
  etc.
$DE2(1)$ = coding to define $\partial^2 y / \partial x_1^2$
$DE2(2)$ = coding to define $\partial^2 y / \partial x_2^2$
  etc.

```
RETURN
END
```

```
      SUBROUTINE MOMENT (CM,N,CC,M)
C.... CALCULATES UP TO FIRST FOUR MOMENTS OF A DEPENDENT VARIABLE
C     FROM MOMENTS OF INDEPENDENT VARIANLES BY TAYLORS SERIES EXPANSION
C.... FUNCTION AND DERIVATIVES ARE DEFINED IN USER WRITTEN SUBROUTINE
C     DERV - SEE MANAL
C
C.... CM(I,J) = ARRAY OF INPUT MOMENTS - J TH MOMENT OF ITH INDEPENDENT
C               VARIABLE - DIMENSION N,M
C     N = NUMBER OF INDEPENDENT VARIABLES
C     CC(I)= MOMENTS OF DEPENDENT VARIABLE - DIMENSION M
C     M = NUMBER OF REQUIRED MOMENTS
      DIMENSION CM(N,M), CC(4), DE1(4), DE2(4)
      DO 1 I=1,N
      CC(I)=CM(I,1)
1     CONTINUE
      CALL DERV (FUN,DE1,DE2,N,CC)
```

```
          CC(1)=FUN
          DO 2 I=2,4
          CC(I)=0.0
2         CONTINUE
          DO 5 I=1,N
          CC(1)=CC(1)+.5*(DE2(I)*CM(I,2))
          IF (M.EQ.1) GO TO 5
          CC(2)=CC(2)+DE1(I)**2*CM(I,2)+DE1(I)*DE2(I)*CM(I,3)
          IF (M.EQ.2) GO TO 5
          CC(3)=CC(3)+DE1(I)**3*CM(I,3)
          IF (M.EQ.3) GO TO 5
          SUM=0.0
          KJ=I+1
          IF (KJ.GT.N) GO TO 4
          DO 3 J=KJ,N
          SUM=SUM+6.*(DE1(I)*DE1(J))**2*CM(I,2)*CM(J,2)
3         CONTINUE
4         CONTINUE
          CC(4)=CC(4)+SUM+DE1(I)**4*CM(I,4)
5         CONTINUE
          RETURN
          END




          SUBROUTINE SMOM(X,M,NSAMP,SM)
C.... CALCULATES SAMPLE CENTRAL MOMENTS
C        X(I) = SAMPLE VALUES,DIMENSION NSAMP
C        M=NUMBER OF MOMENTS DESIRED
C        NSAMP = SAMPLE SIZE
C        SM = VALUE OF MOMENTS, DIMENSION M
          DIMENSION X(1),SM(1)
C.... CALCULATE MEAN
          SUM=0.0
          DO 1 I=1,NSAMP
1         SUM=SUM+X(I)
          SM(1)=SUM/FLOAT(NSAMP)
          IF(M.LT.2)RETURN
C.... CALCULATE VARIANCE
          SUM=0.0
          DO 2 I=1,NSAMP
2         SUM=SUM+(X(I)-SM(1))**2
          SM(2)=SUM/(FLOAT(NSAMP-1))
          IF(M.LT.3)RETURN
C.... CALCULATE HIGHER MOMENTS
          DO 4 I=3,M
          SUM=0.0
          DO 3 J=1,NSAMP
3         SUM=SUM+(X(J)-SM(1))**I
          SM(I)=SUM/FLOAT(NSAMP)
4         CONTINUE
          RETURN
          END




          SUBROUTINE SMOMG(XC,FREQ,N,NSAMP,M,SM)
C.... CALCULATES CENTRAL MOMENTS FOR A GROUPED SAMPLE
C.... INPUT
C        XC(I) = INTERVAL MIDPOINTS, DIMENSION N
C        FREQ(I) = GROUPED FREQ FOR I TH INTERVAL, DIMENSION N
C        N = NUMBER OF INTERVALS OR GROUPS
C        M = NUMBER OF MOMENTS
C        NSAMP = TOTAL SAMPLE SIZE
```

```
C.... OUTPUT
C        SM(I) = VALUES OF CENTRAL MOMENTS, DIMENSION M
         DIMENSION XC(N),FREQ(N),SM(M)
C.... CALCULATE MEAN
         SUM=0.0
         DO 1 I=1,N
1        SUM=XC(I)*FREQ(I)
         SM(1)=SUM/FLOAT(NSAMP)
C.... OBTAIN HIGHER MOMENTS
         DO 2 J=2,M
         SUM=0.0
         DO 3 I=1,N
3        SUM=SUM+(XC(I)-SM(1))**J*FREQ(I)
2        SM(J)=SUM/FLOAT(NSAMP)
         RETURN
         END
```

Subroutine TRANSM (N,M,NMOM,XUPP,XLOW,CC,DELX,XGRID, KVALUE,X)

Purpose

This subroutine calculates any number of moments of a random variable y which is some general function of a set of independent random variables.

$$y = g(x_1, x_2, \ldots, x_n) \tag{1}$$

The user must provide subprograms defining the function in (1), and the joint density function of the x's.

Method

The algorithm is based on integral equations directly relating moments of y to the joint density function of the x's.

Input Variables

M: number of intervals for each x
N: number of independent random variables
NMOM: number of central moments required
XUPP(I): upper bound of x_i, dimensioned N
XLOW(I): lower bound of x_i, dimensioned N

Output Variables

CC(J): array of moments of y, dimensioned with NMOM

Working Arrays

The following working arrays must be dimensioned in the calling program: DELX(N), XGRID(N,M), KVALUE(N), and X(N).

User-Provided Subprograms

FUNCTION FUN(X,N)

DIMENSION X(1)

FUN = coding defining value of y corresponding to a given
set of values for the x's.

RETURN

END

FUNCTION DENSIT(N,X)

DIMENSION X(1)

DENSIT - coding defining value of the joint density function for
the x's corresponding to a given set of values for the x's

RETURN

END

```
      SUBROUTINE TRANSM (N,M,NMOM,XUPP,XLOW,CC,DELX,XGRID,KVALUE,X)
C.... MOMENT TRANSFER BY A DISCRETIZATION ALGORITHM BASED ON INTEGRATION
C     EQUATIONS
C     EQUATIONS
C.... INPUT
C        N = NUMBER OF INDEPENDENT VARIABLES
C        M = GRID SIZE OR NUMBER OF INTERVALS FOR EACH X
C        NMOM = NUMBER OF CENTRAL MOMENTS REQUIRED
C        XUPP(I) = UPPER BOUND OF X(I) - DIMENSION N
C        XLOW(I) = LOWER BOUND OF X(I) - DIMENSION N
C.... OUTPUT
C        CC(J) = ARRAY OF CENTRAL MOMENTS OF Y - DIMENSION NMOM
C.... WORKING ARRAYS - DIMENSION A SHOWN
C        DELX(N),XGRID(N,M),KVALUE(N),X(N)
      DIMENSION XUPP(1), XLOW(1), CC(1), DELX(1), XGRID(N,1), KVALUE(1),
     1 X(1)
C.... INITIALIZATION
      YMEAN=0.0
      SUMP=0.0
      DO 1 I=1,NMOM
      CC(I)=0.0
1     CONTINUE
      CELLA=1.0
      NCELLS=M**N
C.... SET UP GRID
      DO 3 I=1,N
      DELX(I)=(XUPP(I)-XLOW(I))/FLOAT(M)
      XGRID(I,1)=XLOW(I)+DELX(I)/2.
      DO 2 J=2,M
      XGRID(I,J)=XGRID(I,J-1)+DELX(I)
2     CONTINUE
3     CONTINUE
C.... CALCULATE CELL AREA
      DO 5 I=1,N
5     CELLA=CELLA*DELX(I)
C.... SCAN EACH CELL TO GET MEAN
      L=1
6     CONTINUE
C.... IDENTIFY CELL
      CALL CELL (L,N,M,KVALUE)
C.... EVALUATE Y AND JOINT DENSITY FUNCTION OF X'S
      DO 4 I=1,N
```

```
        KX=KVALUE(I)
        X(I)=XGRID(I,KX)
4       CONTINUE
        Y=FUN(X,N)
C....   CALCULATE CELL PROBABILITY
        CELLP=DENSIT(N,X)*CELLA
        SUMP=SUMP+CELLP
C....   CALCULATE MEAN
        YMEAN=YMEAN+Y*CELLP
        L=L+1
        IF(L.LE.NCELLS)GO TO 6
        WRITE (6,7) SUMP
7       FORMAT (* SUM OF CELL PROBABILITIES = *,E12.5,//)
C....   OBTAIN HIGHER MOMENTS
        L=1
11      CONTINUE
        CALL CELL (L,N,M,KVALUE)
        DO 8 I=1,N
        KX=KVALUE(I)
        X(I)=XGRID(I,KX)
8       CONTINUE
        Y=FUN(X,N)
        CELLP=DENSIT(N,X)*CELLA
        DO 10 J=2,NMOM
        CC(J)=CC(J)+(Y-YMEAN)**J*CELLP
10      CONTINUE
        L=L+1
        IF(L.LE.NCELLS)GO TO 11
        CC(1)=YMEAN
        RETURN
        END
        SUBROUTINE CELL (L,N,M,KVALUE)
C....   IDENTIFIES A CELL
        DIMENSION KVALUE(1)
        ISUM=0
        DO 1 I=1,N
        II=N-I+1
        IF (II.EQ.N) GO TO 1
        ISUM=ISUM+KVALUE(II+1)*M**II
1       KVALUE(II)=(L-ISUM-1)/M**(II-1)
        DO 2 I=1,N
        KVALUE(I)=KVALUE(I)+1
2       CONTINUE
        RETURN
        END
```

A.8 ORDER STATISTICS AND EXTREME VALUE DISTRIBUTIONS

```
        FUNCTION CUMRNK(RANK,I,N)
C.....COMPUTES LOCAL VALUES OF CUMULATIVE RANK DISTRIBUTION
C.....INPUT
C       RANK=VALUE OF RANK
C       I=ORDER
C       N=SAMPLE SIZE
        IF(RANK.LT.1.E-10) GO TO 2
        IF(RANK.GT.1.-1.E-10) GO TO 3
        CUMRNK=0.0
        DO 1 K=I,N
        COEF=1.
        IF(N.EQ.K) GO TO 1
```

```
        NK=N-K
        COEF=FACTO(N)/(FACTO(K)*FACTO(NK))
1       CUMRNK=CUMRNK+RANK**K*(1.-RANK)**(N-K)*COEF
        RETURN
2       CUMRNK=0.0
        RETURN
3       CUMRNK=1.0
        RETURN
        END



        FUNCTION CUMXT1(UCHAR,ALPHA,XE,IEXTR)
C.... CALCULATES LOCAL VALUES OF FIRST ASYPTOTIC EXTREME VALUE
C      CUM DIST FUNCTION
C.... INPUT
C           IEXTR = 0, LARGEST EXTREME VALUE DISTRIBUTION
C                 = 1, SMALLEST EXTREME VALUE DISTRIBUTION
C           ALPHA = PARAMETER (EXTREMAL INTENSITY FUNCTION)
C           UCHAR = PARAMETER (CHARACTERISTIC LARGEST VALUE)
C           XE = SPECIFIED VALUE OF EXTREME VALUE RANDOM VARIABLE
C.... OUTPUT
C           CUMXT1 =VALUE OF CUM DIST FUNCT
        Y=ALPHA*(XE-UCHAR)
        IF(IEXTR.EQ.0)CUMXT1=EXP(-EXP(-Y))
        IF(IEXTR.EQ.1)CUMXT1=1.-EXP(-EXP(Y))
        RETURN
        END



        FUNCTION CUMXT2(VCHAR,PARK,XE)
C.... CALCULATES LOCAL VALUES OF SECOND ASYMPTOTIC LARGEST EXTREME
C      VALUE CUM DIST FUNCTION
C.... INPUT
C           PARK = PARAMETER
C           VCHAR = PARAMETER
C           XE = SPECIFIED VALUE OF EXTREME VALUE RANDOM VARIABLE
C.... OUTPUT
C           CUMXT2 = VALUE OF CUM DIST FUNCTION
        IF(XE.LT.1.E-10)GO TO 1
        CUMXT2=EXP(-(VCHAR/XE)**PARK)
        RETURN
1       CUMXT2=0.0
        RETURN
        END



        FUNCTION DEXTR1(UCHAR,ALPHA,XE,IEXTR)
C.... CALCULATES LOCAL VALUES OF FIRST ASYPTOTIC EXTREME VALUE
C      DENSITY FUNCTION
C.... INPUT
C           IEXTR = 0, LARGEST EXTREME VALUE DISTRIBUTION
C                 = 1, SMALLEST EXTREME VALUE DISTRIBUTION
C           ALPHA = PARAMETER (EXTREMAL INTENSITY FUNCTION)
C           UCHAR = PARAMETER (CHARACTERISTIC LARGEST VALUE)
C           XE = SPECIFIED VALUE OF EXTREME VALUE RANDOM VARIABLE
C.... OUTPUT
C           DEXTR1 = VALUE OF DENSITY FUNCTION
        Y=ALPHA*(XE-UCHAR)
        IF(IEXTR.EQ.0)DEXTR1=ALPHA*EXP(-Y-EXP(-Y))
        IF(IEXTR.EQ.1)DEXTR1=ALPHA*EXP(Y-EXP(Y))
        RETURN
        END
```

```
      FUNCTION DEXTR2(VCHAR,PARK,XE)
C.... CALCULATES LOCAL VALUES OF SECOND ASYPTOTIC LARGEST EXTREME
C     VALUE DENSITY FUNCTION
C.... INPUT
C         PARK = PARAMETER
C         VCHAR = PARAMETER
C         XE = SPECIFIED VALUE OF EXTREME VALUE RANDOM VARIABLE
C.... OUTPUT
C         DEXTR2 = VALUE OF DENSITY FUNCTION
      IF(XE.LT.1.E-10)GO TO 1
      TEMP=VCHAR/XE
      DEXTR2=PARK/VCHAR*TEMP**(PARK+1)*EXP(-TEMP**PARK)
      RETURN
1     DEXTR2=0.0
      RETURN
      END


      FUNCTION DORDER(FX,FCUM,NSAMP,IORDER,ORDER)
C.... EVALUATES LOCAL VALUE OF ORDER STATISTIC DENSITY FUNCTION
C.... INPUT
C         NSAMP = SAMPLE SIZE
C         IORDER = ORDER NUMBER COUNTING FROM BOTTOM
C         ORDER = VALUE OF ORDER STATISTIC
C         FX = FUNCTION DEFINING PARENT DENSITY FUNCTION, WRITTEN BY
C              USER AS FX(ORDER) - MUST BE DEFINED EXTERNAL IN CALLING
C              PROGRAM
C         FCUM = FUNCTION DEFINING PARENT CUM DIST FUNCTION, WRITTEN BY
C              USER AS FCUM(ORDER) - MUST BE DEFINED EXTERNAL IN CALL-
C              ING PROGRAM
C.... OUTPUT
C         DORDER = REQUIRED LOCAL VALUE OF ORDER STATISTIC DENSITY
C              FUNCTION
      IF (NSAMP.EQ.1.AND.IORDER.EQ.1)GO TO 3
C.... EVALUATE PARENT CUM DIST FUNCTION
      CUM=FCUM(ORDER)
C.... EVALUATE DENSITY FUNCTION
      TEMP=1.-CUM
C.... AVOID UNDERFLOW
      IF(TEMP.LE.0.)TEMP=1.E-6
      IF(CUM.LE.0.)CUM=1.E-6
      IF(FLOAT(IORDER-1)*ABS(ALOG10(CUM)).GT.12.)GO TO 1
      IF(FLOAT(NSAMP-IORDER)*ABS(ALOG10(TEMP)).GT.12.)GO TO 1
C
      IF(IORDER.EQ.1)GO TO 4
      IF(IORDER.EQ.NSAMP)GO TO 5
      NUM=BFACT(NSAMP,IORDER-1,NSAMP-IORDER)
      COEF=FLOAT(NUM)
      DORDER=COEF*CUM**(IORDER-1)*FX(ORDER)*TEMP**(NSAMP-IORDER)
      RETURN
4     DORDER=NSAMP*FX(ORDER)*TEMP**(NSAMP-IORDER)
      RETURN
5     DORDER=NSAMP*CUM**(IORDER-1)*FX(ORDER)
      RETURN
3     DORDER=FX(ORDER)
      RETURN
1     DORDER=0.0
      RETURN
      END
```

A.9 PLOTTING DENSITY FUNCTIONS AND HISTOGRAMS

PLOTPR plots a density function on a printer when the density function is defined numerically. It uses local system software.

PLOTPL plots a density function on a plotter, using local system software for a Benson-Lehner plotter. However, it could easily be adopted to any plotting software. Axes, scales, and labels are all provided internally. Index mark spacing on the x axis is internally adjusted to give two number integer values which are multiples of 5 or 10. The vertical axis has no scale marks. Multiple-curve plots are possible by successive calls.

Plotting and labeling of axes can be suppressed by setting IFLAG=2. This is done internally after the first call to permit plotting of multiple curves on the same axes. Setting IFLAG=2 externally permits the superposition of the density function curve on other plots— for example, on the histogram from subroutine HIST. Subroutine HIST can be used to plot a histogram; see Sec. A.4.

```
        SUBROUTINE MREX1 (X,NSAMP,NLABEL,LABEL,IPLOT,UCHAR,ALPHA,W)
C.... GENERATES A MEAN RANK PLOT FROM A DATA SAMPLE FOR A TYPE 1
C      EXTREME VALUE DIST
C.... INPUT
C          X(I) = ARRAY OF VALUES IN SAMPLE, NEED NOT BE ORDERED,
C                 DIMENSION NSAMP
C          NSAMP = SIZE OF SAMPLE
C          NLABEL = NUMBER OF CHARACTERS IN LABEL
C          LABEL = NAME ON X AXIS, USE HOLLERITH CODE
C          IPLOT = 0, NO PLOT
C                = 1, PLOTS
C.... OUTPUT
C          UCHAR = PARAMETER
C          ALPHA = PARAMETER
C.... WORKING ARRAY - W(I) - DIMENSION NSAMP
        DIMENSION X(1), XMARK(20), YMARK(20), W(1)
C.... ORDER SAMPLE
        CALL SORT (X,NSAMP)
        BU=X(NSAMP)
        BL=X(1)
C.... DETERMINE REDUCED VARIATE SAMPLE VALUES AND BOUNDS
C      DESIGNATED W(I), YMAX, YMIN
        DO 1 I=1,NSAMP
        W(I)=-ALOG(-ALOG(FLOAT(I)/FLOAT(NSAMP+1)))
1       CONTINUE
        YMAX=W(NSAMP)
        YMIN=W(1)
C.... DETERMINE STRAIGHT LINE FIT USING LEAST SQUARE FIT NORMAL
C      EQUATIONS
C      FIND MEANS
        SUMX=SUMY=0.0
        PROD=SQUAR=0.0
        DO 2 I=1,NSAMP
        PROD=PROD+X(I)*W(I)
        SQUAR=SQUAR+X(I)*X(I)
        SUMX=SUMX+X(I)
        SUMY=SUMY+W(I)
2       CONTINUE
        XMEAN=SUMX/FLOAT(NSAMP)
```

```
            YMEAN=SUMY/FLOAT(NSAMP)
            ALPHA=(PROD-FLOAT(NSAMP)*XMEAN*YMEAN)/(SQUAR-FLOAT(NSAMP)*XMEAN*XM
           1EAN)
            A1=YMEAN-ALPHA*XMEAN
            UCHAR=-A1/ALPHA
            IF (IPLOT.EQ.0) RETURN
      C.... PLOT FIT
      C     SET UP SCALES
            YSCALE=(YMAX-YMIN)/4.5
            XSCALE=(BU-BL)/6.5
            CALL XAXIZ (XSCALE,BL,BU,XMARK,NMARK,RX,RK)
            CALL YAXIZ (YSCALE,YMIN,YMAX,YMARK,NMARKY,QX,QK)
      C.... INITIATE PLOT
            V=-2.*XSCALE+XMARK(1)
            WY=-2.*YSCALE+YMARK(1)
            XMN=V
            YMN=WY
            XMX=XMARK(NMARK)+XSCALE
            YMX=YMAX+2.*YSCALE
            CALL PLTIN (XSCALE,YSCALE,V,WY,XMN,XMX,YMN,YMX)
      C.... PLOT X AXIS AND Y AXIS
            XAXIS=(XMARK(NMARK)-XMARK(1))/XSCALE+2.
            YAXIS=(YMARK(NMARKY)-YMARK(1))/YSCALE+2.
            CALL XAXPL (XMARK,NMARK,XSCALE,RK,RX,XAXIS)
            CALL YAXPL (YMARK,NMARKY,YSCALE,QK,QX,YAXIS)
      C.... LABEL X AXIS
            XLABEL=(XMARK(NMARK)-V)/XSCALE-FLOAT(NLABEL)*.15
            YLABEL=1.4
            CALL LETTER (NLABEL,.15,0.,XLABEL,YLABEL,LABEL)
      C.... LABEL Y AXIS
            CALL LETTER (15,.15,90.,1.2,3.0,15HREDUCED VARIATE)
      C     PLOT POINTS
            DO 3 I=1,NSAMP
            CALL UNITTO (X(I),W(I),XP,YP)
            XPO=XP-.05
            XPE=XP+.05
            YPO=YP-.05
            YPE=YP+.05
            CALL PLOT (XPO,YP,3)
            CALL PLOT (XPE,YP,2)
            CALL PLOT (XP,YPO,3)
            CALL PLOT (XP,YPE,2)
      3     CONTINUE
      C.... PLOT LINE
            XO=(YMIN+ALPHA*UCHAR)/ALPHA
            IF (XO.LT.BL) GO TO 4
            YO=YMIN
            GO TO 5
      4     YO=ALPHA*(BL-UCHAR)
            XO=BL
      5     XE=(YMAX+ALPHA*UCHAR)/ALPHA
            IF (XE.GT.BU) GO TO 6
            YE=YMAX
            GO TO 7
      6     YE=ALPHA*(BU-UCHAR)
            XE=BU
      7     CALL UNITTO (XO,YO,XPO,YPO)
            CALL UNITTO (XE,YE,XPE,YPE)
            CALL PLOT (XPO,YPO,3)
            CALL PLOT (XPE,YPE,2)
      C.... PLOT LETTERING
            CALL LETTER (2,.15,0.0,6.5,3.5,2HY )
            CALL MATH (6.8,3.5,.15,0.0,4)
            CALL GREEK (7.1,3.5,.15,0.0,5HALPHA)
            CALL LETTER (1,.15,0.0,7.25,3.5,1H()
```

```
          CALL LETTER (5,.15,0.0,7.4,3.5,5HX - U)
          CALL LETTER (1,.15,0.0,8.15,3.5,1H))
          CALL GREEK (6.5,3.0,.15,0.0,5HALPHA)
          CALL MATH (6.8,3.0,.15,0.0,4)
          IMFT=7H(E12.5)
          ENCODE (12,IMFT,BCD) ALPHA
          CALL LETTER (12,.15,0.0,7.1,3.0,BCD)
          CALL LETTER (1,.15,0.0,6.5,2.5,1HU)
          CALL MATH (6.8,2.5,.15,0.0,4)
          ENCODE (12,IMFT,BCD) UCHAR
          CALL LETTER (12,.15,0.,7.1,2.5,BCD)
          CALL PLOT (0.0,0.0,999)
          RETURN
          END



          SUBROUTINE MREX2 (X,NSAMP,IPLOT,VCHAR,PARK,W)
C.... GENERATES A MEAN RANK PLOT FROM A DATA SAMPLE FOR A TYPE 2
C     EXTREME VALUE DIST
C.... INPUT
C          X(I) = ARRAY OF VALUES IN SAMPLE, NEED NOT BE ORDERED,
C                 DIMENSION NSAMP
C         NSAMP = SIZE OF SAMPLE
C         IPLOT = 0, NO PLOT
C               = 1, PLOTS
C.... OUTPUT
C         VCHAR = PARAMETER
C          PARK = PARAMETER
C.... WORKING ARRAY - W(I) - DIMENSION NSAMP*2
          DIMENSION X(1), XMARK(20), YMARK(20), W(1)
C.... ORDER SAMPLE
          CALL SORT (X,NSAMP)
C.... DETERMINE TRANSFORMED VARIABLES SAMPLE VALUES
C     DESIGNATED W(I), YMAX, YMIN
          DO 1 I=1,NSAMP
          W(I+NSAMP)=ALOG(X(I))
          W(I)=-ALOG(-ALOG(FLOAT(I)/FLOAT(NSAMP+1)))
1         CONTINUE
          YMAX=W(NSAMP)
          YMIN=W(1)
          BU=W(NSAMP*2)
          BL=W(NSAMP+1)
C.... DETERMINE STRAIGHT LINE FIT USING LEAST SQUARE FIT NORMAL
C     EQUATIONS
C     FIND MEANS
          SUMX=SUMY=0.0
          PROD=SQUAR=0.0
          DO 2 I=1,NSAMP
          PROD=PROD+W(NSAMP+I)*W(I)
          SQUAR=SQUAR+W(NSAMP+I)*W(NSAMP+I)
          SUMX=SUMX+W(I+NSAMP)
          SUMY=SUMY+W(I)
2         CONTINUE
          XMEAN=SUMX/FLOAT(NSAMP)
          YMEAN=SUMY/FLOAT(NSAMP)
          PARK=(PROD-FLOAT(NSAMP)*XMEAN*YMEAN)/(SQUAR-FLOAT(NSAMP)*XMEAN*XME
         1AN)
          A1=YMEAN-PARK*XMEAN
          VCHAR=EXP(-A1/PARK)
          IF (IPLOT.EQ.0) RETURN
C.... PLOT FIT
C     SET UP SCALES
          YSCALE=(YMAX-YMIN)/4.5
          XSCALE=(BU-BL)/6.5
```

```
      CALL XAXIZ (XSCALE,BL,BU,XMARK,NMARK,RX,RK)
      CALL YAXIZ (YSCALE,YMIN,YMAX,YMARK,NMARKY,QX,QK)
C.... INITIATE PLOT
      V=-2.*XSCALE+XMARK(1)
      WY=-2.*YSCALE+YMARK(1)
      XMN=V
      YMN=WY
      XMX=XMARK(NMARK)+XSCALE
      YMX=YMAX+2.*YSCALE
      CALL PLTIN (XSCALE,YSCALE,V,WY,XMN,XMX,YMN,YMX)
C.... PLOT X AXIS AND Y AXIS
      XAXIS=(XMARK(NMARK)-XMARK(1))/XSCALE+2.
      YAXIS=(YMARK(NMARKY)-YMARK(1))/YSCALE+2.
      CALL XAXPL (XMARK,NMARK,XSCALE,RK,RX,XAXIS)
      CALL YAXPL (YMARK,NMARKY,YSCALE,QK,QX,YAXIS)
C.... LABEL X AXIS
      XLABEL=(XMARK(NMARK)-V)/XSCALE-3.6
      YLABEL=1.4
      CALL LETTER (24,.15,0.0,XLABEL,YLABEL,24HTRANSFORMED VARIABLE - Z)
C.... LABEL Y AXIS
      CALL LETTER (24,.15,90.,1.2,3.0,24HTRANSFORMED CUM DIST - Y)
C     PLOT POINTS
      DO 3 I=1,NSAMP
      CALL UNITTO(W(NSAMP+1),W(I),XP,YP)
      XPO=XP-.05
      XPE=XP+.05
      YPO=YP-.05
      YPE=YP+.05
      CALL PLOT (XPO,YP,3)
      CALL PLOT (XPE,YP,2)
      CALL PLOT (XP,YPO,3)
      CALL PLOT (XP,YPE,2)
3     CONTINUE
C.... PLOT LINE
      XO=(YMIN+PARK*ALOG(VCHAR))/PARK
      IF (XO.LT.BL) GO TO 4
      YO=YMIN
      GO TO 5
4     YO=PARK*(BL-ALOG(VCHAR))
      XO=BL
5     XE=(YMAX+PARK*ALOG(VCHAR))/PARK
      IF (XE.GT.BU) GO TO 6
      YE=YMAX
      GO TO 7
6     YE=PARK*(BU-ALOG(VCHAR))
      XE=BU
7     CALL UNITTO (XO,YO,XPO,YPO)
      CALL UNITTO (XE,YE,XPE,YPE)
      CALL PLOT (XPO,YPO,3)
      CALL PLOT (XPE,YPE,2)
C.... PLOT LETTERING
      CALL LETTER (24,.15,0.0,6.5,4.0,24HCUM DIST = EXP(-EXP(-Y)))
      CALL LETTER (17,.15,0.0,6.5,3.5,17HVARIABLE = EXP(Z))
      CALL LETTER (1,.15,0.0,6.5,3.0,1HK)
      CALL MATH (6.8,3.0,.15,0.0,4)
      IMFT=7H(E12.5)
      ENCODE (12,IMFT,BCD) PARK
      CALL LETTER (12,.15,0.0,7.1,3.0,BCD)
      CALL LETTER (1,.15,0.0,6.5,2.5,1HV)
      CALL MATH (6.8,2.5,.15,0.0,4)
      ENCODE (12,IMFT,BCD) VCHAR
      CALL LETTER (12,.15,0.,7.1,2.5,BCD)
      CALL PLOT (0.0,0.0,999)
      RETURN
      END
```

```
      SUBROUTINE MRWEIB (X,NSAMP,IPLOT,ETA,BETA,W)
C.... GENERATES A MEAN RANK PLOT FROM A DATA SAMPLE FOR A WEIBULL
C     DISTRIBUTION
C.... INPUT
C        X(I) = ARRAY OF VALUES IN SAMPLE, NEED NOT BE ORDERED,
C               DIMENSION NSAMP
C        NSAMP = SIZE OF SAMPLE
C        IPLOT = 0, NO PLOT
C              = 1, PLOTS
C.... OUTPUT
C        ETA = POSITION PARAMETER
C        BETA = SHAPE PARAMETER
C.... WORKING ARRAY - W(I) - DIMENSION NSAMP*2
      DIMENSION X(1), XMARK(20), YMARK(20), W(1)
C.... ORDER SAMPLE
      CALL SORT (X,NSAMP)
C.... DETERMINE TRANSFORMED VARIABLES SAMPLE VALUES
C     DESIGNATED W(I), YMAX, YMIN
      DO 1 I=1,NSAMP
      W(I+NSAMP)=ALOG(X(I))
      W(I)=ALOG(ALOG(1./(1.-FLOAT(I)/FLOAT(NSAMP+1))))
1     CONTINUE
      YMAX=W(NSAMP)
      YMIN=W(1)
      BU=W(NSAMP*2)
      BL=W(NSAMP+1)
C.... DETERMINE STRAIGHT LINE FIT USING LEAST SQUARE FIT NORMAL
C     EQUATIONS
C     FIND MEANS
      SUMX=SUMY=0.0
      PROD=SQUAR=0.0
      DO 2 I=1,NSAMP
      PROD=PROD+W(NSAMP+I)*W(I)
      SQUAR=SQUAR+W(NSAMP+I)*W(NSAMP+I)
      SUMX=SUMX+W(I+NSAMP)
      SUMY=SUMY+W(I)
2     CONTINUE
      XMEAN=SUMX/FLOAT(NSAMP)
      YMEAN=SUMY/FLOAT(NSAMP)
      BETA=(PROD-FLOAT(NSAMP)*XMEAN*YMEAN)/(SQUAR-FLOAT(NSAMP)*XMEAN*XME
     1AN)
      A1=YMEAN-BETA*XMEAN
      ETA=EXP(-A1/BETA)
      IF (IPLOT.EQ.0) RETURN
C.... PLOT FIT
C     SET UP SCALES
      YSCALE=(YMAX-YMIN)/4.5
      XSCALE=(BU-BL)/6.5
      CALL XAXIZ (XSCALE,BL,BU,XMARK,NMARK,RX,RK)
      CALL YAXIZ (YSCALE,YMIN,YMAX,YMARK,NMARKY,QX,QK)
C.... INITIATE PLOT
      V=-2.*XSCALE+XMARK(1)
      WY=-2.*YSCALE+YMARK(1)
      XMN=V
      YMN=WY
      XMX=XMARK(NMARK)+XSCALE
      YMX=YMAX+2.*YSCALE
      CALL PLTIN (XSCALE,YSCALE,V,WY,XMN,XMX,YMN,YMX)
C.... PLOT X AXIS AND Y AXIS
      XAXIS=(XMARK(NMARK)-XMARK(1))/XSCALE+2.
      YAXIS=(YMARK(NMARKY)-YMARK(1))/YSCALE+2.
      CALL XAXPL (XMARK,NMARK,XSCALE,RK,RX,XAXIS)
      CALL YAXPL (YMARK,NMARKY,YSCALE,QK,QX,YAXIS)
C.... LABEL X AXIS
      XLABEL=(XMARK(NMARK)-V)/XSCALE-3.6
```

```
        YLABEL=1.4
        CALL LETTER (24,.15,0.0,XLABEL,YLABEL,24HTRANSFORMED VARIABLE - Z)
C.... LABEL Y AXIS
        CALL LETTER (24,.15,90.,1.2,3.0,24HTRANSFORMED CUM DIST - Y)
C       PLOT POINTS
        DO 3 I=1,NSAMP
        CALL UNITTO(W(NSAMP+I),W(I),XP,YP)
        XPO=XP-.05
        XPE=XP+.05
        YPO=YP-.05
        YPE=YP+.05
        CALL PLOT (XPO,YP,3)
        CALL PLOT (XPE,YP,2)
        CALL PLOT (XP,YPO,3)
        CALL PLOT (XP,YPE,2)
3       CONTINUE
C.... PLOT LINE
        XO=YMIN/BETA+ALOG(ETA)
        IF (XO.LT.BL) GO TO 4
        YO=YMIN
        GO TO 5
4       YO=BETA*(BL-ALOG(ETA))
        XO=BL
5       XE=YMAX/BETA+ALOG(ETA)
        IF (XE.GT.BU) GO TO 6
        YE=YMAX
        GO TO 7
6       YE=BETA*(BU-ALOG(ETA))
        XE=BU
7       CALL UNITTO (XO,YO,XPO,YPO)
        CALL UNITTO (XE,YE,XPE,YPE)
        CALL PLOT (XPO,YPO,3)
        CALL PLOT (XPE,YPE,2)
C.... PLOT LETTERING
        CALL LETTER (26,.15,0.0,6.5,4.0,26HCUM DIST = 1-1/EXP(EXP(Y)))
        CALL LETTER (17,.15,0.0,6.5,3.5,17HVARIABLE = EXP(Z))
        CALL LETTER (3,.15,0.0,6.5,3.0,3HETA)
        CALL MATH (7.1,3.0,.15,0.0,4)
        IMFT=7H(E12.5)
        ENCODE (12,IMFT,BCD) ETA
        CALL LETTER (12,.15,0.0,7.4,3.0,BCD)
        CALL LETTER (4,.15,0.0,6.5,2.5,4HBETA)
        CALL MATH (7.25,2.5,.15,0.0,4)
        ENCODE (12,IMFT,BCD) BETA
        CALL LETTER (12,.15,0.,7.55,2.5,BCD)
        CALL PLOT (0.0,0.0,999)
        RETURN
        END



        SUBROUTINE PLOTPL (NCURV,X,FX,M,LABEL,N,BU,BL)
C.... PLOTS A DENSITY FUNCTION ON A PLOTTER
C.... INPUT
C       X(I) = DISCRETE VALUES OF VARIABLES,DIMENSION M
C       FX(I) = DISCRETE VALUES OF DENSITY FUNCTION, DIMENSION M
C       M = NUMBER OF VALUES
C       NCURV = NUMBER OF CURVE THAT IS TO BE PLOTTED. IN ONE EXECUTION
C               ALL CURVES MUST BE PLOTTED ON THE SAME AXES. THE FIRST
C               CALL PLOTS THE AXES AND LABELS. SUBSEQUENT CALLS TO
C               PLOTPL ARE SUPERIMPOSED
C               A FINAL CALL TO PLOTPL WITH NCURV=0 IS REQUIRED TO
C               TERMINATE PLOT
C       LABEL = X AXIS LABEL - INSERT IN A HOLLERITH DATA STATEMENT
C       N = NUMBER OF CHARACTERS IN LABEL
```

```
C           IFLAG = 1, AXES PLOTTED AND LABELED
C                 = 2, CURVE ONLY PLOTTED
C           BU    = UPPER BOUND OF DENSITY FUNCTION
C           BL    = LOWER BOUND
C           ISCALE = 1, YAXIS SCALE IS INTEGERIZED FOR SUPERIMPOSED PLOTTING
C                      OF SIMILAR CURVES - DEFAULTS AT ZERO
C.... BU AND BL SHOULD BE INSIDE INITIAL X(1) TO X(M), FOR NCURV EQUAL 1
C.... IFLAG IS SET IN COMMON/PLOTPL/IFLAG. THE DEFAULT VALUE IS 1 AND
C           IT IS INTERNALLY RESET TO 2 AFTER THE FIRST CALL
            DIMENSION X(1), FX(1), XMARK(20)
            COMMON /PLOTPL/ IFLAG
            COMMON /YSCALE/ISCALE
            DATA IFLAG,ISCALE/1,0/
            IF (NCURV.EQ.0) GO TO 7
            IBND=IFLAG
            IF (IFLAG.GT.1) GO TO 1
            IFLAG=2
C.... DETERMINE BOUNDS ON FX
            CALL BOUNDS (FX,M,FMAX,FMIN)
C.... DETERMINE SCALE OF FX
            YSCALE=FMAX/3.5
C.... DETERMINE SCALE OF X
            RANGE=X(M)-X(1)
C.... XSCALE = DATA UNITS PER INCH
            XSCALE=RANGE/6.5
            CALL XAXIZ (XSCALE,X(1),X(M),XMARK,NMARK,RX,RK)
            IF((XMARK(2)-XMARK(1))/XSCALE.LT..8)XSCALE=XSCALE/1.5
C.... OPTION TO INTEGERIZE YSCALE
            IF(ISCALE.NE.0)CALL YAXIZ(YSCALE,0.0,FMAX,NMARKY,QX,QK)
C.... INITIATE PLOT
            V=-2.*XSCALE+XMARK(1)
            W=-2.*YSCALE
            XMIN=V
            YMIN=W
            XMAX=XMARK(NMARK)+XSCALE
            YMAX=FMAX+2.*YSCALE
            CALL PLTIN (XSCALE,YSCALE,V,W,XMIN,XMAX,YMIN,YMAX)
C.... PLOT X AXIS AND Y AXIS
            XAXIS=(XMARK(NMARK)-XMARK(1))/XSCALE+2.
            YAXIS=6.
            CALL PLOT (2.,2.,3)
            CALL PLOT (2.,YAXIS,2)
            CALL XAXPL (XMARK,NMARK,XSCALE,RK,RX,XAXIS)
C.... LABEL X AXIS
            XLABEL=(XMARK(NMARK)-V)/XSCALE-FLOAT(N)*.15
            YLABEL=1.4
            CALL LETTER (N,.15,0.,XLABEL,YLABEL,LABEL)
C.... LABEL Y AXIS
            CALL LETTER (16,.15,90.,1.85,3.1,16HDENSITY FUNCTION)
C.... PLOT CURVE
1           GO TO (2,3,4,5), NCURV
C.... FIRST CURVE IS SOLID LINE
2           CALL PLTMPL (X,FX,M)
            GO TO 6
C.... SECOND CURVE IS DASHED LINE
3           CALL PLTMPL (X,FX,M)
            GO TO 6
C.... THIRD CURVE IS DOTTED LINE
4           CALL PLTMPL (X,FX,M)
            GO TO 6
C.... FOURTH CURVE IS DOT DASH
5           CALL PLTMPL (X,FX,M)
6           IF (IBND.EQ.2) RETURN
            XBNDU=(BU-V)/XSCALE
            XBNDL=(BL-V)/XSCALE
```

```
         YU=2.5
         TEST=FX(M)/YSCALE+2.
         IF (YU.LT.TEST) YU=TEST
         CALL PLOT (XBNDU,2.0,3)
         CALL PLOT (XBNDU,YU,2)
         YU=2.5
         TEST=FX(1)/YSCALE+2.
         IF (YU.LT.TEST) YU=TEST
         CALL PLOT (XBNDL,2.0,3)
         CALL PLOT (XBNDL,YU,2)
         RETURN
7        CALL PLOT (0.,0.,999)
         RETURN
         END



         SUBROUTINE PLOTPR(X,FX,M)
C....    PLOTS A DENSITY FUNCTION ON A PRINTER
C....    SYSTEM DEPENDENT
C....    PLOTS CANNOT BE SUPERIMPOSED BUT SUCCESSIVE PLOTS CAN BE USED
C....    INPUT
C            X(1) = DISCRETE VALUES OF VARIABLE, DIMENSION M
C            FX(1) = DISCRETE VALUES OF DENSITY FUNCTION, DIMENSION M
C            M = NUMBER OF VALUES
         DIMENSION X(1),FX(1)
         DO 1 I=1,M
         XP=X(I)
         FXP=FX(I)
1        CALL PLOTPT(XP,FXP,9)
         CALL OUTPLT
         RETURN
         END



         SUBROUTINE XAXIZ (XSCALE,BL,BU,XMARK,NMARK,RX,RK)
         DIMENSION XMARK(20)
C....    OBTAIN X INTERVALS
         RX=1.
         RK=.1
C....    XSC= INTERVAL BETWEEN INDEX MARKS IN DATA UNITS
         XSC=XSCALE
         IF (XSC.GT.1.) GO TO 2
1        IF (XSC.GT..1) GO TO 3
         RX=RX/10.
         RK=RX/10.
         XSC=XSC*10.
         GO TO 1
2        IF (XSC.LT.1.) GO TO 3
         RK=RX
         RX=RX*10.
         XSC=XSC/10.
         GO TO 2
3        IF (XSC.LT..25) XSC=.1
         IF (XSC.GE..25.AND.XSC.LT..75) XSC=.5
         IF (XSC.GE..75) XSC=1.
         RF=RX
         IF (XSC.LT..11) RF=RX/10.
         XSC=XSC*RX
C....    DETERMINE PLOTTING ORIGIN
         TEMP=(BL-XSC)/RF
         IZERO=INT(TEMP)
         IF (BL.LT.0.0) GO TO 4
         IF (TEMP-FLOAT(IZERO).GT..5) IZERO=IZERO+1
```

```
          ZERO=RF*FLOAT(IZERO)
          GO TO 5
4         IF (TEMP-FLOAT(IZERO).LT.-.5) IZERO=IZERO-1
          ZERO=RF*FLOAT(IZERO)
5         IF (ZERO+XSC.LT.BL+1.E-6*XSC) ZERO=ZERO+XSC
C.... LOCATE SCALE MARKS
          XMARK(1)=ZERO
          I=1
6         XMARK(I+1)=XMARK(I)+XSC
          IF (XMARK(I+1).GE.BU) GO TO 7
          I=I+1
          GO TO 6
7         NMARK=I+1
          RETURN
          END



          SUBROUTINE XAXPL (XMARK,NMARK,XSCALE,RK,RX,XAXIS)
          DIMENSION XMARK(20)
C.... PLOTS X AXIS
          CALL PLOT (2.,2.,3)
          CALL PLOT (XAXIS,2.,2)
C.... PLOT INDEX MARKS FOR X AXIS
          CALL PLOT (2.,2.,3)
          N1=NMARK-1
          XINVL=(XMARK(2)-XMARK(1))/XSCALE
          DO 1 I=1,N1
          XTEMP=FLOAT(I)*XINVL+2.
          CALL PLOT (XTEMP,2.,3)
          CALL PLOT (XTEMP,2.1,2)
1         CONTINUE
C.... LABEL INDEX MARKS ON X AXIS
          XLET=1.55
          DO 2 I=1,NMARK
          TEMP=XMARK(I)/RK+.001
          IF (XMARK(I).LT.0.0) TEMP=TEMP-.002
          IDATUM=INT(TEMP)
          IF (TEMP-FLOAT(IDATUM).GT..5) IDATUM=IDATUM+1
          IMFT=4H(I4)
          ENCODE (4,IMFT,BCD) IDATUM
          CALL LETTER (4,.15,0.,XLET,1.70,BCD)
          XLET=XLET+XINVL
2         CONTINUE
          IRK=INT(RK)
          IF (RK-FLOAT(IRK).GT..5) IRK=IRK+1
          IF (IRK.EQ.1) RETURN
          XLET=XLET+.7-XINVL
          CALL MATH (XLET,1.70,.1,0.,1HX)
          XLET=XLET+.25
C.... LABEL MULTIPLYING FACTOR
          IF (RK.GT.1.) GO TO 13
          I=1
3         IRK=INT(RX)
          IF (RX-FLOAT(IRK).GT..5) IRK=IRK+1
          IF (IRK.EQ.1) GO TO 4
          I=I+1
          IF (I.GT.7) GO TO 23
          RX=RX*10.
          GO TO 3
4         GO TO (5,6,7,8,9,10,11), I
5         IMFT=6H(F2.1)
          GO TO 12
6         IMFT=6H(F3.2)
          GO TO 12
```

```
7       IMFT=6H(F4.3)
        GO TO 12
8       IMFT=6H(F5.4)
        GO TO 12
9       IMFT=6H(F6.5)
        GO TO 12
10      IMFT=6H(F7.6)
        GO TO 12
11      IMFT=6H(F8.7)
12      NC=I+1
        ENCODE (NC,IMFT,BCD) RK
        GO TO 22
13      IK=INT(RK)
        IF (RK-FLOAT(IK).GT..5) IK=IK+1
        I=1
14      IK=IK/10
        IF (IK.EQ.1) GO TO 15
        I=I+1
        IF (1.GT.5) GO TO 25
        GO TO 14
        GO TO 14
15      GO TO (16,17,18,19,20), I
16      IMFT=4H(I2)
        GO TO 21
17      IMFT=4H(I3)
        GO TO 21
18      IMFT=4H(I4)
        GO TO 21
19      IMFT=4H(I5)
        GO TO 21
20      IMFT=4H(I6)
21      NC=I+1
        IRK=INT(RK)
        IF (RK-FLOAT(IRK).GT..5) IRK=IRK+1
        ENCODE (NC,IMFT,BCD) IRK
22      CALL LETTER (NC,.15,0.,XLET,1.7,BCD)
        RETURN
23      WRITE (6,24)
24      FORMAT (33H X VARIABLE TOO SMALL FOR PROGRAM)
        STOP
25      WRITE (6,26)
26      FORMAT (33H X VARIABLE TOO LARGE FOR PROGRAM)
        STOP
        END



        SUBROUTINE YAXIZ (YSCALE,YMIN,YMAX,YMARK,NMARKY,QX,QK)
        DIMENSION YMARK(20)
        COMMON/YSCALE/ISCALE
C.... OBTAIN Y INTERVALS
        QX=1.
        QK=.1
        YSC=YSCALE
        IF (YSC.GT.1.) GO TO 2
1       IF (YSC.GT..1) GO TO 3
        QX=QX/10.
        QK=QX/10.
        YSC=YSC*10.
        GO TO 1
2       IF (YSC.LT.1.) GO TO 3
        QK=QX
        QX=QX*10.
        YSC=YSC/10.
        GO TO 2
3       IF (YSC.LT..25) YSC=.1
```

```
         IF (YSC.GE..25.AND.YSC.LT..75) YSC=.5
         IF (YSC.GE..25.AND.YSC.LT..75) YSC=.5
         IF (YSC.GE..75) YSC=1.
         QF=QX
         IF (YSC.LT..11) QF=QX/10.
         YSC=YSC*QX
C.... DETERMINE Y PLOTTING ORIGIN
         TEMP=(YMIN-YSC)/QF
         IZERO=INT(TEMP)
         IF (YMIN.LT.0.0) GO TO 4
         IF (TEMP-FLOAT(IZERO).GT..5) IZERO=IZERO+1
         ZERO=QF*FLOAT(IZERO)
         GO TO 5
4        IF (TEMP-FLOAT(IZERO).LT.-.5) IZERO=IZERO-1
         ZERO=QF*FLOAT(IZERO)
5        IF (ZERO+YSC.LT.YMIN+1.E-6*YSC) ZERO=ZERO+YSC
C.... LOCATE SCALE MARKS
         YMARK(1)=ZERO
         I=1
6        YMARK(I+1)=YMARK(I)+YSC
         IF (YMARK(I+1).GE.YMAX) GO TO 7
         I=I+1
         GO TO 6
7        NMARKY=I+1
         IF(ISCALE.NE.0)YSCALE=YSC
         RETURN
         END



         SUBROUTINE YAXPL (YMARK,NMARKY,YSCALE,QK,QX,YAXIS)
         DIMENSION YMARK(20)
C.... PLOT INDEX MARKS FOR Y AXIS
         CALL PLOT (2.,2.,3)
         CALL PLOT (2.,YAXIS,2)
         CALL PLOT (2.,2.,3)
         N1=NMARKY-1
         YINVL=(YMARK(2)-YMARK(1))/YSCALE
         DO 1 I=1,N1
         YTEMP=FLOAT(I)*YINVL+2.
         CALL PLOT (2.,YTEMP,3)
         CALL PLOT (2.1,YTEMP,2)
1        CONTINUE
C.... LABEL INDEX MARKS ON Y AXIS
         YLET=1.92
         DO 2 I=1,NMARKY
         TEMP=YMARK(I)/QK+.001
         IF (YMARK(1).LT.0.0) TEMP=TEMP-.002
         IDATUM=INT(TEMP)
         IF (TEMP-FLOAT(IDATUM).GT..5) IDATUM=IDATUM+1
         IMFT=4H(I4)
         ENCODE (4,IMFT,BCD) IDATUM
         CALL LETTER (4,.15,0.,1.30,YLET,BCD)
         YLET=YLET+YINVL
2        CONTINUE
         IQK=INT(QK)
         IF (QK-FLOAT(IQK).GT..5) IQK=IQK+1
         IF (IQK.EQ.1) RETURN
         YLET=YLET+.5-YINVL
         CALL MATH (1.85,YLET,.1,90.,1HX)
         YLET=YLET+.25
C.... LABEL MULTIPLYING FACTOR
         IF (QK.GT.1.) GO TO 11
         I=1
3        IRK=INT(QX)
         IF (QX-FLOAT(IRK).GT..5) IRK=IRK+1
```

```
        IF (IRK.EQ.1) GO TO 4
        I=I+1
        IF (I.GT.5) GO TO 21
        QX=QX*10.
        GO TO 3
4       GO TO (5,6,7,8,9), I
5       IMFT=6H(F2.1)
        GO TO 10
6       IMFT=6H(F3.2)
        GO TO 10
7       IMFT=6H(F4.3)
        GO TO 10
8       IMFT=6H(F5.4)
        GO TO 10
9       IMFT=6H(F6.5)
10      NC=I+1
        ENCODE (NC,IMFT,BCD) QK
        GO TO 20
11      IK=INT(QK)
        IF (QK-FLOAT(IK).GT..5) IK=IK+1
        I=1
12      IK=IK/10
        IF (IK.EQ.1) GO TO 13
        I=I+1
        IF (I.GT.5) GO TO 23
        GO TO 12
13      GO TO (14,15,16,17,18), I
14      IMFT=4H(I2)
        GO TO 19
15      IMFT=4H(I3)
        GO TO 19
16      IMFT=4H(I4)
        GO TO 19
17      IMFT=4H(I5)
        GO TO 19
18      IMFT=4H(I6)
19      NC=I+1
        IRK=INT(QK)
        IF (QK-FLOAT(IRK).GT..5) IRK=IRK+1
        ENCODE (NC,IMFT,BCD) IRK
20      CALL LETTER (NC,.15,90.,1.85,YLET,BCD)
        RETURN
21      WRITE (6,22)
22      FORMAT (33H Y VARIABLE TOO SMALL FOR PROGRAM)
        STOP
23      WRITE (6,24)
24      FORMAT (33H Y VARIABLE TOO LARGE FOR PROGRAM)
        STOP
        END
```

A.10 PROBABILISTIC ANALYSIS

Subroutine CARLO(N,NSAM,NSTART,NY,NCOMBO,XMIN,XMAX,IDENS, NDENS,NCUM,YCUM,NMOM,IDATA,IRESULT,DENS,YD,FYCUM,YMOM, FEVENT,A,YY,SDENS,X,NEVENT,TDENS,Y,YMAX,YMIN)

Purpose

Probabilistic analysis using Monte Carlo simulation, when combined events may occur. This subroutine provides an estimate of the random nature of a set of dependent variables, y_i, where

$$y_i = g_i(x_1, x_2, \ldots, x_n)$$

and the *probability density functions* of the x's are known. The x's must all be stochastically independent. The user has the option of obtaining the density function, and/or the cumulative relative frequency, and/or the first four moments of the y_i's, and/or the probabilities of combined events associated with the y_i's.

Method

The rejection method is used to obtain a theoretical sample of the x's. A corresponding random value of each y_i is calculated. This is repeated sufficient times to obtain a sample of y_i of the desired size.
The user writes a main program that calls CARLO, which performs the operations noted above. The user must also write subprogram FUN, which defines the set of $g_i(x_i, x_i, \ldots, x_n)$; he or she must write subprogram DENSIT, which defines the density functions of the x's; and must write subprogram LOGIC, which defines the events with the y_i's and how they are to be combined.

How to Use

1. Write the calling program. In its simplest form it is as follows.
    a. *DIMENSION statement.* Check through the list of input and output variables and working arrays. Include all subscripted variables, dimensioning as indicated.
    b. *Define input data.* Include DATA cards, or READ statements, or individual cards such as

IPRINT = 1

so that each variable in the input list is defined.
    c. Include input data for any numerically defined density function. Include the corresponding array in the DIMENSION statement above, and include a labeled COMMON statement in order to transfer the information to SUBROUTINE DENSIT.
    d. Call the subroutine

CALL CARLO (N,NSAM,NSTART,NY,NCOMBO,XMIN,XMAX,NDENS,
    NCUM, YCUM, NMOM, IDATA, IRESULT,DENS,YD,FYCUM,YMOM,
    FEVENT,A,YY,SDENS,X,TDENS,Y,YMAX,YMIN)

    e. Add STOP and END.
2. Write function FUN in accordance with the procedure described to define the function.
3. Write subroutine DENSIT to define the density functions in accordance with the procedure described.

4. Write subroutine LOGIC to define the events associated with the
    y_i's (dependent variables) and how they are to be combined.

Input Variables

N: number of independent variables.

NSAM: sample size to be generated.

NSTART: any integer number except zero, used to start the string
of random numbers.

NY: number of dependent variables, or y_i's.

NCOMBO: number of combined events for which the probabilities must
be estimated.

XMIN(J): lower bounds of the independent variables, dimensioned
with N.

XMAX(J): upper bounds of the independent variables, dimensioned
with N.

IDENS(I):   = 1, a density function for y_i, derived from a smoothed
histogram, is desired,
= 0, a density function is not required,
dimensioned with the value of NY.

NDENS(I): number of points used to define the density function for
y_i set at 0 if IDENS(I) = 0, dimensioned with the value of NY.

NCUM(I): number of points for which it is required to calculate the
cumulative relative frequency of y_i, dimensioned with the value
of NY.

YCUM(I,J): array containing the values for y_i for which the values
of the cumulative relative frequency of y_i are to be calculated,
dimensioned with the value of NY and the maximum value of
NCUM(I).

NMOM(I): number of moments that are to be calculated for y_i from
the sample (maximum 4), set at 0, if no moments are required;
dimensioned with the value of NY. (Note: The first moment will
be the mean, and higher moments will be central moments.)

IDATA = 1, all input data are printed out.
= 0, input data are not printed out.

IRESULT = 1, output data are printed and plotted on printer.
= 0, no output.

Output Variables

DENS(I,J): array defining values for the density function of y_i, dimensioned with the value of NY and the maximum value of
NDENS(I)

YD(I,J): array defining values for y_i for which the density function
is evaluated, dimensioned with the value of NY and the maximum
value of NDENS(I)

FEVENT(L): array giving the estimate of probabilities of the specified combined events, dimensioned with the value of NCOMBO

FYCUM(I,J): array defining the values for the cumulative relative
frequency of y_i, corresponding to the values of YCUM(I,J),
dimensioned with the value of NY and the maximum value of
NCUM(I)

YMOM(I,K): array defining the moments of y_i, dimensioned with the
value of NY and the maximum value of NMON(I)

Working Arrays

A(I): dimensioned with the value of N

X(I): dimensioned with the value of N

YY(I,K): dimensioned with NY, and NSAM if any IDENS(I) = 1;
otherwise, the second subscript dimensioned with 1

SDENS(J): dimensioned with the maximum value of NDENS(I), but
not less than 1

TDENS(J): dimensioned with the maximum value of NDENS(I), but
not less than 1

NEVENT(L): dimensioned with the value of NCOMBO

Y(I): dimensioned with the value of NY

YMAX(I): dimensioned with the value of NY

YMIN(I): dimensioned with the value of NY

Programming Information

CARLO has full variable dimensioning. The calling program must
provide dimensioning as given above. Note that any array not used
should be dimensioned with 1, not zero.

The user must define the functions $g_i(x_1, x_2, \ldots, x_n)$ in subprogram FUNC, the density functions for the x's in subprogram
DENSIT, and the combined events in LOGIC. See below.

If IRESULT is set equal to 1, all output variables are automatically printed except the density functions DENS(I,J), which are plotted only.

Function FUN(X,I)

Purpose

To evaluate all $g_i(x_1, x_2, \ldots, x_n)$.

Method

The functions may be evaluated to any manner, including the
use of additional subprograms, as long as a specific value is returned
for any input point for the x's.

Input Variables

X(J): point at which the function is to be evaluated

I: identifies function required

Output Variables

FUN: function value of $g_i(x_1, x_2, \ldots, x_n)$

How to Set Up FUN

The following cards must be punched by the user:

```
FUNCTION FUN(X,I)
DIMENSION X(1)
GO TO (1,2,3) I
```

1   FUN = arithmetic expression defining the function (or more com-
         plex programming defining the function value)
```
      RETURN
```
2   FUN = etc.
```
      RETURN
```
3   FUN = etc.
```
      RETURN
      END
```

Note that this example is for three y_i's.

Subroutine DENSIT(L,XX,DENSX)

Purpose

To define the density function for the x's.

Method

They may be defined in functional form or by an array of numerical values. In the latter case, the array may be transferred by a
labeled COMMON statement from the main calling program, or by a
DATA declaration in DENSIT.

Input Variables

L: subscript of variable x for which the value of the density function
       is to be returned
XX: value of the particular x for which the corresponding value of
       the density function is to be returned

Output Variables

DENSX: value of density function at the specified value of x

Working Arrays

DENS1(I): array of numerical values defining a density function,
             dimensioned as required

X1(I): random variable corresponding to DENS1(I)
DENS2(I): another similar array to the above
X2(I): another similar array to the above
Etc.

How to Set Up DENSIT

A typical user-written program for DENSIT is given below. In this example, there are two x's, one having a functional representation for the density function, and one having a numerical representation.

```
      SUBROUTINE DENSIT (L,XX,DENSX)
      DIMENSION DENS1(33),X1(33)
C     DENS1(I) IS AN ARRAY DEFINING A DENSITY FUNCTION NU-
          MERICALLY FOR RANDOM VARIABLE X1(I)
      COMMON/AA/DENS1/X1
      GO TO (1,2)L
1     DENSX = .705E-12((XX-.920E6)**1.33*EXP(-.302E-12*(XX-
          .920E6)**2.33)
      RETURN
2     DENSX - FTABLE(X1,DENS1,XX,33)
C     FTABLE IS A LINEAR INTERPOLATION SUBPROGRAM
          RETURN
          END
```

Note that the values in the arrays DENS1(I) and X(I) have been transferred through *labeled* COMMON from the main program. Note also that a COMPUTED GO TO statement, or the equivalent, must be used to designate which density function is being called for.

Subroutine LOGIC(Y,NEVENT)

Purpose

To define the events associated with the y_i's and how they are to be combined. The program is user written.

Method

The function receives a set of trial values of the y_i's, "observes" if they satisfy the events with which they are associated, and then "observes" if the combined events are true or false. The result is returned as 1 or 0 in the array NEVENT(I).

Input Variables

Y(I): array of sample values of dependent variables

Output Variables

NEVENT(I) = 1 if Ith combined event is true
= 0 if Ith combined event is false

How to Set Up LOGIC

A typical program would have the following form:

```
      SUBROUTINE LOGIC (Y,NEVENT)
      DIMENSION Y(1),NEVENT(1),
      DO 1 I = 1,2
1        NEVENT (I) = 0
      IF (Y(1).GT.23000..AND.Y(2).GT.0.)NEVENT(1)=1
      IF (Y(1).GT.12000..AND.Y(1).LE.18000.)NEVENT(2)=1
      RETURN
      END
```

Note that this example is for two combined events.

```
      SUBROUTINE CARLO   (N,NSAM,NSTART,NY,NCOMBO,XMIN,XMAX,IDENS,NDENS,N
     1CUM,YCUM,NMOM,IDATA,IRESULT,DENS,YD,FYCUM,YMOM,FEVENT,A,YY,SDENS,X
     2,NEVENT,TDENS,Y,YMAX,YMIN)
C
      DIMENSION XMIN(1), XMAX(1), IDENS(1), NDENS(1), NCUM(1), YCUM(NY,1
     1), NMOM(1), DENS(NY,1), YD(NY,1), FEVENT(1), FYCUM(NY,1), YMOM(NY,
     21), A(1), X(1), YY(NY), SDENS(1), TDENS(1), NEVENT(1), Y(1), YMAX(
     31), YMIN(1), B(2)
      REWIND 1
C
C     ZERO ARRAYS
C
      DO 1 M=1,NCOMBO
      FEVENT(M)=0.0
1     CONTINUE
      DO 5 I=1,NY
      NM=NMOM(I)
      DO 2 K=1,NM
      YMOM(I,K)=0.0
2     CONTINUE
      NC=NCUM(I)
      DO 3 J=1,NC
      FYCUM(I,J)=0.0
3     CONTINUE
      ND=NDENS(I)
      DO 4 L=1,ND
      DENS(I,L)=0.0
4     CONTINUE
5     CONTINUE
C
C     WRITE INPUT DATA
C
      IF (IDATA.EQ.0) GO TO 9
      WRITE (6,54)
      WRITE (6,63) IDATA
      WRITE (6,64) IRESULT
      WRITE (6,55) N
      WRITE (6,65) NY
```

```
              WRITE (6,56) NSAM
              WRITE (6,57) NSTART
              WRITE (6,66) NCOMBO
              WRITE (6,72) N
              DO 6 I=1,N
              WRITE (6,68) I,XMIN(I)
              WRITE (6,69) I,XMAX(I)
       6      CONTINUE
              WRITE (6,67) NY
              DO 8 I=1,NY
              WRITE (6,73)
              WRITE (6,58) I,IDENS(I)
              WRITE (6,59) I,NDENS(I)
              WRITE (6,60) NCUM(I)
              IF (NCUM(I).EQ.0) GO TO 7
              NC=NCUM(I)
              WRITE (6,61) (YCUM(I,J),J=1,NC)
       7      WRITE (6,62) (NMOM(I))
       8      CONTINUE
       C
       C
       C      DETERMINE MAXIMUM VALUE OF EACH DENSITY FUNCTION
       C
       9      CALL AMAXD (N,A,XMAX,XMIN)
       C
       C      DETERMINE SAMPLE
       C
              CALL FRAND  (B,1,NSTART)
              DO 18 K=1,NSAM
              DO 11 L=1,N
       10     CALL FRAND  (B,2,0)
              XX=B(2)*(XMAX(L)-XMIN(L))+XMIN(L)
              CALL DENSIT (L,XX,DENSX)
              IF (B(1).GT.DENSX/A(L)) GO TO 10
              X(L)=XX
       11     CONTINUE
              DO 16 I=1,NY
              Y(I)=FUN(X,I)
              WRITE (1) Y(I)
              IF (IDENS(I).NE.0) GO TO 12
              IF (NMOM(I).LT.2) GO TO 13
       12     CONTINUE
              IF (K.EQ.1) YMAX(I)=Y(I)
              IF (K.EQ.1) YMIN(I)=Y(I)
              IF (K.EQ.1) GO TO 13
              IF (Y(I).GT.YMAX(I)) YMAX(I)=Y(I)
              IF (Y(I).LT.YMIN(I)) YMIN(I)=Y(I)
       13     IF (NCUM(I).EQ.0) GO TO 15
       C
       C      CALCULATE CUMULATIVE FREQUENCY
       C
              NC=NCUM(I)
              DO 14 J=1,NC
              IF (Y(I).LE.YCUM(I,J)) FYCUM(I,J)=FYCUM(I,J)+1.
       14     CONTINUE
       15     IF (NMOM(I).EQ.0) GO TO 16
       C
       C      CALCULATE THE FIRST MOMENT
       C
              YMOM(I,1)=YMOM(I,1)+Y(I)
       16     CONTINUE
       C
       C      CALCULATE FREQUENCY OF COMBINED EVENTS
       C
              IF (NCOMBO.EQ.0) GO TO 18
              CALL LOGIC (Y,NEVENT)
```

```
         DO 17 M=1,NCOMBO
         FEVENT(M)=FEVENT(M)+FLOAT(NEVENT(M))
17       CONTINUE
18       CONTINUE
         NY1=NY-1
         DO 39 I=1,NY
         IF (IDENS(I).EQ.0) GO TO 29
         REWIND 1
C
C        CALCULATE DENSITY FUNCTION
C
         DEL=(YMAX(I)-YMIN(I))/FLOAT(NDENS(I)-1)
         I1=I-1
         IF (I.EQ.1) GO TO 20
         DO 19 II=1,I1
         READ (1) YY(II)
19       CONTINUE
20       CONTINUE
         DO 25 K=1,NSAM
         READ (1) XX1
         IF (K.EQ.NSAM.OR.NY1.EQ.0) GO TO 22
         DO 21 JJ=1,NY1
         READ (1) YY(JJ)
21       CONTINUE
22       STEP=1.0
23       IF (XX1.LE.(YMIN(I)+DEL*STEP)) GO TO 24
         STEP=STEP+1.0
         GO TO 23
24       KK=IFIX(STEP)
         IF (KK.GT.NDENS(I)) KK=NDENS(I)
         DENS(I,KK)=DENS(I,KK)+1.0
25       CONTINUE
C
C        SMOOTH DENSITY FUNCTION
C
         KDENS=NDENS(I)
         DO 26 J=1,KDENS
         SDENS(J)=DENS(I,J)/FLOAT(NSAM)
26       CONTINUE
         CALL SE15 (SDENS,TDENS,KDENS,IER)
         YD(I,1)=YMIN(I)+DEL/2.0
         DO 28 J=2,KDENS
         YD(I,J)=YD(I,J-1)+DEL
28       CONTINUE
C
C        NORMALIZE DENSITY FUNCTION
C
         RANGE = YMAX(I) - YMIN(I)
         AREA = FSIMP(TDENS,RANGE,KDENS)
         DO 27 J=1,KDENS
         DENS(I,J) = TDENS(J)/AREA
27       CONTINUE
C
C        CALCULATE CUMULATIVE RELATIVE FREQUENCIES
C
29       IF (NCUM(I).EQ.0) GO TO 31
         NC=NCUM(I)
         DO 30 J=1,NC
         FYCUM(I,J)=FYCUM(I,J)/FLOAT(NSAM)
30       CONTINUE
31       IF (NMOM(I).EQ.0) GO TO 39
         YMOM(I,1)=YMOM(I,1)/FLOAT(NSAM)
C
C
C        CALCULATE HIGHER MOMENTS
```

```
C
         REWIND 1
         KMOM=NMOM(I)
         IF (I.EQ.1) GO TO 33
         DO 32 II=1,I1
         READ (1) YY(II)
32       CONTINUE
33       CONTINUE
         DO 37 K=1,NSAM
         READ (1) XX2
         IF (K.EQ.NSAM.OR.NY1.EQ.0) GO TO 35
         DO 34 JJ=1,NY1
         READ (1) YY(JJ)
34       CONTINUE
35       CONTINUE
         DO 36 J=2,KMOM
         YMOM(I,J)=(XX2-YMOM(I,1))**J+YMOM(I,J)
36       CONTINUE
37       CONTINUE
         DO 38 J=2,KMOM
         YMOM(I,J)=YMOM(I,J)/FLOAT(NSAM)
38       CONTINUE
39       CONTINUE
C
C        ESTIMATE PROBABILITIES OF COMBINED EVENTS
C
         IF (NCOMBO.EQ.0) GO TO 41
         DO 40 M=1,NCOMBO
         FEVENT(M)=FEVENT(M)/FLOAT(NSAM)
40       CONTINUE
C
C        OUTPUT
C
41       CONTINUE
         IF (IRESULT.NE.1) RETURN
         WRITE (6,49)
         WRITE (6,53) NSAM
         WRITE (6,70) NY
         DO 48 I=1,NY
         IF (NCUM(I).EQ.0) GO TO 43
         WRITE (6,74) I
         WRITE (6,50)
         KCUM=NCUM(I)
         DO 42 J=1,KCUM
         WRITE (6,51) J,YCUM(I,J),FYCUM(I,J)
42       CONTINUE
43       CONTINUE
         IF (NMOM(I).EQ.0) GO TO 44
         KMOM=NMOM(I)
         WRITE (6,52) (YMOM(I,J),J=1,KMOM)
44       KDENS=NDENS(I)
         IF (KDENS.EQ.0) GO TO 47
C
C        PLOT THE DISTRIBUTION
C
         DELTA=(YD(I,KDENS)-YD(I,1))/50.
         DO 46 J=1,51
         XPLOT=YD(I,1)+FLOAT(J-1)*DELTA
         DO 45 K=1,KDENS
         SDENS(K)=YD(I,K)
         TDENS(K)=DENS(I,K)
45       CONTINUE
         YPLOT=FTABLE(SDENS,TDENS,XPLOT,KDENS)
         CALL PLOTPT (XPLOT,YPLOT,9)
46       CONTINUE
```

```
       CALL OUTPLT
47     CONTINUE
48     CONTINUE
       IF (NCOMBO.EQ.0) RETURN
       WRITE (6,71) ((I,FEVENT(I)),I=1,NCOMBO)
       RETURN
C
C
C
49     FORMAT (1H1,20X,*RESULTS FOR SUBROUTINE CARLO *,/,20X,30(*-*),//)
50     FORMAT (1H0,*     STATION              VARIABLE
      1  CUMULATIVE        *,/56X,*FREQUENCY*,//)
51     FORMAT (I10,15X,E18.9,12X,E18.9,//)
52     FORMAT (//,5X,*THE REQUIRED MOMENTS ARE   *,4E18.9,/)
53     FORMAT (5X,*SAMPLE SIZE = *,I6,//)
54     FORMAT (1H0,//,5X,*INPUT DATA*,/,5X,10(*-*),///)
55     FORMAT (5X,*NUMBER OF INDEPENDENT VARIABLES ....................
      1.....     N=*,I5,//)
56     FORMAT (5X,*SAMPLE SIZE TO BE GENERATED ......................
      1.....   NSAM=*,I5,//)
57     FORMAT (5X,*INTEGER USED TO START THE STRING OF RANDOM NUMBERS....
      1..... NSTART=*,I5,//)
58     FORMAT (5X,*A DENSITY FUNCTION FOR Y(*,I2,*) IS DERIVED FOR IDENS=
      11 ONLY...   IDENS=*,I5,//)
59     FORMAT (5X,*NUMBER OF POINTS USED TO DEFINE THE DENSITY FUNCTION F
      1OR Y(*,I2,*) NDENS=*,I5,//)
60     FORMAT (5X,*THE CUMULATIVE DISTRIBUTION IS REQUIRED AT NCUM POINTS
      1.....   NCUM=*,I5,//)
61     FORMAT (5X,*THE CUMULATIVE DISTRIBUTION IS NEEDED AT POINT ...
      1..... YCUM(I)=*,(4E10.3,/,72X))
62     FORMAT (//,5X,*NUMBER OF MOMENTS REQUIRED  ....................
      1........   NMOM=*,I5,//)
63     FORMAT (5X,*INPUT DATA IS PRINTED OUT FOR IDATA=1 ONLY .........
      1.....   IDATA=*,I5,//)
64     FORMAT (5X,*OUTPUT DATA IS PRINTED OUT AND PLOTTED FOR IRESULT=1 O
      1NLY . IRESULT=*,I5,//)
65     FORMAT (5X,*NUMBER OF DEPENDENT VARIABLES......................
      1.....     NY=*,I5,//)
66     FORMAT (5X,*PROBABILITIES NEEDED FOR NCOMBO COMBINED EVENTS.......
      1..... NCOMBO=*,I5,////)
67     FORMAT (5X,*INPUT DATA FOR(*,I3,*)DEPENDENT VARIABLES*,/,5X,38(*-*
      1),/)
68     FORMAT (5X,*LOWER BOUND OF X(*,I2,*)..........................
      1....     =*,E12.5,//)
69     FORMAT (5X,*UPPER BOUND OF X(*,I2,*)..........................
      1....     =*,E12.5,////)
70     FORMAT (5X,*OUTPUT DATA FOR *,I3,* DEPENDENT VARIABLES*,/,5X,39(*-
      1*),//)
71     FORMAT (//////,5X,* PROBABILITY OF EVENT NO. *,I3,*  = *,E12.5)
72     FORMAT (5X,*INPUT DATA FOR *,I3,* INDEPENDENT VARIABLES*,/,5X,40(*
      1-*),///)
73     FORMAT (//)
74     FORMAT (1H1,//,5X,*VARIABLE NO. *,I3,/,5X,12(*-*),/)
       END


       SUBROUTINE AMAXD (N,A,XMAX,XMIN)
C      CALCULATE MAXIMUM VALUE OF EACH DENSITY FUNCTION USING  GOLDEN
C      SECTION
       DIMENSION A(1), XMAX(1), XMIN(1)
       DO 3 I=1,N
       KOUNT=0
       X1=XMAX(I)-0.618*(XMAX(I)-XMIN(I))
       X2=XMIN(I)+0.618*(XMAX(I)-XMIN(I))
```

```
        AX=XMIN(I)
        BX=XMAX(I)
1       KOUNT=KOUNT+1
        CALL DENSIT (I,X1,F1)
        CALL DENSIT (I,X2,F2)
        IF (ABS(F1-F2).LT.1.E-6*F1) GO TO 3
        IF (KOUNT.GT.500) GO TO 4
        IF (F2.GT.F1) GO TO 2
        BX=X2
        X2=X1
        X1=AX+0.382*(BX-AX)
        GO TO 1
2       AX=X1
        X1=X2
        X2=BX-0.382*(BX-AX)
        GO TO 1
3       A(I)=F1
        RETURN
4       WRITE (6,5) I
        STOP
C
C
C
5       FORMAT (* UNABLE TO FIND A MAXIMUM FOR DENSITY FUNCTION NUMBER  *,
        1I3)
        END



        SUBROUTINE SE15 (Y,Z,NDIM,IER)
        DIMENSION Y(1), Z(1)
        IF (NDIM-5) 3,1,1
1       A=Y(1)+Y(1)
        C=Y(2)+Y(2)
        B=0.2*(A+Y(1)+C+Y(3)-Y(5))
        C=.1*(A+A+C+Y(2)+Y(3)+Y(3)+Y(4))
        DO 2 I=5,NDIM
        A=B
        B=C
        C=0.2*(Y(I-4)+Y(I-3)+Y(I-2)+Y(I-1)+Y(I))
        Z(I-4)=A
2       CONTINUE
        A=Y(NDIM)+Y(NDIM)
        A=0.1*(A+A+Y(NDIM-1)+Y(NDIM-1)+Y(NDIM-2)+Y(NDIM-2)+Y(NDIM-3)+Y(NDI
        1M-1))
        Z(NDIM-3)=B
        Z(NDIM-2)=C
        Z(NDIM-1)=A
        Z(NDIM)=A+A-C
        IER=0
        RETURN
3       IER=-1
        RETURN
        END
```

Subroutine CELLV (N,M,NY,SLOW,SHIGH,XLOW,XUPP,PSPEC, KVALUE,X,XGRID,DELX)

Purpose

This subroutine provides an estimate of the probability of an event occurring associated with the dependent variable or design characteristic y_i, where

$$y_i = g_i(x_1, x_2, \ldots, x_n), \quad i = 1,n \tag{1}$$

and the density function of the x's are known, or the joint density function. The event must be defined by simple specifications limiting the values of the y's, as follows:

$$\ell_i \leqslant y_i \leqslant u_i \tag{2}$$

where ℓ_i and u_i are the upper and lower specifications on y_i.

This subroutine does not provide density functions of the y's.

Method

The algorithm is based on the *independent variable cell technique*.

The user must provide all functions corresponding to (1). The joint density function for the x's must also be defined by the user, plus bounds for all x's.

The functions need not be single valued, nor are derivatives required.

How to Use

1. Write the calling program. In its simplest form it is as follows.
   a. *DIMENSION statement*. Check through the list of input and output variables, and working arrays. Include all subscripted variables, dimensioning as indicated.
   b. *Define input data*. Include DATA statements, or READ statements, or individual lines such as

N = 4

so that each variable in the input list is defined.
   c. Call the subroutine

CALL CELLV(N,M,NY,SLOW,SHIGH,XLOW,XUPP,PSPEC,KVALUE,X, XGRID,DELX)

   d. *Output*. Include any required output statements such as

WRITE(6,10) PSPEC

   e. Add STOP and END.
2. Write function FUN and subroutine DENSIT in accordance with the prescriptions below.

Input Variables

N: number of x variables.
NY: number of y variables.
M: grid size or number of intervals for each x.

SLOW(I): lower specification for y_i; if there is no specification, set at a very low number such as $-1.E10$; dimensioned NY.

SHIGH(I): upper specification for y_i; if there is no specification, set at a very large number such as $1.E10$; dimensioned NY.

XLOW(I): lower bound of X(I), dimensioned N.

XUPP(I): upper bound of X(I), dimensioned N.

Output Variables

PSPEC: probability of specification being satisfied.

Working Arrays

The following working arrays are used: KVALUE(N), X(N), XGRID(N,M), and DELX(N).

Programming Information

CELLV has full variable dimensioning.

Function FUN(X,I,N)

Purpose

To evaluate all y's.

Method

The functions may be evaluated in any manner, including the use of additional subprograms, as long as a specific value is returned for any input point for the x's.

Input Variables

X(J): point at which the function is to be evaluated

I: identifies y_i function required

N: number of variables

Output Variables

FUN: function values of Y(I)

How to Set Up FUN

The following typical coding is provided by the user.

```
      FUNCTION FUN(X,I,N)
      DIMENSION X(1)
      GO TO (1,2,3), I (assuming NY=3)
1     FUN = code defining y₁
      RETURN
```

2   FUN = code defining y_2
      RETURN
3   FUN = code defining y_3
      RETURN
      END

Function DENSIT (N,X)

Purpose

To define the joint density function for the x's.

Method

The joint density function will commonly be the product of independent density functions

$$f(\bar{x}) = f_1(x_1)f_2(x_2) \cdots f_n(x_n)$$

The density functions may be defined by the user in any way as long
as a specific value is returned for any input point for the x's.

Input Variables

N: number of variables
X(I): point at which the function is to be evaluated

Output Variables

DENSIT: value of joint density function

How to Set Up DENSIT

The following subprogram must be supplied by the user.

```
FUNCTION DENSIT(N,X)
DIMENSION X(1)
   coding to define DENSIT
RETURN
END
```

```
      SUBROUTINE CELLV (N,M,NY,SLOW,SHIGH,XLOW,XUPP,PSPEC,KVALUE,X,XGRID
     1,DELX)
C.... PROBABILISTIC ANALYSIS BY INDEPENDENT VARIABLE CELL METHOD
C.... INPUT
C         N = NUMBER OF INDEPENDENT VARIABLES
C         M = GRID SIZE OR NUMBER OF INTERVALS FOR EACH X
C         NY = NUMBER OF DEPENDENT VARIABLES, Y
C         SLOW(I) = LOWER SPECIFICATION FOR Y(I) - IF NO SPEC SET AT
C                   SMALL NUMBER - DIMENSION NY
C         SHIGH(I) = UPPER SPECIFICATION FOR Y(I) - IF NO SPEC SET AT
C                   LARGE NUMBER - DIMENSION NY
```

```
C          XLOW(I) = LOWER BOUND OF X(I) - DIMENSION N
C          XUPP(I) = UPPER BOUND OF X(I) - DIMENSION N
C          YLOW(I) = LOWER BOUND OF Y(I) - DIMENSION NY
C          YUPP(I) = UPPER BOUND OF Y(I) - DIMENSION NY
C.... OUTPUT
C          PSPEC = PROBABILITY OF SPECIFICATIONS BEING SATISFIED
C.... WORKING ARRAYS - DIMENSIONING SHOWN
C          KVALUE(N),X(N),XGRID(N,M),DELX(N)
           DIMENSION SLOW(1), SHIGH(1), XLOW(1), XUPP(1)
           DIMENSION KVALUE(1), XGRID(N,1), DELX(1), X(1)
           COMMON /TCHECK/ TLIMIT
C.... INITIALIZE
           DATA TLIMIT/1.E10/
           NCELLS=M**N
           PSPEC=0.0
C.... SET UP GRID
           DO 2 I=1,N
           DELX(I)=(XUPP(I)-XLOW(I))/FLOAT(M)
           XGRID(I,1)=XLOW(I)+DELX(I)/2.
           DO 1 J=2,M
           XGRID(I,J)=XGRID(I,J-1)+DELX(I)
1          CONTINUE
2          CONTINUE
C.... SCAN EACH CELL
           L=1
10         CONTINUE
C.... IDENTIFY CELL
           ISUM=0.0
           DO 3 I=1,N
           II=N-I+1
           IF (II.EQ.N) GO TO 3
           ISUM=ISUM+KVALUE(II+1)*M**II
3          KVALUE(II)=(L-ISUM-1)/M**(II-1)
           DO 4 I=1,N
           KVALUE(I)=KVALUE(I)+1
4          CONTINUE
C.... EVALUATE X FOR EACH CELL
           DO 5 I=1,N
           KX=KVALUE(I)
           X(I)=XGRID(I,KX)
5          CONTINUE
C.... CHECK IF CELL CONTAINS AN EVENT OUTSIDE SPECIFICATIONS
           DO 6 I=1,NY
           Y=FUN(X,I,N)
           IF (Y.LT.SLOW(I).OR.Y.GT.SHIGH(I)) GO TO 7
6          CONTINUE
           GO TO 9
C.... DETERMINE PROBABILITY FOR CELL
7          PYCELL=DENSIT(N,X)
           DO 8 I=1,N
           PYCELL=PYCELL*DELX(I)
8          CONTINUE
           PSPEC=PSPEC+PYCELL
9          CALL SECOND (T)
           IF (T.GT.TLIMIT) GO TO 11
           L=L+1
           IF(L.LE.NCELLS)GO TO 10
           PSPEC=1.-PSPEC
           RETURN
11         WRITE (6,12) L
12         FORMAT (* TIME LIMIT AT CELL *,I8)
           STOP
           END
```

Subroutine MANAL (N,CM,YMIN,YMAX,NYP,YP,AL,CUM)

Purpose

This subroutine provides an estimate of the probability density function for y, where

$$y = g(x_1, x_2, \ldots, x_n) \tag{1}$$

and the first four moments of the x's are known. The density function for y has the form

$$f(y) = \exp (\lambda_1 + \lambda_2 y + \lambda_3 y^2 + \lambda_4 y^3 + \lambda_5 y^4)$$

The subroutine output provides the values of the parameters. The cumulative distribution function of y can also be obtained.

Method

A truncated Taylor's series expansion of (1) is used as a basis for obtaining the moments of y as a function of the moments of the x's; see subroutine MOMENT. The user must supply a subroutine defining $g(x_1, x_2, \ldots, x_n)$, $\partial y / \partial x_i$, and $\partial^2 y / \partial x_i^2$; see subroutine MOMENT documentation. A maximum entropy distribution is generated by y using MEP1.

The bounds for y require special consideration. The user has the default option of inputting these bounds, or using internally set bounds based on the standard deviation (the square root of the second moment) and the skewness of y.*

How to Use

The calling program in its simplest form is set up as follows.

a. *DIMENSION statement.* Check through the list of input and output variables and include all subscripted variables.
b. *COMMON statement.* If nondefault values of the parameters are desired, a labeled COMMON statement must be added. See the section on programming information below.
c. *Define input data.* This may be done by simple arithmetic statements, DATA declarations, or READ statements. Each variable in the input list below must be included unless the default value is used.
d. Call subroutine MANAL.
e. Call the output subroutine or write output statements. Library output subroutines available from MEP1 can be used.
f. Add STOP and END.

---

*This logic feature was suggested by T. Pal.

Input Variables

Variables marked with an asterisk have internally set default values.

N: number of independent variables.
CM(I,J): array containing the first moment about the origin and the next three central moments of the independent variables, dimensioned exactly (N, 4).
YMIN: lower bound of the dependent variable.
YMAX: upper bound of the dependent variable.
*NBND: = 1, bounds of dependent variable are set internally.
= 0, bounds are provided by user—default value.
NYP: number of points for which it is required to calculate the cumulative distribution function.
YP(I): array containing the values of y for which the values of the cumulative distribution function are to be calculated, dimensioned with the value of NYP.
*IPRINT: prints results every IPRINT cycle, set = 0 for no intermediate output. [Note that intermediate results are related to the entropy maximization method, and have no direct relevance to probability analysis (see subroutine MEP1).]
*IDATA: = 1, all input data are printed out.
= 0, input data are not printed out—default value.
KSTART: sets the starting method used in MEP1; use a value of 1 for first trials; see MEP1 documentation.

Output Variables

AL(I): array containing the Lagrangian multipliers or λ;s, dimension at 6. (Note that although there are five λ's, the sixth subscript is used internally.)
CUM(I): array containing the values of the cumulative distribution function of f(y) corresponding to YP(I), dimensioned with the value of NYP.

Programming Information

MANAL has full variable dimensioning. The calling program must provide dimensioning as given above. The user must define the function $g(x_1, x_2, \ldots, x_n)$ and the first and second partial derivatives. See the documentation for MOMENT.

If the user wishes to change default values of input parameters, the following COMMON statement must be included in the calling program:

COMMON /MANAL/IPRINT,IDATA,NBND

The user may change parameters used in MEP1 by adding the statement

COMMON /MEP1/KPRINT ,TOL,MAXFN

See MEP1 documentation.

```
      SUBROUTINE MANAL(N,CM,YMIN,YMAX,NYP,YP,KSTART,AL,CUM)
      DIMENSION CM(N,4),CC(4),YP(1),CUM(1),AL(1)
C.... DEFAULT DATA
      COMMON/MANAL/IPRINT,IDATA,NBND
      DATA IPRINT,IDATA,NBND/3*0/
      IF(IDATA.EQ. 0) GO TO 2
      WRITE(6,11)
      WRITE(6,12) IDATA
      WRITE(6,13) IPRINT
      WRITE(6,15) N
      WRITE(6,16)
      DO 1 I=1,N
      WRITE(6,17) I,(CM(I,J),J=1,4)
1     CONTINUE
2     CONTINUE
      CALL MOMENT (CM,N,CC,4)
      SQR= SQRT(CC(2))
      BETA1= CC(3)/(SQR**3)
      IF (NBND.EQ.0) GO TO 3
      S1=3.5
      IF (BETA1.LT. -.5) S1=4.
      YMIN=CC(1)-S1*SQR
3     IF(NBND.EQ.0) GO TO 4
      S2=3.5
      IF(BETA1.GT. .5) S2=4.
      YMAX=CC(1)+S2*SQR
4     CONTINUE
      IF(IDATA.EQ.0)GO TO 5
      WRITE(6,18) YMIN,YMAX
5     CALL MEP1(4,CC,YMIN,YMAX,NYP,YP,KSTART,0,AL,CUM)
      RETURN
11    FORMAT (1H1,//,20X,*INPUT DATA FOR SUBROUTINE MANAL*,/,20X,31(*-*)
     1,//)
12    FORMAT(* INPUT DATA IS PRINTED OUT FOR IDATA =1 ONLY .... IDATA =*
     1,I18,/)
13    FORMAT(* INTERMEDIATE OUTPUT EVERY IPRINT(TH) CYCLE .... IPRINT =*
     1,I18,/)
15    FORMAT(* NUMBER OF INDEPENDENT VARIABLES.....................N =*
     1,I18,/)
16    FORMAT(//,* VARIABLE     FIRST MOMENT      SECOND MOMENT      THIRD
     1MOMENT      FOURTH MOMENT*,//)
17    FORMAT(1X,I3,6X,4E18.9,/)
18    FORMAT(* BOUNDS ARE ....*,2E18.6)
      END
```

Subroutine MOMINT (N,M.NMOM,XUPP,XLOW,YMIN,YMAX,NYP,YP, KSTART,AL,CUM,DELX,XGRID,KVALUE,X)

Purpose

This subroutine provides an estimate of the probability density function for y, where

$$y = g(x_1, x_2, \ldots, x_n) \tag{1}$$

and the first m moments of the x's are known. The density function for y has the form

$$f(y) = \exp \left(\lambda_1 + \lambda_2 y + \lambda_3 y^2 + \lambda_4 y^3 + \lambda_5 y^4 + \cdots + \lambda_{m+1} y^m \right) \tag{2}$$

The subroutine output provides the values of the λ parameters. The cumulative distribution function of y can also be obtained.

Method

An integral equation is used as a basis for obtaining the moments of y as a function of the density functions of the x's; see subroutine TRANSM. The user must supply a subroutine defining $g(x_1, x_2, \ldots, x_n)$, and the density functions of the x's; see subroutine TRANSM documentation. The maximum entropy distribution is generated for y using MEP1.

The bounds for y require special consideration. The user has the default option of inputting these bounds, or using internally set bounds based on the standard deviation (the square root of the second moment) and the skewness of y.*

How to Use

The calling program in its simplest form is set up as follows.

a. *DIMENSION statement.* Check through the list of input and output variables and include all subscripted variables.
b. *COMMON statement.* If nondefault values of the parameters are desired, a labeled COMMON statement must be added. See the section on programming information below.
c. *Define input data.* This may be done by simple arithmetic statements, DATA declarations, or READ statements. Each variable in the input list below must be included unless the default value is used.
d. Call subroutine MOMINT.
e. Call the output subroutine or write output statements. Library output subroutines available from MEP1 can be used.
f. Add STOP and END.

Input Variables

Variables marked with an asterisk have internally set default values.

---

*This logic feature was suggested by T. Pal.

N: number of independent variables.

M: grid size or number of intervals for each x.

XUPP(I): upper bound of Ith independent variable, dimensioned N.

XLOW(I): lower bound of Ith independent variable, dimensioned N.

NMOM: number of moments to be used.

YMIN: lower bound of the dependent variable.

YMAX: upper bound of the dependent variable.

*NBND: = 1, bounds of dependent variable are set internally.

        = 0, bounds are provided by user—default value.

NYP: number of points for which it is required to calculate the cumulative distribution function.

YP(I): array containing the values of y for which the values of the cumulative distribution function are to be calculated, dimensioned with the value of NYP.

*IPRINT: prints results every IPRINT cycle, set = 0 for no intermediate output. [Note that intermediate results are related to the entropy maximization method, and have no direct relevance to probability analysis (see subroutine MEP1).]

*IDATA: = 1, all input data are printed out.

        = 0, input data are not printed out—default value.

KSTART: Sets the starting method used in MEP1; use a value of 1 for first trials; see MEP1 documentation.

Output Variables

AL(I): array containing the Lagrangian multipliers of λ's, dimensioned at NMOM+2

CUM(I): array containing the values of the cumulative distribution function of $f(y)$ corresponding to YP(I), dimensioned with the value of NYP

Working Arrays

The following working arrays must be dimensioned as shown: DELX(N), XGRID(N,M), KVALUE(N), and X(N).

Programming Information

MOMINT has full variable dimensioning. The calling program must provide dimensioning as given above. The user must define the function $g(x_1, x_2, \ldots, x_n)$, and the density functions of the x's. See the documentation for TRANSM.

If the user wishes to change default values of input parameters, the following COMMON statement must be included in the calling program:

COMMON /MOMINT/ IPRINT, IDATA, NBND

The user may change parameters used in MEP1 by adding the statement

COMMON/MEP1/KPRINT,TOL,MAXFN

See MEP1 documentation.

```
       SUBROUTINE MOMINT(N,M,NMOM,XUPP,XLOW,YMIN,YMAX,NYP,YP,KSTART,AL,
      1CUM,DELX,XGRID,KVALUE,X)
       DIMENSION XUPP(1),XLOW(1),DELX(1),XGRID(N,1),KVALUE(1),X(1)
       DIMENSION CC(8),YP(1),CUM(1),AL(1)
       COMMON/MOMINT/IPRINT,IDATA,NBND
C.... DEFAULT DATA
       DATA IPRINT,IDATA,NBND/3*0/
       IF(IDATA.EQ. 0) GO TO 2
       WRITE(6,11)
       WRITE(6,12) IDATA
       WRITE(6,13) IPRINT
       WRITE(6,15) N
       WRITE(6,16)M
       WRITE(6,17)NMOM
       WRITE(6,19)(XUPP(I),I=1,N)
       WRITE(6,20)(XLOW(I),I=1,N)
1      CONTINUE
2      CONTINUE
       CALL TRANSM(N,M,NMOM,XUPP,XLOW,CC,DELX,XGRID,KVALUE,X)
       WRITE(6,100)(CC(I),I=1,NMOM)
100    FORMAT(//,* MOMENTS ARE   *,5E12.5)
       SQR= SQRT(CC(2))
       BETA1= CC(3)/(SQR**3)
       IF (NBND.EQ.0) GO TO 3
       S1=3.5
       IF (BETA1.LT. -.5) S1=4.
       YMIN=CC(1)-S1*SQR
3      IF(NBND.EQ.0) GO TO 4
       S2=3.5
       IF(BETA1.GT. .5) S2=4.
       YMAX=CC(1)+S2*SQR
4      CONTINUE
       IF(IDATA.EQ.0)GO TO 5
       WRITE(6,18) YMIN,YMAX
5      CALL MEP1(NMOM,CC,YMIN,YMAX,NYP,YP,KSTART,0,AL,CUM)
       RETURN
11     FORMAT(1H1,//,20X,*INPUT DATA FOR SUBROUTINE MOMINT*,/,20X,32(*-*)
      1,//)
12     FORMAT(* INPUT DATA IS PRINTED OUT FOR IDATA =1 ONLY .... IDATA =*
      1,I18,/)
13     FORMAT(* INTERMEDIATE OUTPUT EVERY IPRINT(TH) CYCLE .... IPRINT =*
      1,I18,/)
15     FORMAT(* NUMBER OF INDEPENDENT VARIABLES......................N =*
      1,I18,/)
16     FORMAT(* NUMBER OF INDEPENDENT VAR. INTERVALS .............. M =*
      1,12X,I6,/)
17     FORMAT(* NUMBER OF MOMENTS USED ......................... NMOM =*
      1,12X,I6,/)
19     FORMAT(* UPPER BOUNDS OF X VARIABLES ................. XUPP(I) =*
      1,//,(5E16.8),/)
20     FORMAT(* LOWER BOUNDS OF X VARIABLES ................. XLOW(I) =*
      1,//,(5E16.8),//)
18     FORMAT(* BOUNDS ARE ....*,2E18.6)
       END
```

Subroutine TRANSF (N,M,ISPEC,SLOW,SHIGH,NDENS,NDIM,MINT,
    DELX,XLOW,XUPP,YLOW,YUPP,PSPEC,YDENS,FY,DIFF,Y,
    ISHIGH,ISLOW,KVALUE,X,W1,W2,W3,DELY,FREQ,YI,FREQI,
    DENS,FDENS)

Purpose

This subroutine provides an estimate of the probability of an event occurring associated with the dependent variables or design characteristic y_i, where

$$y_i = g_i(x_1, x_2, \ldots, x_n), \ i = 1, n \tag{1}$$

and the density functions of the x's are known, or the joint density function. The event must be defined by simple specifications limiting the values of the y's, as follows*:

$$\ell_i \leq y_i \leq u_i \tag{2}$$

where ℓ_i and u_i are either the lower and upper bounds on y_i, or upper and lower specifications. The subroutine will also provide the marginal density functions for the y_i's in numerical form.

It may be noted that the formulation requires that there be the same number of y's as there are x's. This will not normally be the case in practice; usually, there are fewer y's. In this event the user must formulate arbitrary additional y's to make up the required number of n. These *auxiliary variables* can simply be made equal to one of the independent variables. Thus, if we have two real dependent variables and four independent variables, the formulation would be

$$y_1 = g_1(x_1, x_2, x_3, x_4) \tag{3}$$

$$y_2 = g_2(x_1, x_2, x_3, x_4) \tag{4}$$

$$y_3 = x_j \tag{5}$$

$$\left. \begin{array}{c} \\ \end{array} \right\} \text{auxiliary variables}$$

$$y_4 = x_k \tag{6}$$

Method

The algorithm is based on the *transformation of variable* method. The user must provide all functions corresponding to (1), including auxiliary variables. The joint density function for the x's must also

_____

*This is a limitation of the algorithm, not an inherent limitation of the method.

be defined by the user, plus bounds for all x's and y's. The user must provide a subroutine that inverts (3), (4), (5), and (6), either directly or numerically.

The theory is based on the assumptions that the y's are single-valued functions and the first derivatives are all continuous. The latter are required for the Jacobian, which is used in the transformation of the density function. These derivatives are calculated numerically in subroutine JACOBI, using simple forward differences.

The bounds on the y's are important. If it is not possible to determine them explicitly, they must be found by trial or by optimization. It is very helpful to plot the marginal density function for the y in question.

How to Use

1. Write the calling program. In its simplest form it is as follows.
   a. *DIMENSION statement.* Check through the list of input and output variables and working arrays. Include all subscripted variables dimensioning as indicated.
   b. *Define input data.* Include DATA statements or READ statements, or individual lines such as

NDIM = 2

   so that each variable in the input list is defined.
   c. Call the subroutine

CALL TRANSF(N,M,ISPEC,SLOW,SHIGH,NDENS,NDIM,MINT,DELX, XLOW,XUPP,YLOW,YUPP,YSPEC,YDENS,FY,DIFF Y,ISHIGH, ISLOW,KVALUE,X,W1,W2,W3,DELY,FREQ,YI,FREQI,DENS, FDENS)

   d. *Output.* Include any required output statement.
   e. Add STOP and END.
2. Write function FUN, subroutine DENSIT, and subroutine INVRT in accordance with the prescriptions below.

Input Variables

N: number of variables.
M: grid size, or number of intervals for each y.
ISPEC: = 1, specification probability required, all specifications must be simple bounds.
   = 0, otherwise.
SLOW(I): lower specification for Y(I), if there is no specification set at lower bound of Y(I), dimensioned N.

SHIGH(I): upper specification for Y(I), if there is no specification
    set at upper bound of Y(I); dimensioned N.
NDENS: number of density functions required.
NDIM: = NDENS but not less than 1.
MINT: number of stations defining the required marginal density
    functions.
DELX: finite-difference factor for numerical differentiation; suggest-
    ed value is 1.E−3.
XLOW(I): lower bound of X(I); dimensioned N.
XUPP(I): upper bound of X(I); dimensioned N.
YLOW(I): lower bound of Y(I), may have to be set by trial; dimen-
    sioned N.
YUPP(I): upper bound of Y(I), may have to bet set by trial;
    dimensioned N.

It is essential that the Y(I)'s be ordered so that the first NDENS are
those requiring density functions.

Output Variables

PSPEC: probability of specifications being satisfied
YDENS(I,J): station values of J defining Y(I): dimension (NDENS,
    MINT), or if NDENS is zero dimension (1, 1)
FY(I,J): discrete values at J for marginal density function of Y(I):
    dimension (NDENS,MINT), or if NDENS is zero dimension (1, 1)

Working Arrays

The following working arrays are used: DIFF(N), Y(N,M),
ISHIGH(N), ISLOW(N), KVALUE(N), X(N), DELY(N), FREQ(NDIM,
N), YI(M+1), FREQI(M), DENS(MINT), FDENS(MINT), W1(M),
W2(M), and W3(N,M).

Programming Information

TRANSF has variable dimensioning, but N has an upper limit of
20.

Function FUN(X,I,N)

Purpose

To evaluate all y's.

Method

The functions may be evaluated in any manner, including the use
of addition subprograms, as long as a specific value is returned for
any input point for the x's.

Input Variables

X(J): point at which the function is to be evaluated
I: identifies y_i function required
N: number of variables

Output Variables

FUN: function value of Y(I)

How to Set Up FUN

The following typical coding must be provided by the use.

```
     FUNCTION FUN(X,I,N)
     DIMENSION X(1)
     GO TO (1,2,3), I (assuming that N=3)
1    FUN = code defining y₁
     RETURN
2    FUN = code defining y₂
     RETURN
3    FUN = code defining y₃
     RETURN
     END
```

Function DENSIT(N,X)

Purpose

To define the joint density function for the x's.

Method

The joint density function will commonly be the product of independent density functions

$$f(\bar{x}) = f_1(x_1)f_2(x_2) \cdots f_n(x_n)$$

The density functions may be defined by the user in any way as long as a specific value is returned for any input point for the x's.

Input Variables

N: number of variables
X(I): point at which the function is to be evaluated

Output Variables

DENSIT: value of joint density function

How to Set Up DENSIT

The following subprograms must be set up by the user.

```
FUNCTION DENSIT(N,X)
DIMENSION X(1)
   Coding to define DENSIT
RETURN
END
```

Subroutine INVRT(L,N,X)

Purpose

To invert

$$y_i = g_i(x_1, x_2, \ldots, x_n), \quad i = 1, n$$

to the form

$$x_i = h_i(y_1, y_2, \ldots, y_n), \quad i = 1, n$$

Method

Commonly, it will be possible to invert explicitly in analytical form, because of the simple form of the auxiliary functions. However, in some cases a solution of implicit functions will be necessary.

Input Variables

L: unused
N: number of variables
YCELL(I): values of y's for which inversions are to be made

Output Variables

X(I): values of x's resulting from inversion

How to Set Up INVRT

The following subroutine must be provided by the user.

```
SUBROUTINE INVRT(L,N,X)
DIMENSION X(1)
COMMON/IN/NVAR,YCELL(20)
   coding to define X(1), X(2), ..., X(N)
RETURN
END
```

```
      SUBROUTINE TRANSF (N,M,ISPEC,SLOW,SHIGH,NDENS,NDIM,MINT,DELX,XLOW,
     1XUPP,YLOW,YUPP,PSPEC,YDENS,FY,DIFF,Y,ISHIGH,ISLOW,KVALUE,X,W1,W2,W
     23,DELY,FREQ,YI,FREQI,DENS,FDENS)
C.... PROBABILISTIC ANALYSIS BY TRANSFORMATION OF VARIABLES
C     AND DISCRETIZATION
C.... INPUT
C        N = NUMBER OF INDEPENDENT VARIABLES
C        M = GRID SIZE OR NUMBER OF INTERVALS FOR EACH Y
C        ISPEC = 1, SPECIFICATION PROBABILITY DESIRED - SPECS MUST ALL
C                BE SIMPLE BOUNDS
C              = 0, OTHERWISE
C        SLOW(I) = LOWER SPECIFICATION FOR Y(I) - IF NO SPEC SET AT
C                  LOWER BOUND - DIMENSION N
C        SHIGH(I) = UPPER SPECIFICATION FOR Y(I) - IF NO SPEC SET AT
C                   UPPER BOUND - DIMENSION N
C        NDENS = NUMBER OF DENSITY FUNCTIONS REQUIRED
C        NDIM = NDENS BUT NOT LESS THAN ONE
C NOTE -- Y'S SHOULD BE ORDERED SO FIRST NDENS REQUIRE DENSITY FUNCTIONS
C        MINT = NUMBER OF STATIONS DEFINING REQUIRED DENSITY FUNCTIONS
C        DELX = FINITE DIFFERENCE FACTOR FOR NUMERICAL DIFFERENTIATION
C        XLOW(I) = LOWER BOUND OF X(I) - DIMENSION N
C        XUPP(I) = UPPER BOUND OF X(I) - DIMENSION N
C        YLOW(I) = LOWER BOUND OF Y(I) - DIMENSION N
C        YUPP(I) = UPPER BOUND OF Y(I) - DIMENSION N
C.... OUTPUT
C        PSPEC = PROBABILITY OF SPECIFICATIONS BEING SATISFIED
C        YDENS(I,J) = STATION VALUES OF Y(I) TO DETERMINE MARGINAL DENSITY
C                     FUNCTION - DIMENSION (NDENS,MINT) OR IF NDENS = 0
C                     DIMENSION (1,1)
C        FY(I,J) = DISCRETE VALUES FOR DENSITY FUNCTION AT Y(I)
C                  - DIMENSION (NDENS,MINT) OR IF NDENS = 0 DIMENSION
C                  (1,1)
C.... WORKING ARRAYS - DIMENSIONING SHOWN
C        DIFF(N),Y(N,M),ISHIGH(N),ISLOW(N),KVALUE(N),X(N),
C        DELY(N),FREQ(NDIM,M),YI(M+1),FREQI(M),DENS(MINT),FDENS(MINT),
C        W1(M),W2(M),W3(N,N)
      DIMENSION SLOW(1), SHIGH(1), YDENS(NDIM,1)
      DIMENSION FY(NDIM,1), DIFF(1), Y(N,1), ISHIGH(1), ISLOW(1), KVALUE
     1(1)
      DIMENSION XLOW(1), XUPP(1), YLOW(1), YUPP(1), X(1)
      DIMENSION DELY(1), FREQ(NDIM,1), YI(1), FREQI(1), XS(1)
      DIMENSION DENS(1), FDENS(1), W1(1), W2(1), W3(N,1)
      COMMON /IN/ NVAR,YCELL(20)
      COMMON /HIST/ ICALL
      COMMON /TCHECK/ TLIMIT
C.... INITIALIZE
      DATA TLIMIT/1.E10/
      NVAR=N
      DO 1 I=1,N
      DIFF(I)=DELX*(XUPP(I)-XLOW(I))
1     CONTINUE
      NCELLS=M**N
      PSPEC=0.0
      ICALL=1
      IF (NDENS.EQ.0) GO TO 3
      DO 2 I=1,NDENS
      DO 2 J=1,M
2     FREQ(I,J)=0.0
C.... SET UP GRID
3     DO 5 I=1,N
      DELY(I)=(YUPP(I)-YLOW(I))/FLOAT(M)
      Y(I,1)=YLOW(I)+DELY(I)/2.0
      DO 4 J=2,M
      Y(I,J)=Y(I,J-1)+DELY(I)
```

```
4       CONTINUE
5       CONTINUE
C....   CONVERT SPECIFICATIONS TO INTEGERS
        DO 10 I=1,N
        DO 6 J=1,M
        IF (SLOW(I).LT.Y(I,J)) GO TO 7
6       CONTINUE
7       ISLOW(I)=J
        DO 8 J=1,M
        IF (SHIGH(I).LT.Y(I,J)) GO TO 9
8       CONTINUE
        ISHIGH(I)=M
        GO TO 10
9       ISHIGH(I)=J-1
10      CONTINUE
C....   SCAN EACH CELL
        L=1
22      CONTINUE
C....   IDENTIFY CELL
        ISUM=0.0
        DO 11 I=1,N
        II=N-I+1
        IF (II.EQ.N) GO TO 11
        ISUM=ISUM+KVALUE(II+1)*M**II
11      KVALUE(II)=(L-ISUM-1)/M**(II-1)
        DO 12 I=1,N
        KVALUE(I)=KVALUE(I)+1
12      CONTINUE
C....   CHECK IF CELL CONTAINS AN EVENT INSIDE SPECIFICATIONS
        KEVENT=1
        IF (ISPEC.EQ.0) GO TO 14
        DO 13 I=1,N
        KI=0
        IF (KVALUE(I).GE.ISLOW(I).AND.KVALUE(I).LE.ISHIGH(I)) KI=1
        KEVENT=KEVENT*KI
13      CONTINUE
14      IF (NDENS.GT.0) GO TO 15
        IF (KEVENT.EQ.0) GO TO 22
15      CONTINUE
C....   INVERT ALL FUNCTIONS TO GET VALUES OF X'S
C       DEFINE CELL VALUE OF Y'S
        DO 16 I=1,N
        KY=KVALUE(I)
        YCELL(I)=Y(I,KY)
16      CONTINUE
        CALL INVRT (L,N,X)
C....   OBTAIN DETERMINANT OF JACOBIAN
        CALL JACOBI (N,X,DIFF,DETERM,W3)
        IF (DETERM.LT.1.E8) GO TO 18
        WRITE (6,17) L
17      FORMAT (52H WARNING - CALCULATION OF DETERMINANT FAILED AT CELL,I8
       1,/33H PROBABILITY FOR CELL SET AT ZERO)
C....   EVALUATE JOINT DENSITY FUNCTION OF X'S AND CORRESPONDING DENSITY
C       FUNCTION FOR Y'S
18      PYCELL=DENSIT(N,X)/DETERM
        DO 19 I=1,N
        PYCELL=PYCELL*DELY(I)
19      CONTINUE
        IF (KEVENT.EQ.0) GO TO 20
        PSPEC=PSPEC+PYCELL
20      IF (NDENS.EQ.0) GO TO 22
C....   EVALUATE RELATIVE FREQUENCIES FOR MARGINAL DENSITY FUNCTIONS
        DO 21 I=1,NDENS
        J=KVALUE(I)
        FREQ(I,J)=FREQ(I,J)+PYCELL
```

```
21      CONTINUE
        CALL SECOND(T)
        IF(T.GT.TLIMIT)GO TO 26
        L=L+1
        IF(L.LE.NCELLS)GO TO 22
C.... GENERATE DENSITY FUNCTIONS
        IF (NDENS.EQ.0) RETURN
        DO 25 I=1,NDENS
        DO 23 K=1,M
        YI(K)=Y(I,K)-DELY(I)/2.
        FREQI(K)=FREQ(I,K)
23      CONTINUE
        YI(M+1)=YI(M)+DELY(I)
        CALL HIST (XS,NS,M,YUPP(I),YLOW(I),1,0,3,3HVAR,MINT,YI,FREQI,DENS,
       1FDENS,W1,W2)
        DO 24 J=1,MINT
        YDENS(I,J)=DENS(J)
        FY(I,J)=FDENS(J)
24      CONTINUE
25      CONTINUE
        RETURN
26      WRITE(6,27)L
27      FORMAT(* TIME LIMIT AT CELL *,I8)
        STOP
        END



        FUNCTION ADET(N,A,EPS)
C.... CALCULATES ABSOLUTE VALUE OF DETERMINANT OF MATRIX
C.... INPUT
C        A(I,J) = MATRIX ELEMENTS - DIMENSION AS A SINGLE ARRAY N*N
C                 - N CANNOT EXCEED 20
C        N = SIZE OF MATRIX
C        EPS = MINIMUM PIVOT MAGNITUDE
C.... OUTPUT
C        ADET = ABSOLUTE VALUE OF DETERMINANT - IF MINIMUM PIVOT IS
C               LESS THAN EPS, A VALUE OF 1./EPS IS RETURNED
        DIMENSION A(N,1),IROW(20),JCOL(20)
        ADET=1.
        DO 18 K=1,N
        KM1=K-1
        PIVOT=0.
        DO 11 I=1,N
        DO 11 J=1,N
        IF(K.EQ.1)GO TO 9
        DO 8 ISCAN=1,KM1
        DO 8 JSCAN=1,KM1
        IF(I.EQ.IROW(ISCAN))GO TO 11
8       IF(J.EQ.JCOL(JSCAN))GO TO 11
9       IF(ABS(A(I,J)).LE.ABS(PIVOT))GO TO 11
        PIVOT=A(I,J)
        IROW(K)=I
        JCOL(K)=J
11      CONTINUE
        IF(ABS(PIVOT).GT.EPS)GO TO 13
        ADET=1./EPS
        RETURN
13      IROWK=IROW(K)
        JCOLK=JCOL(K)
        ADET=ADET*PIVOT
        DO 14 J=1,N
14      A(IROWK,J)=A(IROWK,J)/PIVOT
        A(IROWK,JCOLK)=1./PIVOT
        DO 18 I=1,N
```

```
      AIJK=A(I,JCOLK)
      IF(I.EQ.IROWK)GO TO 18
      A(I,JCOLK)=-AIJK/PIVOT
      DO 17 J=1,N
17    IF(J.NE.JCOLK)A(I,J)=A(I,J)-AIJK*A(IROWK,J)
18    CONTINUE
      ADET=ABS(ADET)
      RETURN
      END



      SUBROUTINE JACOBI (N,X,DIFF,DETERM,W)
C.... CALCULATES LOCAL VALUE OF DETERMINANT OF JACOBIAN MATRIX -
C     DERIVATIVES BY FORWARD DIFFERENCES
C.... INPUT
C        N=NUMBER OF VARIABLES
C        X(I) = ARRAY OF VALUES OF INDEPENDENT VARIABLES
C        DELX = FACTOR DEFINING INCREMENT FOR ESTIMATE OF DERIVATIVES
C        DIFF(I) = ARRAY DEFINING X(I) INCREMENTS
C
C.... OUTPUT
C        DETERM = VALUE OF DETERMINANT
C.... WORKING ARRAY - W(N*N)
C
      DIMENSION X(1), DIFF(1), W(N,1)
C     OBTAIN JACOBIAN
      DO 1 I=1,N
      GXO=FUN(X,I,N)
      DO 1 J=1,N
      X(J)=X(J)+DIFF(J)
      GXP=FUN(X,I,N)
      X(J)=X(J)-DIFF(J)
1     W(I,J)=(GXP-GXO)/DIFF(J)
C.... OBTAIN DETERMINANT
      DETERM=ADET(N,W,1.E-8)
      RETURN
      END
```

A.11 OPERATIONS ON PROBABILITY FUNCTIONS

This section provides a set of subroutines which enables the user to convert between the following types of probability functions of a continuous random variable: density function, cumulative distribution function, hazard function, and reliability function. The cumulative hazard function can also be calculated from the hazard function.

Subprograms are also provided to determine the maximum value of a density function and the mode, and the value of the cumulative distribution function at any point when the density function is defined numerically.

```
      SUBROUTINE AMAX(FD,XMIN,XMAX,A,X1)
C.... CALCULATES MAXIMUM VALUE OF A DENSITY FUNCTION DEFINED IN
C     FUNCTION SUBPROGRAM FD - USES GOLDEN SECTION
C.... INPUT
C        XMIN = LOWER BOUND
C        XMAX = UPPER BOUND
C.... OUTPUT
```

```
C          A    =MAXIMUM VALUE OF DENSITY FUNCTION
C          X1   =LOCATION OF MAXIMUM VALUE
           KOUNT=0
           X1=XMAX-.618*(XMAX-XMIN)
           X2=XMIN+.618*(XMAX-XMIN)
           AX=XMIN
           BX=XMAX
1          KOUNT=KOUNT+1
           F1=FD(X1)
           F2=FD(X2)
           IF(ABS(X1-X2).LT.1.E-6*(XMAX-XMIN))GO TO 3
C.... CHECK FOR LOSS OF SIGNIFICANT FIGURES IN DIFFERENCE BETWEEN F1 AND
C          F2
           IF(X2.LT.X1)GO TO 3
           IF(KOUNT.GT.50)GO TO 4
           IF(F2.GT.F1)GO TO 2
           BX=X2
           X2=X1
           X1=AX+.382*(BX-AX)
           GO TO 1
2          AX=X1
           X1=X2
           X2=BX-.382*(BX-AX)
           GO TO 1
3          A=F1
           RETURN
4          WRITE(6,5)
5          FORMAT(* UNABLE TO FIND A MAXIMUM FOR DENSITY FUNCTION*)
           STOP
           END



           SUBROUTINE CUMTOF (DISTF,DENSI,RANGE,M)
C....    CONVERTS A DISTRIBUTION FUNCTION TO A DENSITY FUNCTION
C        NOTE THAT DISTF MAY BE SLIGHTLY CHANGED BY SMOOTHING
C        DISTF(I)= DISTRIBUTION FUNCTION ARRAY
C        DENSI(I)= DENSITY FUNCTION ARRAY
C        RANGE= RANGE OF VARIABLE
C        M= NUMBER OF STATIONS, MUST BE ODD AND GT 6
         DIMENSION DISTF(1), DENSI(1)
C....    MAKE SURE INPUT IS SMOOTH
         CALL SMOOTH (DISTF,DENSI,M)
C....    PUT DATA BACK IN DISTF
         DO 1 I=1,M
         DISTF(I)=DENSI(I)
1        CONTINUE
C....    USE STANDARD NUMERICAL DIFFERENTIATION SUBROUTINE
         CALL DERIV (DISTF,DENSI,RANGE,M)
C....    SMOOTH AND NORMALIZE
         CALL SMOOTH (DENSI,DENSI,M)
C        MAKE SURE ALL VALUES ARE POSITIVE
         DO 2 I=1,M
         IF (DENSI(I).LT.0.0) DENSI(I)=0.0
2        CONTINUE
         CALL FNORM (DENSI,RANGE,M)
         RETURN
         END



         SUBROUTINE CUMTOH (DISTF,HAZ,RANGE,M)
C....    CONVERTS A CUMULATIVE DISTRIBUTION FUNCTION TO A HAZARD FUNCTION
C        DISTF(I)= DISTRIBUTION FUNCTION ARRAY
C        HAZ(I)= HAZARD FUNCTION ARRAY
```

```
C           RANGE= RANGE OF VARIABLE
C           M= NUMBER OF STATIONS, MUST BE ODD AND GT 6
            DIMENSION DISTF(1), HAZ(1)
C....       OBTAIN DENSITY FUNCTION
            CALL CUMTOF(DISTF,HAZ,RANGE,M)
C           OBTAIN HAZARD FUNCTION
            DO 3 I=1,M
            IF (1.-DISTF(I).LT.1.E-10) GO TO 1
            HAZ(I)=HAZ(I)/(1.-DISTF(I))
            GO TO 3
C           EXTRAPOLATE TO GET HAZ(I) BECAUSE REL(I) NEAR ZERO
C           CHECK FOR NO CHANGE IN LAST TWO VALUES
1           IF (ABS(HAZ(I-1)-HAZ(I-2)).LT.1.E-10) GO TO 2
C           EXTRAPOLATE BY FITTING QUADRATIC TO PREVIOUS THREE POINTS
C           SEE PAGE 303 OF CARNAHAN AND WILKES
            HAZ(I)=HAZ(I-3)-3.*HAZ(I-2)+3.*HAZ(I-1)
            GO TO 3
2           HAZ(I)=HAZ(I-1)
3           CONTINUE
            RETURN
            END




            SUBROUTINE CUMTOR (DISTF,REL,RANGE,M)
C...        CONVERTS A DISTRIBUTION FUNCTION TO A RELIABILITY FUNCTION
C           DISTF(I)= DISTRIBUTION FUNCTION ARRAY, DIMENSION M
C           REL(I)= RELIABILITY FUNCTION ARRAY, DIMENSION M
C           RANGE= RANGE OF VARIABLE
C           M= NUMBER OF STATIONS, MUST BE ODD AND GT 6
            DIMENSION DISTF(1), REL(1)
            DO 1 I=1,M
            REL(I)=1.-DISTF(I)
1           CONTINUE
            RETURN
            END




            FUNCTION CUM1(VAR,DENSI,N,X,MINT,W)
C....       DETERMINES VALUE OF CUMULATIVE DISTRIBUTION FUNCTION AT ANY
C           GIVEN SINGLE POINT WHEN DENSITY FUNCTION IS DEFINED NUMERICALLY
C....       INPUT
C           VAR(I) = ARRAY OF VALUES OF VARIABLE, NEED NOT BE EQUALLY
C                    SPACED - DIMENSION N
C           DENSI(I) = ARRAY OF VALUES OF DENSITY FUNCTION - DIMENSION N
C           N = NUMBER OF STATIONS DEFINING DENSITY FUNCTION
C           MINT = NUMBER OF INTEGRATION STATIONS TO BE USED TO OBTAIN
C                    SOLUTION
C           X = GIVEN POINT
C....       OUTPUT
C           CUM1 = REQUIRED VALUE OF CUMULATIVE DISTRIBUTION FUNCTION
C....       WORKING ARRAY
C           W(I) = DIMENSION MINT
C
            DIMENSION VAR(1),DENSI(1),W(1)
C....       CHECK THAT X IS WITHIN BOUNDS
            IF(X.LE.VAR(N).AND.X.GE.VAR(1))GO TO 2
            WRITE(6,1)
1           FORMAT(/,34H SPECIFIED VALUE IS OUTSIDE BOUNDS)
            CALL EXIT
C....       OBTAIN EQUALLY SPACED VALUES OF DENSI OVER REQUIRED INTERVAL OF
C           INTEGRATION - PUT IN W(I)
2           RANGE=X-VAR(1)
            DEL=RANGE/FLOAT(MINT-1)
```

```
          XTEMP=VAR(1)
          W(1)=DENSI(1)
          DO 3 I=2,MINT
          XTEMP=XTEMP+DEL
3         W(I)=FTABLE(VAR,DENSI,XTEMP,N)
C.... INTEGRATE TO GET CUM1
          CUM1=FSIMP(W,RANGE,MINT)
          RETURN
          END




          FUNCTION FINVRT(XA,DISTF,M,CUM)
C.... INVERTS A CUMULATIVE DISTRIBUTION FUNCTION DEFINED NUMERICALLY
C     BY MEANS OF CUBIC INTERPOLATION
C.... INPUT
C        XA(I) = ARRAY OF VALUES OF RANDOM VARIABLE, DIMENSION M
C        DISTF(I) = DISTRIBUTION FUNCTION ARRAY, DIMENSION M
C        M = SIZE OF ARRAYS, OR NUMBER OF STATIONS DEFINING FUNCTIONS
C        CUM = GIVEN VALUE OF CUMULATIVE DISTRIBUTION FUNCTION
C     OUTPUT
C        FINVRT = CORRESPONDING VALUE OF RANDOM VARIABLE
          DIMENSION XA(1),DISTF(1)
C.... LOCATE INTERVAL
          DO 1 I=2,M
          IF(CUM.LT.DISTF(I))GO TO 2
1         CONTINUE
2         IF(I.EQ.2)GO TO 3
          IF(I.GT.M-1)GO TO 4
          K=I-2
          GO TO 5
C.... CUM IN FIRST INTERVAL
3         K=1
          GO TO 5
C.... CUM IN LAST INTERVAL
4         K=M-3
5         FINVRT=FCUBIN(K,DISTF,XA,CUM,M)
          RETURN
          END




          SUBROUTINE FNORM (DENSI,RANGE,M)
C....    NORMALIZES A DENSITY FUNCTION
C        DENSI(I)= INPUT IS ORIGINAL DENSITY FUNCTION VALUES, OUTPUT IS
C        NORMALIZED VALUES
C        RANGE= RANGE OF ARGUMENT
C        M=NUMBER OF STATIONS, MUST BE ODD
          DIMENSION DENSI(1)
          AREA=FSIMP(DENSI,RANGE,M)
          DO 1 I=1,M
          DENSI(I)=DENSI(I)/AREA
1         CONTINUE
          RETURN
          END




          SUBROUTINE FTOCUM (DENSI,DISTF,RANGE,M)
C.... CONVERTS A DENSITY FUNCTION TO A DISTRIBUTION FUNCTION
C        DENSI(I) =   DENSITY FUNCTION ARRAY - DIMENSION M
C        DISTF(I) = DISTRIBUTION FUNCTION ARRAY - DIMENSION M
C        RANGE    = RANGE OF ARGUMENT
C        M        = NUMBER OF STATIONS
C
```

```
       DIMENSION DENSI(1), DISTF(1)
       DEL=RANGE/FLOAT(M-1)
       DISTF(1)=0.0
       AUX1=DENSI(1)+3.875*(DENSI(2)+DENSI(5))+2.625*(DENSI(3)+DENSI(4))+
      1DENSI(6)
       AUX2=DENSI(4)+4*DENSI(5)+DENSI(6)
       AUX3=DENSI(2)+4*DENSI(3)+DENSI(4)
       DISTF(2)=DEL/3.*(AUX1-AUX2-AUX3)
       DISTF(3)=DEL/3.*(DENSI(1)+4.*DENSI(2)+DENSI(3))
       DISTF(4)=DEL/3.*(AUX1-AUX2)
       DISTF(5)=DEL/3.*(DENSI(3)+4.*DENSI(4)+DENSI(5))+DISTF(3)
       DISTF(6)=DEL/3.*AUX1
       NN=0
       SUM=DISTF(5)
       DO 2 J=7,M
       IF (NN.EQ.0) GO TO 1
C      USE NEWTONS 3/8 RULE
       DISTF(J)=SUM+.375*DEL*(DENSI(J-3)+3.*(DENSI(J-2)+DENSI(J-1))+DENSI
      1(J))
       SUM=DISTF(J-2)
       SUM=DISTF(J-1)
       NN=0
       GO TO 2
C
C      USE SIMPSONS RULE
1      DISTF(J)=SUM+DEL/3.*(DENSI(J-2)+4.*DENSI(J-1)+DENSI(J))
       SUM=DISTF(J-2)
       NN=1
2      CONTINUE
       WRITE (6,3) DISTF(M)
3      FORMAT (1X,38HLAST VALUE OF DISTRIBUTION FUNCTION IS,F12.8/)
C      NORMALIZE LAST VALUE TO 1
       DO 4 I=1,M
       DISTF(I)=DISTF(I)/DISTF(M)
4      CONTINUE
       RETURN
       END



       SUBROUTINE FTOH (DENSI,HAZ,RANGE,M)
C....  CONVERTS A DENSITY FUNCTION TO A HAZARD FUNCION
C      DENSI(I)= DENSITY FUNCTION ARRAY, DIMENSION M
C      HAZ(I)= HAZARD FUNCTION ARRAY, DIMENSION M
C      RANGE= RANGE OF VARIABLE
C      M= NUMBER OF STATIONS, MUST BE ODD AND GT 6
       DIMENSION DENSI(1), HAZ(1)
C....  OBTAIN RELIABILITY FUNCTION
       CALL FTOR (DENSI,HAZ,RANGE,M)
C....  CALCULATE HAZARD FUNCTION
       DO 3 I=1,M
       IF (HAZ(I).LT.1.E-10) GO TO 1
       HAZ(I)=DENSI(I)/HAZ(I)
       GO TO 3
C....  EXTRAPOLATE TO GET HAZ(I) BECAUSE REL(I) NEAR ZERO
C      CHECK FOR NO CHANGE IN LAST TWO VALUES
1      IF (ABS(HAZ(I-1)-HAZ(I-2)).LT.1.E-10) GO TO 2
C....  EXTRAPOLATE BY FITTING A QUADRATIC TO PREVIOUS THREE POINTS
C      SEE PAGE 303 OF CARNAHAN AND WILKES
       HAZ(I)=HAZ(I-3)-3.*HAZ(I-2)+3.*HAZ(I-1)
       GO TO 3
2      HAZ(I)=HAZ(I-1)
3      CONTINUE
       RETURN
       END
```

```
      SUBROUTINE FTOR (DENSI,REL,RANGE,M)
C.... CONVERTS A DENSITY FUNCTION TO A RELIABILITY FUNCTION
C     DENSI(I) = DENSITY FUNCTION ARRAY - DIMENSION M
C     REL(I)   = RELIABILITY FUNCTION ARRAY - DIMENSION M
C     RANGE    = RANGE OF VARIABLE
C     M        = NUMBER OF STATIONS, MUST BE ODD AND GT 6
      DIMENSION DENSI(1), REL(1)
      CALL FTOCUM (DENSI,REL,RANGE,M)
      DO 1 I=1,M
      REL(I)=1-REL(I)
1     CONTINUE
      RETURN
      END



      SUBROUTINE HTOCMH (HAZ,CUMH,RANGE,M)
C....    CONVERTS A HAZARD FUNCTION TO A CUMULATIVE HAZARD FUNCTION
C        HAZ(I)=  HAZARD FUNCTION ARRAY DIMENSION M
C        CUMH(I) = CUMULATIVE HAZARD FUNCTION ARRAY - DIMENSION M
C        RANGE = RANGE OF ARGUMENT
C        M= NUMBER OF STATIONS
C
      DIMENSION HAZ(1), CUMH(1)
      DEL=RANGE/FLOAT(M-1)
      CUMH(1)=0.0
      AUX1=HAZ(1)+3.875*(HAZ(2)+HAZ(5))+2.625*(HAZ(3)+HAZ(4))+HAZ(6)
      AUX2=HAZ(4)+4*HAZ(5)+HAZ(6)
      AUX3=HAZ(2)+4*HAZ(3)+HAZ(4)
      CUMH(2)=DEL/3.*(AUX1-AUX2-AUX3)
      CUMH(3)=DEL/3.*(HAZ(1)+4.*HAZ(2)+HAZ(3))
      CUMH(4)=DEL/3.*(AUX1-AUX2)
      CUMH(5)=DEL/3.*(HAZ(3)+4.*HAZ(4)+HAZ(5))+CUMH(3)
      CUMH(6)=DEL/3.*AUX1
      NN=0
      SUM=CUMH(6)
      DO 2 J=7,M
      IF (NN.EQ.0) GO TO 1
C        USE NEWTON,S 3/8 RULE
      CUMH(J)=SUM+.375*DEL*(HAZ(J-3)+3.*(HAZ(J-2)+HAZ(J-1))+HAZ(J))
      SUM=CUMH(J-1)
      NN=0
      GO TO 2
C
C        USE SIMPSONS RULE
1     CUMH(J)=SUM+DEL/3.*(HAZ(J-2)+4.*HAZ(J-1)+HAZ(J))
      SUM=CUMH(J-2)
      NN=1
2     CONTINUE
      RETURN
      END



      SUBROUTINE HTOCUM (HAZ,DISTF,RANGE,M)
C....    CONVERTS A HAZARD FUNCTION TO A DISTRIBUTION FUNCTION
C        HAZ(I)= HAZARD FUNCTION ARRAY, DIMENSION M
C        DISTF(I)= DISTRIBUTION FUNCTION ARRAY, DIMENSION M
C        RANGE=RANGE OF VARIABLE
C        M= NUMBER OF STATIONS MUST BE ODD AND GT 6
      DIMENSION HAZ(1), DISTF(1)
C        OBTAIN RELIABILITY FUNCTION
      CALL HTOR (HAZ,DISTF,RANGE,M)
      DO 1 I=1,M
      DISTF(I)=1-DISTF(I)
```

```
1       CONTINUE
C          NORMALIZE
        DO 2 I=1,M
        DISTF(I)=DISTF(I)/DISTF(M)
2       CONTINUE
        RETURN
        END



        SUBROUTINE HTOF (HAZ,DENSI,RANGE,M)
C....      CONVERTS A HAZARD FUNCTION TO A DENSITY FUNCTION
C          HAZ(I)= HAZARD FUNCTION ARRAY, DIMENSION M
C          DENSI(I)= DENSITY FUNCTION ARRAY, DIMENSION M
C          RANGE= RANGE OF VARIABLE
C          M= NUMBER OF STATIONS, MUST BE ODD AND GT 6
        DIMENSION HAZ(1), DENSI(1)
C....      OBTAIN CUMULATIVE HAZARD FUNCTION
        CALL HTOCMH (HAZ,DENSI,RANGE,M)
C....      OBTAIN DENSITY FUNCTION
        DO 1 I=1,M
        DENSI(I)=HAZ(I)*EXP(-DENSI(I))
1       CONTINUE
C....      NORMALIZE
        CALL FNORM (DENSI,RANGE,M)
        RETURN
        END



        SUBROUTINE HTOR (HAZ,REL,RANGE,M)
C....      CONVERTS A HAZARD FUNCTION TO A RELIABILITY FUNCTION
C          HAZ(I)= HAZARD FUNCTION ARRAY, DIMENSION M
C          REL(I)= RELIABILITY FUNCTION ARRAY, DIMENSION M
C          RANGE= RANGE OF VARIABLE
C          M= NUMBER OF STATIONS, MUST BE ODD AND GT 6
        DIMENSION HAZ(1), REL(1)
C....      OBTAIN DENSITY FUNCTION
        CALL HTOF (HAZ,REL,RANGE,M)
C....      OBTAIN RELIABILITY FUNCTION
        DO 2 I=1,M
        IF (REL(I).LT.1.E-10.AND.HAZ(I).LT.1.E-10) GO TO 1
        REL(I)=REL(I)/HAZ(I)
        GO TO 2
1       REL(I)=1.
2       CONTINUE
C....      NORMALIZE
        DO 3 I=1,M
        REL(I)=REL(I)/REL(1)
        IF (REL(I).GT.1.) REL(I)=1.0
3       CONTINUE
        CALL SMOOTH (REL,REL,M)
        DO 4 I=1,M
        IF (REL(I).GT.1.) REL(I)=1.0
4       CONTINUE
        RETURN
        END



        SUBROUTINE RTOCUM (REL,DISTF,RANGE,M)
C....      CONVERTS A RELIABILITY FUNCTION TO A DISTRIBUTION FUNCTION
C          DISTF(I)= DISTRIBUTION FUNCTION ARRAY, DIMENSION M
C          REL(I)= RELIABILITY FUNCTION ARRAY, DIMENSION M
C          RANGE= RANGE OF VARIABLE
```

```
C         M= NUMBER OF STATIONS , MUST BE ODD AND GT 6
          DIMENSION REL(1), DISTF(1)
          DO 1 I=1,M
          DISTF(I)=1.-REL(I)
1         CONTINUE
          RETURN
          END




          SUBROUTINE RTOF (REL,DENSI,RANGE,M)
C....     CONVERTS A RELIABILITY FUNCTION TO A DENSITY FUNCTION
C         REL(I)=RELIABILITY FUNCTION ARRAY, DIMENSION M
C         DENSI(I)=DENSITY FUNCTION ARRAY, DIMENSION M
C         RANGE=RANGE OF VARIABLE
C         M=NUMBER OF STATIONS, MUST BE ODD AND GT 6
          DIMENSION REL(1), DENSI(1)
C....     OBTAIN DISTRIBUTION FUNCTION
          CALL RTOCUM (REL,DENSI,RANGE,M)
C....     OBTAIN DENSITY FUNCTION
          CALL CUMTOF (DENSI,DENSI,RANGE,M)
          RETURN
          END




          SUBROUTINE RTOH (REL,HAZ,RANGE,M)
C....     CONVERTS A RELIABILITY FUNCTION TO A HAZARD FUNCTION
C         REL(I)=RELIABILITY FUNCTION ARRAY, DIMENSION M
C         HAZ(I)=HAZARD FUNCTION ARRAY, DIMENSION M
C         RANGE=RANGE OF VARIABLE
C         M=NUMBER OF STATIONS, MUST BE ODD AND GT 6
          DIMENSION REL(1), HAZ(1)
C         OBTAIN DENSITY FUNCTION
          CALL RTOF (REL,HAZ,RANGE,M)
C         OBTAIN HAZARD FUNCTION
          DO 3 I=1,M
          IF (REL(I).LT.1.E-10) GO TO 1
          HAZ(I)=HAZ(I)/REL(I)
          GO TO 3
C         EXTRAPOLATE TO GET HAZ(I) BECAUSE REL(I) NEAR ZERO.
C         CHECK FOR NO CHANGE IN LAST TWO VALUES
1         IF (ABS(HAZ(I-1)-HAZ(I-2)).LT.1.E-10) GO TO 2
C         EXTRAPOLATE BY FITTING QUADRATIC TO PREVIOUS THREE POINTS
C         SEE PAGE 303 OF CARNAHAN AND WILKES
          HAZ(I)=HAZ(I-3)-3.*HAZ(I-2)+3.*HAZ(I-1)
          GO TO 3
2         HAZ(I)=HAZ(I-1)
3         CONTINUE
          RETURN
          END
```

A.12 PROBABILITY LAWS

These subroutines are designed for us in solving probability law
problems using basic probabilities. The basic probabilities are generated in subroutine BASIC. Joint probabilities are expressed in
terms of basic probabilities in subroutine JOINT. Combined probabilities, where each term is a simple probability, are expressed in terms
of basic probabilities in subroutine UNION. The joint probabilities,

known equal to zero because of mutually exclusive events, are generated in subroutine COMBO. Factorials are calculated in function FACTO.

$A(I)$ is represented by a number code, $A(I)=I$, $ACOMP(I)=I+N$

$KC(I,J)$: Jth event of the Ith basic probability

$NEVENT(L)$: number of simple events in Lth known compound probability.

$KCOMP(L,K)$: Kth simple probability of Lth known compound probability, in number code.

$COEF(L,I)$: coefficient of the Ith basic probability in the expression for the Lth known compound probability.

$NEQUAT$: counter for number of equations in terms of basic probabilities.

The following example may assist in using these subroutines. Three simple events are defined: A_1, A_2, and A_3. The following probabilities are known:

$$P(A_1|\overline{A}_2) = 0.35 \qquad P(A_3|A_1\overline{A}_2) = 0.35$$

$$P(A_2|\overline{A}_1) = 0.35 \qquad P(A_3|\overline{A}_1A_2) = 0.35$$

$$P(A_1|A_2) = 0.73 \qquad P(A_3|A_2A_2) = 0.66$$

$$P(A_3|\overline{A}_1\overline{A}_2) = 0$$

We require $P(\overline{A}_3)$.

```
C.... EXAMPLE OF USING BASIC PROBABILITIES
      DIMENSION CEQ(8,8),RIGHT(8),W(8)
      COMMON/BASIC/N,KCOMP(64,6),M,COEF(64,64),KC(64,6),NEVENT(64)
C.... INPUT
      DATA (NEVENT(L),L=1,11)/2,1,2,1,2,1,4*3,1/
      DATA (KCOMP(I,1),I=1,11)/1,5,4,4,1,2,1,4,1,4,6/
      DATA (KCOMP(I,2),I=1,11)/5,0,2,0,2,0,5,2,2,5,0/
      DATA (KCOMP(I,3),I=1,11)/6*0,4*3,0/
      N=3
C.... GENERATE BASIC EVENTS
      CALL BASIC
C.... GENERATE COEFFICIENTS
      DO 1 L=1,10
1     CALL JOINT(L)
C.... SET UP EQUATIONS
C     CEQ(I,J) = J TH COEF OF I TH EQUATION
      DO 2 J=1,M
2     CEQ(1,J)=1.0
C.... SET UP EQUATIONS 8 TO 14
      DO 3 J=1,M
      CEQ(2,J)=COEF(1,J)-0.35*COEF(2,J)
      CEQ(3,J)=COEF(3,J)-0.35*COEF(4,J)
      CEQ(4,J)=COEF(5,J)-0.73*COEF(6,J)
      CEQ(5,J)=COEF(7,J)-0.35*COEF(1,J)
```

```
          CEQ(6,J)=COEF(8,J)-0.35*COEF(3,J)
          CEQ(7,J)=COEF(9,J)-0.66*COEF(5,J)
3         CEQ(8,J)=COEF(10,J)
C.... SET UP RIGHT HAND SIDES
          RIGHT(1)=1.0
          DO 4 I=2,M
4         RIGHT(I)=0.0
C.... SOLVE EQUATIONS USING ISML LIBRARY SUBROUTINE
          CALL LEQT1F(CEQ,1,8,8,RIGHT,0,W,IER)
C.... PRINT BASIC PROB VALUES
          WRITE(6,11)(RIGHT(I),I=1,8)
11        FORMAT(* BASIC PROBABILITIES ARE*,/,(4E12.5))
C.... CALCULATE COMPLEMENT OF A3
          CALL JOINT(11)
          PROB=0.0
          DO 5 I=1,8
5         PROB=PROB+COEF(11,I)*RIGHT(I)
          WRITE(6,10)PROB
10        FORMAT(* REQUIRED PROBABILITY IS*,E12.5)
          STOP
          END



          SUBROUTINE BASIC
C.... GENERATES SET OF BASIC PROBABILITIES
          COMMON/BASIC/N,KCOMP(64,6),M,COEF(64,64),KC(64,6),NEVENT(64)
          KOUNT=0
          M=2**N
          DO 7 J=1,N
          MA=M/2**J
          DO 9 I=1,M
          KOUNT=KOUNT+1
          IF(KOUNT.LE.MA)GO TO 8
          KC(I,J)=J+N
          IF(KOUNT.EQ.2*MA)KOUNT=0
          GO TO 9
8         KC(I,J)=J
9         CONTINUE
7         CONTINUE
          RETURN
          END



          SUBROUTINE JOINT(L)
C.... EXPRESSES ANY JOINT PROBABILITY IN TERMS OF BASIC PROBABILIIES
          COMMON/BASIC/N,KCOMP(64,6),M,COEF(64,64),KC(64,6),NEVENT(64)
          NN=NEVENT(L)
          DO 11 I=1,M
          MATCH=0
          DO 13 K=1,NN
          DO 10 J=1,N
          IF(KC(I,J).NE.KCOMP(L,K))GO TO 10
          MATCH=MATCH+1
          GO TO 13
10        CONTINUE
13        CONTINUE
          IF(MATCH.EQ.NN)GO TO 12
          COEF(L,I)=0.0
          GO TO 11
12        COEF(L,I)=1.
11        CONTINUE
          RETURN
          END
```

```
         SUBROUTINE UNION(L)
C....    EXPRESSES ANY COMBINED PROBABILITY WHERE EACH TERM IS A SIMPLE
C        PROBABILITY IN TERMS OF BASIC PROBABILITIES
         COMMON/BASIC/N,KCOMP(64,6),M,COEF(64,64),KC(64,6),NEVENT(64)
         NN=NEVENT(L)
         DO 11 I=1,M
         DO 13 K=1,NN
         DO 10 J=1,N
         IF(KC(I,J).EQ.KCOMP(L,K))GO TO 12
10       CONTINUE
13       CONTINUE
         COEF(L,I)=0.0
         GO TO 11
12       COEF(L,I)=1.
11       CONTINUE
         RETURN
         END



         SUBROUTINE COMBO(NEQUAT)
C....    EXPRESSES ALL POSSIBLE COMBINATIONS OF NEVENTS TAKEN N AT A TIME,
C        AND N-1 AT A TIME, ETC.
C....    NEQUAT = COUNTER FOR COMBINATIONS, INPUT VALUE IS NUMBER OF
C                 LAST EQUATION GENERATED IN BASIC PROBABILITIES, OUTPUT
C                 VALUE IS NUMBER OF NEXT EQUATION TO BE GENERATED IN
C                 BASIC PROBABILITIES
C        NPROD = NUMBER OF SUBSETS OF COMBINATIONS
C        NA = NUMBER OF SIMPLE PROBABILITIES IN A COMBINATION
C        NB = NUMBER OF COMBINATIONS IN A SUBSET
C        KODE(J) = INTERNAL VARIABLE USED TO GENERATE PROBABILITY CODE
C                  VALUES
         DIMENSION KODE(6)
         COMMON/BASIC/N,KCOMP(64,6),M,COEF(64,64),KC(64,6),NEVENT(64)
         NPROD=N-1
         DO 7 L=1,NPROD
         NA=2-L+NPROD
         DO 1 J=1,NA
1        KODE(J)=J
         NB=FACTO(N)/(FACTO(NA)*FACTO(N-NA))
         DO 2 I=1,NB
         NEQUAT=NEQUAT+1
         DO 3 J=1,NA
3        KCOMP(NEQUAT,J)=KODE(J)
         NEVENT(NEQUAT)=NA
         IF(I.EQ.NB)GO TO 2
         DO 4 K=1,NA
         IF(KODE(K).EQ.K+N-NA)GO TO 5
         GO TO 4
5        KODE(K-1)=KODE(K-1)+1
         KK=K
         DO 6 KL=KK,NA
6        KODE(KL)=KODE(KL-1)+1
         GO TO 2
4        CONTINUE
         KODE(NA)=KODE(NA)+1
2        CONTINUE
7        CONTINUE
         NEQUAT=NEQUAT+1
         RETURN
         END
```

A.13 SAMPLE ANALYSIS

The following subprograms are in this category but are found elsewhere.

BOUNDS: determines upper and lower bounds of a sample (see Sec. A.6)
SMOM: calculates central moments from a sample (see Sec. A.7)
SORT: sorts a sample from low to high (see Sec. A.6)

A.14 ALPHABETICAL INDEX OF PROGRAMS BY NAME

A.15 MACHINE-DEPENDENT PLOTTING SUBROUTINES

Subroutine GREEK

Purpose

To plot a Greek letter.

Usage

CALL GREEK (X,Y,SCALE,THETA,NCH)

Description of Parameters

X: X coordinate in plotter units of the lower left corner of the
    character
Y: Y coordinate in plotter units of the lower left corner of the
    character
SCALE: character height in plotter units
THETA: angle of plotting measured counterclockwise in degree (0 to
    360) with respect to positive X axis
NCH: integer or Hollerith string specifying the character to be
    plotted

Hollerith strings and their integer equivalents are listed below. Either may be used in calling subroutine GREEK.

1 5HALPHA	9 4HIOTA	17 3HRHO
2 4HBETA	10 5HKAPPA	18 5HSIGMA
3 5HGAMMA	11 6HLAMBDA	19 3HTAU
4 5HDELTA	12 2HMU	20 6HUPSLON
5 6HEPSLON	13 2HNU	21 3HPHI
6 4HZETA	14 2HXI	22 3HCHI
7 3HETA	15 6HOMCRON	23 3HPSI
8 5HTHETA	16 2HPI	24 5HOMEGA

Subroutine LETTER

Purpose

T plot a block of N alphanumeric characters with input coordinates in plotter units.

Usage

CALL LETTER (N,SCALE,ANGLE,XP,YP,BCD)

Description of Parameters

N: number of characters to be plotted.
SCALE: character height in plotter units (minimum 0.06 plotter units).
ANGLE: angle of plotting measured counterclockwise in degrees (0 to 360) with respect to positive X axis.
XP: X coordinate in plotter units of the lower left corner of the first character.
YP: Y coordinate in plotter units of the lower left corner of the first character.
BCD: variables or array containing the characters left justified. (Note that this argument may also be a Hollerith constant, e.g., 6HMYPLOT.)

Subroutine MATH

Purpose

To plot a mathematical symbol.

Usage

CALL MATH (X,Y,SCALE,THETA,NCH)

Description of Parameters

X: X coordinate in plotter units of the lower left corner of the
     symbol
Y: Y coordinate in plotter units of the lower left corner of the
     symbol
SCALE: symbol height in plotter units
THETA: angle of plotting measured counterclockwise in degrees (0
     to 360) with respect to positive X axis
NCH: integer or Hollerith string specifying the symbol to be plotted

    Hollerith strings and their integer equivalents are listed below.
Either may be used in calling subroutine MATH.

1	+	4HPLUS	19	ψ	3HPSI
2	−	5HMINUS	20	Ω	5HOMEGA
3	±	6HPLSMNS	21	√	4HSQRT
4	=	5HEQUAL	22	∫	5HINTEG
5	/	5HSLASH	23	[	5HLBRKT
6	X	1HX	24	]	5HRBRKT
7	·	5HPOINT	25	(	6HLPAREN
8	←	6HLARROW	26	)	6HRPAREN
9	·	3HDOT	27	{	6HLBRACE
10	<	5HLTHAN	28	}	6HRBRACE
11	>	5HGTHAN	29	π	2HPI
12	≥	6HGTHEQL	30	α	5HALPHA
13	≤	6HLTHEQL	31	τ	5HTHETA
14	≈	6HCONSIM	32	φ	3HPHI
15	∞	5HINFIN	33	*	4HSTAR
16	Σ	5HSIGMA	34	≠	6HNEQUAL
17	Γ	5HGAMMA	35	e	1HE
18	Δ	5HDELTA	36	→	6HRARROW

Subroutine PLOT

Purpose

    Basic subroutine for all plotting: choose plotter units as in. or
cm, axis translation, end current plot, end all plotting, control pen
position on or off the paper, obtain current pen coordinates, specify
boundary limits, and so on.

Usage

CALL PLOT (X,Y,N)

Description of Parameters

N : integer to control the pen mode as follows:
  = 9 X and Y are output arguments containing the coordinates (in plotter units) of the present pen position relative to the plotter origin.
  = 1 to plot with pen in present mode (which may be up or down) to the point (X,Y).
  = 2 to plot with pen down to the point (X,Y).
  = 3 to plot with pen up to the point (X,Y).
  = 999 to end all plotting and write an end-of-file mark on the plot tape. X and Y are dummy arguments for this call.

X : X coordinate in plotter units.

Y : Y coordinate in plotter units.

Subroutine PLTIN

Purpose

Plot initialization routine to provide factors and boundary values for X and Y in data units.

Usage

CALL PLTIN (XSCALE,YSCALE,V,W,XMIN,XMAX,YMIN,YMAX)

Description of Parameters

XSCALE: scale along X axis in data units per plotter unit
YSCALE: scale along Y axis in data units per plotter unit
V: value of X in data units corresponding to X=0.0 in plotter units (i.e., corresponding to the Y axis of the plotter)
W: value of Y in data units corresponding to Y=0.0 in plotter units (i.e., corresponding to the X axis of the plotter)
XMIN: minimum boundary value of X in data units
XMAX: maximum boundary value of X in data units
XMIN: minimum boundary value of Y in data units
YMAX: maximum boundary value of Y in data units

Subroutine PLTMPL

Purpose

To plot a series of straight lines between a set of data points with input coordinates in data units.

Usage

CALL PLTMPL (XD,YD,M)

Description of Parameters

XD: array of dimension M containing the X coordinates in data units
of the data points
YD: array of dimension M containing the Y coordinates in data units
of the data points
M: number of data points

Subroutine UNITTO

Purpose

To convert from data units to plotter units.

Usage

CALL UNITTO (XD,YD,XP,YP)

Description of Parameters

XD: X coordinate in data units
YD: Y coordinate in data units
XP: X coordinate in plotter units
YP: Y coordinate in plotter units

appendix B
Combinatorial Theory*

Combinatorial analysis is used for determining a priori probabilities, based on known possible combinations of simple events, each having equal probabilities of occurring. Various rules can be developed for calculating the total possible number of arrangements (permutations) and combinations of sets of items. The rules will be summarized here without proof, which may be found in introductory books in probability theory.†

1. The number of perceivable arrangements of n objects that can be distinguished as separate is n!.
2. The number of perceivable arrangements of n items, x of which are indistinguishable of one type, and $(n - x)$ of which are indistinguishable of a second type, is

$$\frac{n!}{x!(n - x)!}$$

This is called the xth binomial coefficient and is commonly designated by the symbol

---

*The need for this theory was first encountered in developing the binomial distribution in Sec. 3.3.2.
†See, for example, C. Derman, L. J. Gleser, and I. Olkin (1973), *A Guide to Probability Theory and Applications*, Holt, Rinehart and Winston, New York.

$$\binom{n}{x}$$

The items are commonly events or trials or tests, x of which occur in one way and $(n - x)$ of which occur in another. The rule gives the number of ways in which this can happen.

3. The rule above can be generalized to accommodate n trials with k different possible outcomes, n_1 of type 1, n_2 of type 2, ..., n_k of type k. The number of perceivable arrangements is

$$\binom{n}{n_1, n_2, \ldots, n_k} = \frac{n!}{n_1! n_2! \cdots n_k!}$$

4. The number of perceivable ordered arrangements of x items selected from a total of n items is $n!/(n - x)!$.

5. The number of unordered arrangements (combinations) of x items selected from a total of n items is

$$\binom{n}{x}$$

appendix C

Confidence Limits

We saw in Chapter 5 how parameters of an analytical distribution could be estimated by various methods. This is commonly called *point estimation*. There is obviously uncertainty associated with such an estimate, since it is based on a sample of values, and a different sample of the same size would yield different results.

A major topic of classical statistics, called *interval estimation*, is an attempt to measure this uncertainty of parameter estimates by establishing *confidence limits*, or *confidence intervals*, for the parameter, rather than a point estimate.

We shall represent the estimated value of the parameter by the random variable θ. The most common parameter of concern is the sample mean, used to estimate the population mean. The classical approach requires that we define a quantity θ^*, the true parameter for the population density function, which is an unknown constant. The random variable θ need not in general be a specific density function parameter; it can be any quantity that is a function of the observed sample.

We next imagine that we have obtained a large number of samples of the primary random variable x, each sample having a specified size n. We can represent the sample by x_1, x_2, ..., x_n. If it is unsorted, and the sample is truly randomly drawn from a stable population, each value x_i in the sample can be though of as being a random variable with the same probability distribution as x.

If we made a point estimate for θ for each sample, we could generate a probability density function for θ, designated $f(\theta, n)$, and represented in Fig. C.1. It shape would clearly depend on n. We

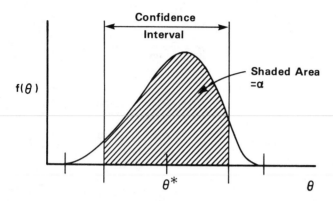

FIG. C.1 Illustration of confidence limits for a parameter θ.

cannot obtain $f(\theta, n)$ in this way, so it must somehow be developed theoretically.

We next must carefully define the confidence interval as that interval in θ that will contain the unknown value $\theta*$ with a probability α, where α is called the *confidence level* or *confidence coefficient*. How can this be written as a probabilistic statement? We are *not* saying that the event of interest is θ having a value between specified bounds, since this would not require that $\theta*$ lie inside the interval. We can achieve our requirement by using a variable transformation to obtain a new random variable.

$$\theta' = \theta* - \theta \tag{C1}$$

We now define our event as

$$P(z_L < \theta' < z_U) = \alpha \tag{C2}$$

where z_L and z_U are bounds on θ'.

The corresponding bounds on $\theta*$ are

$$P(z_L + \theta < \theta* < z_U + \theta) = \alpha \tag{C3}$$

This probability must be derived from $f(\theta, n)$, and we shall see that this can only be done under very limited circumstances.

The archtypical example is a normally distributed random variable for which we wish to determine the confidence interval in estimating the population mean, μ. The sample mean is calculated by

$$\bar{x} = \frac{n}{x_1 + x_2 + \cdots x_n} \tag{C4}$$

We demonstrated in Sec. 6.4 that the variance of $\bar{x}$ is

$$\sigma_m^{\;2} = \frac{\sigma^2}{n} \tag{C5}$$

where σ^2 is the population variance. It can easily be shown that $f(\bar{x})$ is normal if x is normal, using (C4). We next assume that we "know" the population variance, so we now "know" the distribution of $\bar{x}$.

$$f(\bar{x}) = \left(\frac{n}{2\pi\sigma^2}\right)^{1/2} \exp\left[-(\bar{x} - \mu)^2 \frac{n}{2\sigma^2} \right] \tag{C6}$$

This is our required theoretical $f(\theta, n)$.

It is characteristic of the method that μ, the unknown true value of the parameter, appears in the density function. This feature is used to derive (C3). We first convert $\bar{x}$ to a standard normal variate.

$$\bar{x}_s = \frac{\bar{x} - \mu}{\sigma_m} \tag{C7}$$

It is also characteristic of the method that the distribution must be made independent of $\theta*$ (in this case μ) by some type of variable transformation like that used above.

The event of $\bar{x}_s$ being inside bounds $-t$ and t has the probability

$$P\left(-t < \frac{\bar{x} - \mu}{\sigma_m} < t \right) = \alpha \tag{C8}$$

or

$$P(\bar{x} - \sigma_m t < \mu < \bar{x} + \sigma_m t) = \alpha \tag{C9}$$

Thus we have achieved the form of (C3). Given α, the value of t can be calculated from the standard normal distribution.*

It may be noted that, in going from (C8) to (C9) we have altered the random nature of the statement. The bounds have become random and the central quantity has become fixed. We must, therefore, reword our definition of the confidence interval so as to acknowledge this fact. It is that random interval that will enclose $\theta*$ with a probability of α, or in frequency terms, in a large number of sample sets of size n, the fraction of times that the calculated bounds will enclose $\theta*$ will tend to be α. These bound will vary from sample to sample because θ varies.

The first difficulty one has with this sample is the question of "knowing" the population variance. There may be evidence to sug-

---

*The standard normal distribution is a normal distribution with zero mean and unit variance.

gest that the current population is closely related to another very
well defined population, so that the variances can be taken as the
same, even though the means are not. But in practice there will always be some degree of uncertainty about this. The distribution of x
is also a source of uncertainty; it may not be truly normal. If n is
large, the distribution of $\bar{x}$ will tend to be normal based on the central limit theorem, even if x is not normal. But to what degree this
is true in any given instance is always uncertain. The method thus
presents, in this example, a rather comforting illusion that we have
adequately coped with uncertainty by deriving confidence limits, when
in fact there are other uncertainties that are simply ignored.

There are a limited number of other special cases that can be
solved in a similar manner, and with the same uncertainty superimposed on them.

An interesting example of the engineering use of confidence limits can be found in Kececioglu and Lamarre.* Their approach is typical and the comments that follow are not intended as criticisms of the
authors. They are concerned with the problem of calculating the
probability of no failure, or the strength reliability of a structural
member when the probability density functions for stress and
strength are assumed to be normal, and the distribution parameters
(mean and standard deviation) from a sample of specified but different size for each of the stress and strength. More specifically, they
wish to determine the lower one-sided confidence limit, R_ℓ, on the reliability R, with confidence level α. By some ingenious mathematics,
the authors develop an effective distribution for R that depends only
on the samples of stress and strength and the sample sizes, and obtain from this the required one-sided confidence limit. One inherent
uncodified approximation is the estimate of an equivalent sample size
for margin of strength, in terms of the sample sizes for stress and
strength. The important question also arises of how realistic the assumption of normal distribution is. Reference to the *Metals Handbook*†
would suggest that such an assumption would be quite unrealistic
for the yield, ultimate, and fatigue strengths of most steels. The approximation due to the deviation from normality is, again, uncodified.

A further practical objection to the classical method is its lack
of generality. Only a limited number of theoretical distributions are
amenable to this analysis. Theory for simultaneous confidence intervals for more than one parameter is practically nonexistent.‡

---

*D. Kecedioglu and G. Lamarre (1977), Mechanical Reliability Confi-
dence Limits, *ASME Paper 77-WA/DE-12*.
†*Metals Handbook*, American Society for Metals, Metals Park, Ohio.
‡See M. Kendall and A. Stuart (1979), *The Advanced Theory of
Statistics*; Vol. 2: *Inference and Relationship*, 4th ed., Griffin,
London.

It is typical of the subject of statistics that there is more than one theoretical approach for solving a problem and no obvious way of choosing between them. A second approach to the estimation of intervals is *fiducial intervals.** This approach gives essentially the same *appearing* results as confidence intervals, and the distinction between the methods is rather subtle. It can perhaps be best illustrated by reference to the first illustration above. The derived distribution in (C6) is now considered a subjective density function for μ rather than $\bar{x}$, and is thus

$$f(\mu) = \left(\frac{n}{2\pi\sigma^2} \right)^{1/2} \exp\left[-(\bar{x} - \mu)^2 \frac{n}{2\sigma^2} \right] \tag{C10}$$

This is called the *fiducial distribution*, and to quote Kendall and Stuart: "It may be regarded as a distribution of probability in the sense of degrees of belief."† The fiducial limits, corresponding to confidence limits, are calculated by the same equation (C8), but are now interpreted as those limits that we are 100α percent sure of being correct in the current instance. Although this approach to interval estimates is more compatible with the subjective interpretation of probability advocated in this book, the objections to using limits as a measure of uncertainty remain the same.

A third, and more significantly different approach to interval estimation is the *Bayesian method*. It is somewhat similar to the fiducial method in that the parameter is considered a random variable. The procedure for obtaining its distribution was given in Sec. 3.6. The method requires that a subjective estimate of the prior distribution of the parameter be made, and that it be updated by successive application of a kind of recurrence formula:

$$f(\theta_i | x_i) = \frac{f(x_i | \bar{\theta}_{i-1}) f(\theta_{i-1})}{\int_R f(x_i | \bar{\theta}_{i-1}) f(\theta_{i-1}) \, d\theta_{i-1}} \tag{C11}$$

where

$$\theta_i = \text{ith updated value of the parameter } \theta$$

$$x_i = \text{ith value of } x \text{ in a sample}$$

$$\bar{\theta}_{i-1} = \text{mean of the } (i-1)\text{th update of } \theta$$

$$f(\theta_0 | x_0) = \text{prior distribution of } \theta \text{ using no sample information}$$

---

*Ibid.
†Ibid.

There is a tendency in the literature to use some convenient theoretical distribution for the prior density function. It is also necessary that an assumed theoretical distribution be used for $f(x)$, containing of course the parameter θ. The confidence limits are obtained directly from the final updated distribution for θ, and they must be interpreted in the same way as are fiduciary limits.

The arguments against the Bayesian method of "evolving" an estimate of a density function were given in some detail in Sec. 3.6, and it was suggested that the concept of a subjective prior distribution for a parameter is quite illusory. It would seem that the concept of having confidence limits associated with a parameter having for its roots a subjective distribution is equally illusory. What uncertainty is being codified?

It thus shares with the other methods the characteristic of attempting to codify uncertainty about the magnitude of a distribution parameter without including all the components of uncertainty.

Above this objection there are philosophical difficulties with the whole concept of confidence limits. There is the lack of universality; the methods will only handle a limited number of theoretical distribution; and they will not do so simultaneously for more than one parameter. Finally, the most basic philosophical objection, from an engineering viewpoint, is that there is no judgmental basis for selecting the confidence level. In physical terms the confidence level has no real meaning to an engineer; he or she cannot relate it in any direct way to previous experience. Intuition tells us nothing about the significance of a 0.90 confidence level with relation to risk of failure of a design, nor when one should use 0.95 or 0.99 or 0.999 rather than the almost standardized 0.90 level. It seems to this author that engineering intuition must be available as a basic tool in risk decision making.

appendix D

Transformation of Domains
for the Maximum Entropy Distribution

It is useful when working with this distribution to be able to transform from the original region, $\ell \leqslant x \leqslant u$, to the region $\ell' \leqslant x' \leqslant u'$. The density functions are designated as follows in the two domains.

$$\ell \leqslant x \leqslant u, \quad f(x) = \exp\left(\lambda_0 + \sum_{i=1}^{m} \lambda_i x^i\right) \tag{D1}$$

$$\ell' \leqslant x' \leqslant u', \quad f'(x') = \exp\left[\lambda'_0 + \sum_{i=1}^{m} \lambda'_i (x')^i\right] \tag{D2}$$

It is convenient to define the following quantities.

$$S = \frac{u' - \ell'}{u - \ell} \tag{D3}$$

$$A = \frac{S\ell - \ell'}{S} \tag{D4}$$

The variable transformation is

$$x = \frac{(x' - \ell')}{S} + \ell = A + \frac{x'}{S} \tag{D5}$$

And the density functions are related by using Eq. (6.3.3).

$$f(x) = Sf'(x') \tag{D6}$$

Substituting these transformations into Eq. (D1) gives

$$Sf'(x') = \exp\left[\lambda_0 + \sum_{i=1}^{m} \lambda_i\left(A + \frac{x'}{S}\right)^i\right] \tag{D7}$$

Transferring S from the left side to the right, and using the binomial expansion, we get

$$f'(x') = \exp\left\{- \ln S + \lambda_0 \right.$$

$$\left. + \sum_{i=1}^{m} \lambda_i\left[\sum_{j=0}^{i} \frac{i!}{(i-j)!j!} A^{i-j}\left(\frac{x'}{S}\right)^j\right]\right\} \tag{D8}$$

It is possible to rearrange terms in order to have the following form:

$$f'(x') = \exp\left\{-- \ln S + \lambda_0 + \sum_{i=1}^{m} \lambda_i A^i \right.$$

$$\left. + \sum_{i=j}^{m} \sum_{j=i}^{m} \frac{j! A^{i-j} \lambda_j}{(j-i)!i! S^i} x^i\right\} \tag{D9}$$

Comparing this equation with (D2), we can match corresponding terms, and abstract the following relationships:

$$\lambda'_0 = -\ln S + \lambda_0 + \sum_{i=1}^{m} A^i \lambda_i \tag{D10}$$

$$\lambda'_i = \sum_{j=1}^{m} \frac{j! A^{i-j}}{(j-i)!i! S^i} \lambda_j, \quad i = 1, m \tag{D11}$$

appendix E

Proof That the Maximum Entropy Solution Is Based on a Global Optimum

The basic formulation is found in (5.5.28) to (5.5.30), repeated below.

$$S = - \int_R f(x) \ln [f(x)] \; dx = \text{maximum} \tag{E1}$$

$$\int_R f(x) \; dx = 1 \tag{E2}$$

$$\int_R x^i f(x) \; dx = m_i, \quad i = 1, m \tag{E3}$$

It was shown that $f(x)$ must have the form

$$f(x) = \exp \left(\lambda_0 + \sum_{i=1}^{m} \lambda_i x^i \right) \tag{E4}$$

We now assume $f(x)$ has the form satisfying these expressions, and define $g(x)$ as any function replacing $f(x)$ that satisfies (E2) and (E3), with entropy expressed as

---

*This proof follows M. Tribus (1969), *Rational Descriptions, Decisions and Designs*, Pergamon, New York.

$$S' = - \int_R g(x) \ln [g(x)] \, dx \tag{E5}$$

and

$$\int_R g(x) \, dx = 1 \tag{E6}$$

$$\int_R x^i g(x) \, dx = m_i, \quad i = 1, m \tag{E7}$$

We combine (E1) and (E5)

$$S - S' = - \int_R f(x) \ln [f(x)] \, dx + \int_R g(x) \ln [f(x)] \, dx \tag{E8}$$

and put this in the form

$$S - S' = \int_R [g(x) - f(x)] \ln [f(x)] \, dx$$

$$+ \int_R g(x) \ln \left[\frac{g(x)}{f(x)} \right] dx \tag{E9}$$

We now use (E4) in the form

$$\ln [f(x)] = \lambda_0 + \sum_{i=1}^{m} \lambda_i x^i \tag{E10}$$

We use this in the first term of (E9) and do some rearranging.

$$S - S' = - \lambda_0 \int_R f(x) \, dx + \lambda_0 \int_R g(x) \, dx$$

$$+ \sum_{i=1}^{m} \lambda_i \left[\int_R g(x) x^i \, dx - \int_R f(x) x^i \, dx \right]$$

$$+ \int_R g(x) \ln \left[\frac{g(x)}{f(x)} \right] dx \tag{E11}$$

When we apply the constraints to (E11) and all terms vanish but the last.

$$S - S' = \int_R g(x) \ln \left[\frac{g(x)}{f(x)} \right] dx \tag{E12}$$

We next define H for convenience as

$$H = g(x) \ln \left[\frac{g(x)}{f(x)} \right] - g(x) + f(x) \tag{E13}$$

and note that

$$S - S' = \int_R H \, dx \tag{E14}$$

Taking the derivatives of H gives

$$\frac{\partial H}{\partial g} = \ln [g(x)] - \ln [f(x)] \tag{E15}$$

$$\frac{\partial^2 H}{\partial g^2} = \frac{1}{g(x)} \tag{E16}$$

The first derivative vanishes at $g(x)$ equal to $f(x)$, and the second derivative is always positive. Thus, since H has a global minimum of zero for all values of x, S must be greater than S' for all values of $g(x)$ other than $f(x)$, and S represents a global maximum for the entropy.

Author Index

Subject Index